Civil Liability for Bunker Oil Pollution Damage

Internationalrechtliche Studien

Beiträge zum Internationalen Privatrecht, zum Einheitsrecht und zur Rechtsvergleichung

Herausgegeben von Ulrich Magnus und Peter Mankowski

Band 57

PETER LANG

Frankfurt am Main · Berlin · Bern · Bruxelles · New York · Oxford · Wien

Dan Malika Gunasekera

Civil Liability for Bunker Oil Pollution Damage

PETER LANG
Internationaler Verlag der Wissenschaften

Bibliographic Information published by the Deutsche Nationalbibliothek
The Deutsche Nationalbibliothek lists this publication in the Deutsche Nationalbibliografie; detailed bibliographic data is available in the internet at http://dnb.d-nb.de.

Zugl.: Hamburg, Univ., Diss., 2010

D 18
ISSN 0947-0395
ISBN 978-3-631-60460-1

www.peterlang.de

I dedicate this piece of work to
my beloved parents,
my wife Shiluka, daughter Kushagree, son Dunith
and Drs. C. P. Burger

Contents

Preface 13

Abbreviations 15

Introduction 19
Public International Law v. private interests 21
Role of the IMO and the historical development of the area of concern .. 23
'Run' to the 'Bunkers' 26
Interpretation of the text 28
Introduction to the relevant Chapters 28

Chapter 1: The Convention 29
The Need 31
The Underlying Doctrines 33
Polluter Pays Principle 34
Imposition of Strict Liability 35
The prevailed tools 36
The evolution of the concept 40
The Instrument 45
"Bunker Oil" 46
"Ship" 52

Chapter 2: Scope of Application 59
(a) Geographical Scope: *jurisdiction ratione loci* 59
(b) Pollution Damage and Preventive Measures: *jurisdiction ratione materiae* 66
Components of Damage 68
The important characteristics: Externality 70
Contamination 71
Escape or discharge 74
"Preventive measures" 76
The notion of 'reasonableness' 78
The right to act 80
The appropriate instance 81

Further loss or damage by Preventive measures 82
Exclusions 84
Situation under the CLC 84
State immunity and its effects 86

Chapter 3: Status Quo – Liable and Aggrieved Parties 89
Personal Scope of Application: *jurisdiction ratione personae* 89
The Liable Party 89
"Ship Owner" 90
The ultimate compromise 95
Legal implication on responsible corporate entity 96
(a) The owner / registered owner 99
Application of '*alter ego*' 100
The role of Mortgagee and Financial Security Provider 102
Situations of cessation and abandonment of ownership 103
(b) Bareboat charterer 104
(c) Manager 106
(d) Operator 109
Joint and Several Liability 110
Multiplicity of ships 111
Multiplicity of 'shipowners' 113
Convention being the supreme law 113
Implication of liability on related parties and 'responder immunity' 115
The Victim 119

Chapter 4: Jurisdiction 123
Jurisdiction as proclaimed 123
The causative link 124
The geographical link 124
The collective link 127
Suing the appropriate defendant 129
Direct actions 130
Prescription 133
Institution of Actions 135
Matters relating to non-States Parties 136
Fairness Rule 138
The Forum 139
The Common Law Systems' Strategy 140
The Civil Law Systems' Strategy 142
Jurisdiction under the OPA 1990 143

Claims for natural resource damages 145
Status of environmental organizations and pressure groups 147

Chapter 5: Requirements of Liability and Admissibility of Claims 149
The Requirements 150
Operation of the ship 150
Causation 152
Factual Causation 153
Legal Causation 154
(a) *conditio sine qua non* as the sole cause 155
(b) proximate cause test 155
(c) adequate causation test 156
(d) flexible approach test 157
Alternative approaches used by States 157
Causation in events of concurrent tragedies and in recurrence 159
Concurrent Causes 159
Recurring cause 160
Uncertain cause 161
Existence of Loss or Damage 161
Types of damage 162
Direct and indirect damage 162
Actual and future damage 163
Pecuniary and Non-Pecuniary Damage 164
Damage to Property 166
Damage to Person 167
Physical Damage 167
Non-Material Personal Damage 167
Pure Economic Loss 168
Varying claims on Pure Economic Loss and their Admissibility 172
Claims by Commercial Fishermen 172
Claims other than by Commercial Fishermen 173
(a) Pure loss of use 173
(b) Pure loss of earning capacity 173
(c) Pure loss of holiday and free time 174
(d) Costs of un-repairable things 174
Environmental Damage, Costs of Restoration and Reparation 175
Need of Prevention 177
Shipowner's Response 178
Costs incurred by Governmental Authorities 179
Fixed and Additional Costs 180

Costs incurred by Independent Contractors 181
Salvage of the Ship 181
Oil Recovery Process 182
Spraying of Dispersants 183
Rehabilitation of Wildlife 183
Costs incurred by Non-Governmental Agencies, Groups and Interested Parties 184
Miscellaneous Costs 185
Litigation costs including Professional Fees 186
Plaintiffs' Standard of Proof 186
Identification of the Respondent 187
Certainty as to Causation 187
Certainty as to Damage 188
Necessity and Reasonableness in taking Preventive Measures 189

Chapter 6: Available Defences and Exoneration from Liability 190
The defence of "act of war" 193
The defence of "act of civil war" 197
The defence of "act of hostilities" 299
The defence of "act of insurrection" 200
Specific shortcomings 201
The defence of "a natural phenomenon" 204
Narrowing the possibilities 207
Malicious act of a third party 208
Act of piracy as a defence 209
Act of barratry as a defence 211
Acts of strangers and guests 212
The gravity of the onus 213
State liability 214
The obligation 216
Nature of proof 217
Proportionate Liability 219
Malicious conduct of the Victim 220
Contributory negligence 221
The existence 222
Proving contributory negligence 222
Causation 223
Duty of care 223
Standard of care 223

Apportionment of Damage ... 224
Specific problems ... 225

Chapter 7: Limitation of Liability ... 227
Classical Approaches ... 227
Concerns over the issue in the Legal Committee ... 228
1957 Brussels Limitation Convention ... 230
Insurance as a Basic Element ... 231
The Final Settlement ... 233
The Effect and Application of the Relevant Provision ... 234
Applicable Law ... 235
Limitation of Liability under LLMC 1976 ... 236
Scope of Application I ... 237
Persons Entitled to Limit Liability ... 237
(a) Shipowners ... 237
(b) Liability Insurers and Financial Security Providers ... 238
(c) Exempted Persons ... 238
Scope Application II ... 242
Category of Ships – Inclusions and Exclusions ... 243
Scope of Application III ... 245
Claims Subject to Limitation ... 245
(A) Claims under Article 2.1(a) ... 246
(B) Claims under Article 2.1(c) ... 248
(C) Claims under Article 2.1(d) and (f) ... 249
Claims Excepted by Limitation ... 252
Conduct Barring Limitation ... 252
Concept of Actual Fault or Privity ... 253
Restriction as to Article 4 ... 254
The Person Liable ... 255
His Personal Act or Omission ... 255
Loss and Such Loss ... 256
Intent to Cause such loss ... 256
Recklessly and with Knowledge that such loss would probably result ... 256
Onus of Proof ... 257
Limits of Liability ... 258
Priorities of Claims ... 260
The Procedural Aspects of Limitation ... 261
Invocation without the constitution of a Limitation Fund ... 263
Invocation upon the constitution of a Limitation Fund ... 264
The volume of the Fund and the competent forum ... 265

Distribution of the Fund and safeguards 267
Governing law 267

Conclusion 269
Saving the Respondent 270
Avoiding "forum shopping" 271
Limitation v. Adequate Compensation 272
The absence of a supplementary Fund 272
Concluding Remarks 273

Bibliography 275

Preface

International commercial maritime has developed immensely throughout the history of maritime activities in the world facilitating needs of international trade more than any other aspect of transportation. This development has taken related areas of concern such as the law to very great heights identifying the importance of regulating harmonized international legal structure. Since the very inception of the International Maritime Consultative Committee that later became the International Maritime Organization, the said harmonization of varying laws of States has brought effective outcomes for the smooth functioning of the maritime industry thus creating a somewhat fair level field on which different actors could take on their respective roles in maintaining growth. This book focuses on a most recent development in the field of maritime commerce that applies to the entire operation of ships, especially upon its entering into force.

The said work would not only provide an opportunity for international as well as respective national practitioners in applying the area concern in their general legal practice, but also widen knowledge of all interested parties, especially academicians, students and morefully persons involved in the maritime business. The subject of Bunker Oil Pollution Damage and related liabilities have well been discussed in this work understanding the importance of providing a wide coverage of the application and implementation of the provisions dealt in the International Convention on Civil Liability for Bunker Oil Pollution Damage, 2001.

The Max Planck Research School for Maritime Affairs inspired me in taking up this challenge upon been granted its fellowship few years ago, which in my personal belief paved way to complete a successful piece of work of this nature. Multi faceted opportunities at the said research school gave me immense courage to further my knowledge while obtaining relevant understandings in various fields from law to science and geological aspects associated with ocean and maritime affairs. Having been able to gather varying diversified knowledge as aforesaid, I truly admire the efforts of the said research school in lending its fullest co-operation to me and my fellow researchers who have published important pieces of work in different related areas, in furthering our aspirations in the distribution of research and knowledge.

It is with great sincere and gratitude I recall the assistance of Prof. Dr. Jurgen Basedow and his team of experts at the said research school, and especially my PhD supervisor Prof. Dr. Ulrich Magnus who was an inspiration behind the success of this work. Further, I wish to thank my former lecturer at University of Utrecht, Prof. B. Kwiatkowska and my late senior counsel Mr. Chula de Silva, BCL (Oxn.), Presidents' Counsel who taught me the first lessons in admiralty law during my early years in the practice.

Dr. Malika Gunasekera
Colombo, Sri Lanka

January, 2010

Abbreviations

ASTM: American Society for Testing Materials
Brussels Convention 1968: Brussels Convention on Jurisdiction and the Enforcement of Judgments in Civil and Commercial Matters
Brussels Limitation Convention: International Convention for the Unification of Certain Rules relating to the Limitation of Liability of the Owners of Sea-going Vessels, Brussels,1924
Bunkers Convention: International Convention on Civil Liability for Bunker Oil Pollution Damage, 2001
CERCLA: Comprehensive Environmental Response, Compensation, and Liability Act
CLC 69: International Convention on Civil Liability for Oil Pollution Damage, 1969
CLC 92: Protocol to the International Convention on Civil Liability for Oil Pollution Damage, 1992
CLC Protocol 1984: Protocol to the International Convention on Civil Liability for Oil Pollution Damage
CMI: Comite Maritime International
CRISTAL: Contract Regarding an Interim Supplement to Tanker Liability for Oil Pollution, 1971
EC: European Commission
EEC: European Economic Community
EEZ: Exclusive Economic Zone
EU: European Union
FSU: Floating Storage Units
FUND 71: International Convention on the Establishment of an International Fund for Oil Pollution Damage, 1971
FUND 92: Protocol to the International Convention on the Establishment of an International Fund for Compensation for Oil Pollution Damage, 1992
Oil Pollution Damage
Harv, L. Rev.: Harvard Law Review
HELCOM: Baltic Marine Environment Protection Commission also known as HELSINKI Commission
HNSC 96: International Convention on Liability and Compensation for Damage in connection with the HNS Fund: Hazardous and Noxious Substances Fund

Carriage of Hazardous and Noxious Substances by Sea, 1996
ICS: International Chamber of Shipping
IHO: International Hydrographic Organization
IJECL: International Journal of Estuarine and Coastal Law
IMCO: International Maritime Consultative Committee
IMDG Code: International Maritime Dangerous Goods Code
IMO: International Maritime Organization
Intervention Convention: International Convention Relating to Intervention on the High Seas in Cases of IOPF: International Oil Pollution Fund
ISM Code: International Safety Management Code
ITLOS: International Tribunal for Law of the Sea
ITOPF: International Tanker Owners Pollution Federation Limited
IUCN: International Union for Conservation of Nature Oil Pollution Casualties, 1969
JEL: Journal of Environmental Law
JIML: Journal of International Maritime Law
J. Mar. L & Com.: Journal of Maritime Law and Commerce LEG: Legal Committee, International Maritime Organization
LLMC 76: Convention on Limitation of Liability for Maritime Claims, 1976
LLMC 96: Protocol to Amend the Convention on Limitation of Liability for Maritime Claims, 1976
LMCLQ: Lloyd's Maritime and Commercial Law Quarterly
LOSC: Law of the Sea Convention, 1982
Lugano Convention 1968: Convention on Jurisdiction and the Enforcement of Judgments in Civil and Commercial Matters
MARPOL: Int'l Conv., for the Prevention of Oil Pollution from Ships, 1973 as modified by Protocol of 1978
MEPC: Marine Environmental Protection Committee
MODU: Mobile Off-shore Drilling Units
MOU: Mobile Off-shore Units
MSA: Merchant Shipping Act
OECD: Organization for Economic Co-operation and Development
OILPOL: International Convention for the Prevention of Pollution of the Sea by Oil, 1954
OPA 1990: Oil Pollution Act, United States of America
Organizations or between International Organizations, 1986
OSPAR: Oslo-Paris Convention for the Protection of Marine Environment of the North and East Atlantic also known as OSPAR Commission
P & I Clubs: Protection and Indemnity Clubs
SOLAS: International Convention for the Safety of Life at Sea , 1974

TEU: Treaty on European Union, 1992
Tonnage Convention: International Convention on Tonnage Measurement of Ships, 1969
TOVALOP: Tanker Owners Voluntary Agreement Concerning Liability for Oil Pollution, 1969
Tul. Mar. L. J.: Tulane Maritime Law Journal
UN: United Nations
UNCED: United Nations Conference on Environment and Development, 1992
UNCLOS: United Nations Conference on Law of the Sea
UNCLOS III: Third United Nations Conference on Law of the Sea
UNCLOS 82: United Nations Convention on Law of the Sea, 1982
UNCTAD: United Nations Conference on Trade and Development, 1964
UNEP: United Nations Environmental Programme
UK: United Kingdom
US: United States
Vienna Convention: Vienna Convention on the Law of Treaties between States and International
Vir. L. Rev.: Virginia Law Review
Wreck Removal Convention: Nairobi International Convention on the Removal of Wrecks, 2007

Introduction

In 1995 the waters of the sea off the coast of Tasmania experienced one of the most disastrous effects of bunker oil spill.[1] The *Iron Baron*, bulk carrier which ran aground spilling three hundred tonnes of bunker fuel oil[2] affecting fifteen kilometres of shoreline and large number of living and non-living resources of the nature set a huge response cost of ten million Australian Dollars.[3] This dramatic incident opened up the eyes of the Government of Australia and especially the international legislators dealing with liability and compensation for maritime pollution, in taking prompt initiatives to fill the gap that existed in international maritime law aftermath the entry-into-force of the International Convention on Civil Liability for Oil Pollution Damage 1992[4]. Subsequently, two years later, a fuel tank of the *Kure*[5], wood chip carrier broke out in Humboldt Bay, California spilling one hundred and five barrels of bunker fuel oil costing a total of twenty million US Dollars, marking a record for the most expensive oil spill ever in terms of dollars per barrel.[6] These incidents are only a few to name among massive bunker spills that have occurred in different parts of the oceans and seas up to this date.

Comparing to oil spills from tankers that have taken centre stage in national and international contexts, one might consider oil spills resulting from the escape or discharge of fuel oil from the bunkers[7] of dry cargo ships as a negligible event.[8] Although, it is reasonable to ascertain, the real effects of such incidents would not allow such thought to proceed longer. In justification, it can be shown that many bulk carriers and container ships which carry more than ten thousand tonnes of

1 Australian Government, Australian Maritime Safety Authority, Marine Environment Protection (July 10, 1995), available at http://www.amsa.gov.au/Marine_Environment_Protection/Major_oil_spills_in_australia/Iron_Baron/index.asp
2 See Fuel Oil, http://en.wikipedia.org/wiki/Fuel_oil
3 For details see, Lloyds' Casualty Reports of 21 July – 11 August, and December 1995
4 See http://www.imo.org/Conventions/contents.asp?doc_id=660&topic_id=256
5 http://www.dfg.ca.gov/ospr/spill/nrda/nrda_kure.html
6 See *Trade Winds*, 3 April 1998
7 The term "bunkers" is used in modern days due to its usage in the past with regard to coal-burned steam turbines that have been replaced by fuel engines operating with marine fuels carried in the so-called bunkers of a ship in its bottom part.
8 See Regulation Impact Statement, http://www.aph.gov.au/House/committee/jsct/28march2006/treaties/bunkers_ris.pdf

bunker fuel form much bigger quantity than many tankers carry as cargo. Considering the types of modern merchant ships that are in use in maritime activities at present, one may not disregard even minor spills of bunker oil that can pose considerable danger to the marine environment.[9] Especially, concerning residual oils, which have high viscosities and persistent natures, could cause immense damage.[10]

According to various sources such as the International Oil Pollution Fund (IOPF)[11] and P & I Clubs[12], majority of oil pollution and compensation claims concern bunker spills, and some claims even rise up to around forty percent of the insured value.[13] Potentially, bunker spills could occur in many instances, such as during bunkering operations, i.e. oil transferring process from shore-to-ship and ship-to-ship, in maritime accidents and wilful and operational discharges. One of the major problems encountered around the issue of facilitating compensation for bunker spills, is the lack of international uniform rules in this area of concern. Apart from compensating damage caused by bunker oil used in tankers in events of tanker disasters via CLC regime, an international regime for pure bunker oil pollution damage was a highlighted absentee for a long period of time. The only tools in addressing such issue were available at national level, and these differing laws created much of the set-back for a common effective strategy prior to the entering into force of the International Convention on Civil Liability for Bunker Oil Pollution Damage, 2001[14] which came into operation on November 21, 2008.

Prior to such international instrument, many jurisdictions used different strategies at respective domestic level in addressing the subject. The first category of such States can be found having no distinct rules or regime to tackle the problem, rather considering bunker oil pollution incidents within the legal regime on torts and remedial actions for tortuous liability. They seem to apply different tools depending on the type of legal system, i.e. either common law or civil law which is applicable to such State, and opt to apply some tool which provides much satisfactory ends. The second category of States tend to extend the application of the CLC regime to bunkers as well with some changes or modifications. The English

9 See Int'l Maritime Org. [IMO], Maritime Knowledge Centre, International Shipping and World Trade, Facts and Figures, (updated November 2008), http://www.imo.org/includes/blastDataOnly.asp/data_id%3D23754/InternationalShippingand-WorldTrade-factsandfigures..pdf, pp 23 – 27

10 An extensive illustration is provided in Chapter 1.

11 http://www.iopcfund.org

12 See http://www.igpandi.org

13 See UK Clubs Analysis of Major Claims, 1993, available at http://www.ukpandi.com/ukpandi/infopool.nsf/html/LP_Init_AnalyMajorClaims

14 See http://www.imo.org/Conventions/contents.asp?topic_id=256&doc_id=666

Merchant Shipping Act[15] is a quite an example for a domestic law of that nature.[16] The third category of States is the jurisdictions which have a distinct legal instrument at national level to address the issue, such as the Oil Pollution Act 1990 of the United States. Having a diversified situation to address one important area of concern such as this would in no way bring fruitful results. Therefore, it is important to create a harmonized situation in achieving positive outcomes for a much needed issue at hand that affects the whole international legal order.

Public International Law v. private interests

The issues concerning the protection of marine environment as well as determination of liability at international context derive primarily from public international law through customary rule *pacta sunt servanda*. Although, the latter is much closer to private international law due to its relationship with private maritime interests, its emergence is closely linked to international customary law on sea related matters as well as the rules laid down by the United Nations Convention on Law of the Sea 1982 (UNCLOS)[17] and other various instruments that deals with environmental aspects. At present, marine environmental law in general consists of all these traditional sources, namely, customs, treaties, and general principles of law, case laws and works of prominent legal experts. As a result many both hard-law and soft-law instruments have been designed at international level for the protection of the marine environment as well as to facilitate the smooth functioning of private maritime interests. In order to protect both these aspects, large numbers of treaties have come into existence with the initiatives of the International Maritime Organization (IMO)[18] and under the Regional Seas Programme of the United Nations Environmental Programme (UNEP)[19]. Apart from such multilateral approach, regional organizations such as OSPAR[20] and HELSINKI[21] Commissions have contributed immensely towards the subject matter.

Customary international law on the other hand has helped evolve the development of the area concerned. Especially, the Third United Nations Conference

15 See http://www.opsi.gov.uk/ACTS/acts1995/ukpga_19950021_en_1
16 See Merchant Shipping Act (1995), Chaps. II – IV
17 United Nations Convention on the Law of the Sea, Dec 10, 1982 available at http://www.un.org/Depts/los/convention_agreements/texts/unclos/unclos_e.pdf
18 Int'l Maritime Org. [IMO], Home Page, http://www.imo.org
19 For more details see http://www.unep.org/regionalseas
20 See OSPAR Commission [OSPAR], Home Page, http://www.ospar.org
21 See Helsinki Commission [HELCOM], home Page, http://www.helcom.fi

on Law of the Sea[22] outlined the importance of marine environmental protection and urged States to carry out their undertakings in good faith and further develop binding rules on important issues such as responsibility and liability, in this context, through specialized international and regional arrangements. Furthermore, customary law has evolved to create rule-making concepts through State practice and *opinio juris*. In doing so, it requires States to execute due diligence by way of taking necessary steps to protect the natural environment and minimize harm.[23] Such steps could only be effective when minimum standards are guaranteed and enforcement is carried out accordingly, for example by way of port-State controls, issuance of certificates to ships and co-operation with other Sates. In addition, customary law was instrumental in creating new concepts such as "precautionary principle"[24] and "polluter-pays principle"[25], which operate as primary foundations of the present subject matter.

General principles of law forms another important aspect, which upholds the idea of State responsibility, which is a vital element concerning liability issues both at public as well as private levels. Hence, it paves way for the acceptance of the idea of reparation for damage including the principle of *restitutio in integrum*[26] and more modern concepts of cleaner seas and oceans[27].

These aforementioned sources of law have immensely contributed towards the development of both public and private international laws concerning the protection of the marine environment and especially the effects posed by private maritime interests. In addressing issues of bunker oil pollution damage with the view of creating international rules on the subject, it is highly relevant to pay much attention to the aforesaid. Upon being derived from public law perspec-

22 For more reference on the work of the third United Nations Conference on Law of the Sea and related work, see B. H. Oxman, The Third United Nations Conference on the Law of the Sea: The Ninth Session (1980) available at http://www.questia.com/PM.qst?a=o&d=79271683 and G. Plant, The Third United Nations Conference on the Law of the Sea and the Preparatory Commission: Models for United Nations Law Making, Int'l and Comparative Law Quarterly (1987), 36:525–558 Cam. Uni. Press available at http://journals.cambridge.org/action/displayAbstract;jsessionid=00FF63431474917003A43416528AEF9D.tomcat1?fromPage=online&aid=1512780

23 Much detailed provisions have been made by the UNCLOS in facilitating the desired objectives in its Part XII

24 See L. Gundling, "*The Status in International Law of the Principle of Precautionary Action*", IJECL (1990) pp. 23–29.

25 See OECD: OECD and the Environment, Paris 1986, at p. 24. See also OECD Recommendation of 14 November 1974 on the Implementation of the Polluter Pays Principle, C (74) 223

26 See also *Chorzow Factory Case* (Indemnity) (Jurisdiction), PCIJ Ser. A, No. 8/9, (1927), p.31.

27 See *Supra* n.18 at 4

tives, the IMO which chiefly concerns shipping interests, has taken into consideration the importance of the protection of the marine environment and has created effective rules on liability and compensation for pollution damage under its competence throughout the past decades. Before turning to the main topic of this work, it would be interesting to analyse some of its past work in this regards.

Role of the IMO and the historical development of the area of concern

Since it inception as the specialized organization attending to maritime affairs, the IMO as it is known today and its predecessor IMCO have carried out tremendous piece of work. Initially, the Organizations' primary concerns were connected to matters relating to safety of shipping and the prevention of marine pollution.[28] But since the event of *Torrey Canyon* disaster[29] in 1967, it has paid huge amount of attention towards liability and compensation issues concerning various kinds of damage including that of pollution damage, caused by ships. Its first initiatives on oil pollution regulations date back to 1954, during which the International Convention for the Prevention of Pollution of the Sea by Oil (OILPOL) was adopted.[30] This Convention which imposed regulations dealing with prevention of pollution by oil was effectively managed through its Maritime Safety Committee followed by a Subcommittee on Oil Pollution. Understanding the effects of crude oil washing process involved in routine ship operations of most ships, the provisions of the Convention prohibited the dumping of oily waste within a certain distance from land and in 'special areas' where the danger to the environment was especially acute. The destructive consequences of the *Torrey Canyon* disaster and growing interests of commercial exploitation of the shipping industry led to the belief of many States as well the IMO that the OILPOL

28 For more about the relevant work of the IMO see *Supra* n. 18 at 4 then follow "ABOUT IMO" hyperlink

29 See also S. Casey, *Set phases on stun, and other true tales of design, technology, and human error* (1998), Santa Barbara, Aegean Publishing Co., 2nd edition, pp 40–58; R.A. Cahill, *Strandings and their causes*, London, Fairplay Publications Inc., (1985), and also available at http://www.lboro.ac.uk/departments/hu/ergsinhu/aboutergs/lasttrip.htm; R. Petrow, *In the wake of the Torrey Canyon*, New York, David Mackay Co, (1968); R.A. Cahill, *Strandings and their causes*, London, Fairplay Publications Inc., (1985), and also available at http://www.lboro.ac.uk/departments/hu/ergsinhu/aboutergs/lasttrip.htm

30 The Conference for the adoption of this Convention was organized by the United Kingdom together with the Conference for the institution of the IMO itself. The OILPOL Convention came into force in 1958. A brief analysis is available at http://www.imo.org/Conventions/contents.asp?doc_id=678&topic_id=258#3

is inadequate to handle pollution incidents any longer. As a result, the existing Subcommittee was given a wider scope of powers to attend to marine pollution in general, but later transformed it to the Marine Environment Protection Committee (MEPC) with a mandate to deal with all matters relating to marine pollution.

In 1973, the Diplomatic Conference of the IMO incorporated much of the OILPOL 1954 provisions and its amendments with certain modifications as an annexure to a newly constructed convention, named, International Convention for the Prevention of Pollution from Ships, commonly known as MARPOL 1973. The creators of this Convention took into consideration of the growing industrial concerns, increasing sizes of tankers used for transportation, chemicals that were carried at sea and more specially the growing concern over global environment. Due to certain technical problems as to the ratification of the incorporation of OILPOL into MARPOL, many States showed lack of desire to ratify and accept this Convention as binding on them. During this period between 1976 and 1977, there existed of practical difficulties in enforcing MARPOL, including the series of tanker accidents that took place in adjacent waters of the United States. These occurrences led to the reconsidering of stringent rules to combat accidental and operational oil pollution incidents.[31] The 1978 Protocol to MARPOL was a more successful attempt, which absorbed the parent Convention and made stringent requirements for tankers to be upgraded to levels of protecting environment from being polluted.[32] These measures were further supplemented through the adoption of the 1978 Protocol to the International Convention for the Safety of Life at Sea (SOLAS) 1974. The importance of the adoption of the parent document and the Protocol were not felt greatly by States concerned till the lessons learned after the stranding of *Amoco Cadiz*[33], which was one of the gravest oil spills that affected the French coastal and marine environment.[34] As a result of all such incidents, MARPOL was finally ratified by the sufficient number of States by October 1982, upon which the MARPOL 1973/78 Convention later entered into force on October 2, 1983.

Paralleled to the aforementioned process at the IMO, rules on liability and compensation issues were developing at a greater pace. In order to attend to legal

---

31 For example the Agro Merchant ran aground off Massachusetts in December 1976 carrying 27, 000 tons of oil which caused a huge public concern over the threat of oil being flowed to the fishing fields of New England and Georges Bank.

32 Full texts of the Conventions and Protocols available at *Supra* n. 1 at 1, http://www.amsa.gov.au/Marine_Environment_Protection/Revision_of_Annexes_I_and_II_of_MARPOL_73–78/index.asp

33 Details of the incident is available at http://www.incidentnews.gov/incident/6241

34 Amoco Cadiz ran aground off Brittany losing its entire cargo consisting of 223,000 tons of crude oil leaving 130 beaches under a 30 cm. thickness

issues arising out of the *Torrey Canyon* incident, the Organization created an *ad hoc* Legal Committee which soon became a permanent subsidiary organ of its Council.[35] It was entrusted with the functions of determining international legal rules on liability and compensation for oil pollution damage. Establishing legal regimes with regard to the prevention of environmental pollution caused by oil during transportation by maritime means was alone incapable to address the liability and compensation issues that were erupted after many disasters. One may correctly point out this situation as such oil pollution incidents not only affects the losses to the respective ship and cargo, but ultimately becoming responsible for third party claims. This was the primary issue at the face of the Legal Committee when the matter was taken up for discussion for the creation of a new legal regime for liability and compensation. Apart from the general question as to who is liable in such an incident, there existed of number of other sensitive issues that had to be resolved prior to the drafting process, *inter alia* the basis of liability, level of compensation, quantification of damages and limits of liability. Considering the enormous damages that took place in several disasters, the prevailing international law on maritime claims could not be taken as a method for settlement of claims as to their incapability to deal in such situations and insufficiency as to the maximum compensation that could be awarded under such regimes. As a result, the IMO in 1969 adopted a convention dealing with the civil liability of the shipowner for pollution damage caused by such incident, known as the International Convention on Civil Liability for Oil Pollution Damage, 1969 (CLC 69).[36] The foremost objective of this Convention was to ensure that adequate compensation is paid to victims. Furthermore, a supplementary convention on settlement of claims which exceeds limits under the previously mentioned convention was created, known as the International Convention on the Establishment of an International Fund for Oil Pollution Damage, 1971 (FUND 71).[37] The Fund is made up of contributions from oil importers, who import oil to their designated destination by ships, although they are not in anyway responsible for the incidents as in the case of the shipowner. Both these Conventions were further amended with the primary intention of increasing limits of compensation due to the growth of the industry and rapid increase of the financial character of damages that caused to oil pollution victims in 1992, came into being as the CLC 92 and FUND 92. The

35 At present the IMO consists of six main bodies. The Assembly and the Council are the main organs with four main Committees, namely, the Maritime Safety Committee, Marine Environment Committee, Legal Committee and the Facilitation Committee.

36 A brief introduction of the convention and its adoption is available at http://www.imo.org/Conventions/mainframe.asp?topic_id=256&doc_id=660

37 For a summary of the convention and its amendments see http://www.imo.org/Conventions/mainframe.asp?topic_id=256&doc_id=661

new regimes substituted their predecessors after coming into force in 2003, and further led to the denunciation of the said former regimes.

The IMO was also responsible for the creation of the Convention on Limitation of Liability for Maritime Claims, 1957, 1976 and Protocol of 1996.[38] This was an essential move to introduce limits of liability on the part of the shipowner as a matter of protecting commercial interests of the trade and subsequent amendments ensured an up to date revision of financial limits based on the international economic structure and ever increasing costs of response and clean-up.[39] Apart from above mentioned treaties, the IMO also adopted the International Convention on Liability and Compensation for Damage in Connection with the Carriage of Hazardous and Noxious Substances 1996[40] (HNSC 96) dealing with compensation for accidents involving substances so mentioned in that Convention.

'Run' to the 'Bunkers'

Understanding the gravity of the problems connected with bunker oil pollution and its adverse effects that lead to damage, the IMO persuaded chiefly by the Government of Australia and several others convened the Legal Committee to draft a new international liability and compensation regime to attend to such damage. As a result, in 2001 it adopted the International Convention on Civil Liability for Bunker Oil Pollution Damage, 2001[41] (Bunkers Convention) consisting of 19 Articles together with an Annex.[42] It also adopted three resolutions dealing with connected issues that are assumed to be of vital importance for the effectiveness of the main regime.[43] The Convention is basically modelled on the CLC 69/92 and has cited certain provisions with reference to the HNSC 96. Unbelievably, the Convention was ratified by the required number of States understanding its importance and came into force with overwhelming support on November 21, 2008 opening new horizons in liability and compensation concerns.

38 For a summary of the convention and its amendments see http://www.imo.org/Conventions/contents.asp?doc_id=664&topic_id=256

39 See also P Griggs, "Limitation of Maritime Claims: The Search for International Uniformity", Lloyds Maritime and Commercial Law..., 1997 – Lloyds of London Press

40 A brief introduction of this convention is available at http://www.imo.org/Conventions/contents.asp?topic_id=256&doc_id=665

41 For a summary of the convention see http://www.imo.org/Conventions/contents.asp?topic_id=256&doc_id=666

42 The Annex deals with "Certificate of Insurance or other Financial Security in respect of Civil Liability for Bunker Oil Pollution Damage"

43 LEG 83/11, 6 September 2001

The foregoing work of this research concentrates on relevant aspects that have been covered in this new instrument with respect to 'liability issues',[44] but before illustrating those discussed contents in introduction, it would be beneficial to emphasise one of the main important strategies used in the on-going discussion in analysing substance of the Convention and relevant principles that are involved.

Interpretation of the text

Pursuant to the provisions laid down in the Vienna Convention on the Law of the Treaties 1969[45], the present Convention in question is expected to be interpreted at first sight. This would mean that both international public law and international private law rules play an important role in interpretation of this Convention, as they form the foundation for the establishment of the rules of the present Convention. In substance, the relevant Convention would be interpreted according to Article 31 of the Law of the Treaties dealing with general rule of interpretation at first sight. Secondly, approaches towards supplementary means would take effect pursuant to Article 32 of the said law. In doing so, the *travaux prėparatoires* and other related aspects would be dealt with, as the most appropriate and applicable tools. Using such an approach is inevitable in a Convention of this nature, which is indeed modelled or referred to as under another similar regime, such as in this case, the CLC and HNSC regimes.

Introduction to the relevant Chapters

The basic idea being the evaluation of the effectiveness of civil liability as established under the Bunkers Convention, the foregoing research is based on its application in maritime liability and compensation regime with regard to its implementation as gap filler. Therefore, the present study is mainly focused on liability issues, while especially concentrating on its practical aspects towards attributing liability on the responsible party and ways of obtaining compensation. In view of attaining these goals, the Chapter 2 deals with the basic elements of the Con-

44 The relevant work of the author does not deal with the insurance aspect. This is purely confined to liability issues and establishment of liability under the Bunkers Convention as a means of gap filler to the present context of liability and compensation for oil pollution damage, as the title suggests

45 United Nations, *Treaty Series*, vol. 1155, p 331, Vienna, May 23, 1969, entered into force January 27, 1980. Full text available at http://untreaty.un.org/ilc/texts/instruments/english/conventions/1_1_1969.pdf

vention and its importance in the entire regime, while Chapter 3 focuses on its scope of application.

Chapter 4 outlines the status of liable parties and victims, covering a broad spectrum of entities that fall within the ambit of the conventional meaning. Once these aspects have been properly discussed, the foregoing work concentrates on issues pertaining to institution of action and mode of adjudication. Though, it might sound strange for some readers who expect to find the relevant analysis at the latter stage, it is the idea of the author to include that aspect in Chapter 5 due to its importance in a discussion based on practical aspects, as mentioned before. Upon the realization of jurisdictional competency through the said Chapter, the discussion on requirements for the establishment of liability would be taken up in the next. Thereafter, Chapter 7 deals with issues concerning available defences on behalf of the liable party accompanied with limitation aspect that is dealt in Chapter 8.

Chapter 1: The Convention

The international maritime law for liability and compensation opened up a new chapter with the adoption of the International Convention on Civil Liability for Bunker Oil Pollution Damage, 2001[46] under the auspices of the International Maritime Organization, recognizing formally the importance of imposing liability for bunker oil pollution damage caused by ships and compensating victims of such damage in consequence. The much awaited need for an independent exclusive regime ended up successfully due to great knowledge, expertise and highly controversial legislative procedure that existed in the Assembly and in the Legal Committee[47] in general with tremendous encouragement, support and commitment of participated States, governmental and non-governmental organizations and interested entities.[48]

Persuasive efforts of the Government of Australia led to propose the importance of having a free standing treaty for bunker oil pollution damage having understood its urgency in international maritime activities, after being victimized from a spill of three hundred tons of heavy fuel oil by a bulk carrier resulted near its own coast.[49] This disastrous occurrence appeared as one of its most expensive oil spill response ever faced by Australia,[50] and furthermore, the lack of international legislative instruments to deal with the matter together with insufficient insurance covers, compelled it to forward this proposal. Number of States made submissions in favour of the Australian proposal forwarding numerous evidences as to the gravities of damages that bunker oil originating from ships could cause on and around the marine environment adjacent to coasts of such States. In sup-

---

46 *Supra* n. 46

47 In line with the work carried out aftermath the *Torrey Canyon* incident the Legal Committee was established to deal with legal issues and was made a permanent organ under the supervision of the IMO Council and has been entrusted with the formulation relevant work pertaining to liability and compensation.

48 Done at London, 23 March 2001, see also Status of Multilateral Conventions and Instruments in Respect of which the International Maritime Organization or its Secretary General Performs Depository or Other Functions, as at December 31, 2008 , L:\LED\LO\MS-Depositary\STATUS OF MULTILATERAL CONVENTIONS\2008 & 2009\Status – 2008.doc, p 445 available at http://www.imo.org/includes/blastDataOnly.asp/data_id%3D25891/Status-2008.pdf

49 LEG 73 agenda item 12, 12 October 1995. Also *Supra* n. 1 at 1

50 *Supra* Introduction, p. 1

port, the United Kingdom submitted its view alleging that such damage is of high magnitude as of tanker disasters for which the remedial instruments of the CLC 92 and complimentary Fund regimes are in existence. In view, the initial idea that prevailed among the delegation was uncertain whether to pursue with the subject as adopting this proposed instrument as of free standing nature or as an amendment or a protocol to the CLC, amendment to the LLMC Convention or to the MARPOL. These suggestions were turned down during the past activities of the Organization including the rejection of inserting bunkers in the HNS Convention, and therefore felt necessary at this initiative of having a free-standing instrument. In spite, a formal draft was presented by a group of States consisting Australia, Finland, the Netherlands, Norway, Ireland, South Africa and the United Kingdom.[51] The main aim for such motivation was to ensure that prompt and adequate compensation[52] be paid for damage caused by ships' bunkers.[53] The presented draft stood on four main principles on which the intended final outcome should rest, namely;

(a) imposition of strict liability for damage caused by ships' bunkers;
(b) application of this instrument to all ships, including those on innocent passage, subject to the compliance with the LLMC Convention, CLC or applicable national law on limitation of liability and to the UNCLOS;
(c) liability under which is to be backed up by a compulsory insurance; and
(d) the new regime to be of no conflictions with either the LLMC Convention, CLC or applicable national law on limitation of liability.[54]

During the early stages of deliberations, many views were forwarded by both State and commercial concerns, especially with regard to the main objectives and desired outcomes. Although international law was in absence of the subject, respective national laws proved to be somewhat satisfactory despite jurisdictional complications and slow recovery process of compensation.[55] The International Chamber of Shipping (ICS)[56] was of the view that a need for such an extensive co-

51 LEG 74/4/1, 9 August 1996
52 See Ling Zhu, *Can the Bunkers Convention Ensure Adequate Compensation for Pollution Victims*, Journal of Maritime Law and Commerce, April 2009
53 See *Supra* n. 5 at 1
54 *Supra* n. 49 at para. 3
55 Y. K. Lee, A Brief Introduction to the International Convention on Liability and Compensation for Bunker Oil Pollution Damage (The Bunkers Convention), available at http://www.seatransport.org/seaview_doc/SV_84/ICLC%20for%20Bunker%20Oil%20Pollution%20Damage.pdf
56 Int'l Chamber of Shipping and Int'l Federation of Shipping, Home Page, http://www.marisec.org

verage was inevitable, but creating another instrument itself would unnecessarily multiply the numbers thus prevailing. Accordingly, it was of the view that shipowners' liabilities arising from bunker spills are insured internationally with respective underwriters,[57] and the important characteristic of a new agenda should be to include all types of ships such as to include uninsured fish factory ships,[58] within a prevailing regime. The P & I Clubs held somewhat similar view opposing to an instrument of such nature citing the fact that it would pose enormous administrative burdens, and proposed that the Legal Committee should attend to practical problems involved in the process of issuing certificate of entry as a gap-filler,[59] in combating the present situation. The Greenpeace International[60] on the other side promoted the idea of a free-standing instrument supplemented by the customary law principle of 'polluter pays'[61], emphasizing the importance of marine environmental protection. It further proposed certain criteria for adoption that would best support the objectives of the Convention, namely, unlimited and absolute liability, State liability, broader coverage for environmental damages, non-preemption, and effective enforcement mechanisms. These propositions would be discussed in detail for justification in appropriate circumstance of the foregoing piece of work.

The Need

The ultimate requirement for a regime of this nature would initially rest on the severe impacts that have caused towards the urgent practical need, rather than setting up legal rules based on certain concerns or idealisms. Therefore, it is important to unveil the growing concern for the need of a regime, before addressing the important characteristics of a legal framework. Many delegations at the Legal Committee were of the opinion that the need for an international regime had not been demonstrated by the proponents to a considerable extent.[62] But however, many delegations represented their willingness towards a new regime, and some of them tabled convincing information to that end. Among them was the Australian delegation which sponsored the approach as a precautionary measure in compliance with the IMO Guidelines on incorporation of the Precautionary

57 Although 90% of the world merchant fleet is insured, the remaining uninsured fleet could have at least 90,000 to 180,000 tonnes of bunker oil on board at any time.
58 Note by the International Chamber of Shipping, LEG 74/4/4, 13 September 1996.
59 See LEG 76/3, document submitted by the P & I Clubs.
60 Greenpeace Int'l, Home Page, http://www.greenpeace.org/international
61 See *infra* pp. 17–19
62 Draft Report of the Work at LEG 75/WP.1 24 April 1997, para 43.

Approach[63] for matters relating to environmental protection. Annual figures in this area of study have shown drastic pollution claims arising out of bunker spills amounting to almost half of the entire incidents in certain parts of the world involving dry cargo ships[64], and compelling notes of the ITOPF on uncertainty of reimbursement by respective government agencies and others, the expenditure it bares for clean-up costs for bunker spills, prove negative outcomes. According to it, the world wide statistics have proven that non-tanker spills amount to a number greater than that of tanker spills, based on Oil Spill Intelligence Report in the USA.[65] On the other hand, an absence of an international regime would leave a huge vacuum in adjudication of liability in several respects, mainly, the inability of instituting direct action against the insurer due to its non-undertaking claims for breach of contract by ship-owners', inability to identify the location of ship-owners' assets, and the absence of a roll-back system arising out of the victims' inability to claim from the insurer. These deficiencies in international law lead to controversial issues of detentions of ships under arrest procedure by respective States for liability claims on *res*, increasing the associated risks for shipowners while been engaged in highly competitive international maritime business.[66]

Some of the sponsoring delegations tabled their own experiences with regard to the handling of bunker spills in their respective jurisdictions, during the meetings of the Legal Committee that attended to the matters concerning the importance of creating this conventional regime. According to the Australian delegation, 83% of oil spills occurred in Australian waters since 1975 had been by non-tankers for which an international regime on liability is not in existence. Further, its clean-up costs concerning 12 largest spills have amounted to an aggregate value of five times greater for non-tanker spills than that of tanker spills.[67] Major spills in Norway in the recent past have recorded a cost of more than US$13,500,000 out of which, a large amount being unable to recover due to the absence of compulsory insurance requisite, mode of direct actions against the insurer, ship-owner's identification and jurisdictional complications.[68] South Africa had bourn nearly US$ 10,000,000 since 1991 whilst facing similar obstacles

63 MEPC.67(37).

64 See *Supra* n. 8 at 2

65 International Oil Spill Statistics: 1994 – Oil Spill Intelligence Report published by Cutter Information Corporation, 37 Broadway, Arlington MA 02174, USA.

66 For example, detention of ships in Australia between 1992 and 1996 were 61, 72, 153, 244 and 248 respectively, while detentions in Norway during the same period being 22, 9, 37, 22 and 23, and in Sweden there were 41, 44, 77, 60 and 54. UK record high numbers such as 124, 174, 212, 206 and 184

67 Clean-up costs for tanker spills amounted to US$ 115,000 while for non-tankers an amount of US$ 625,000. See LEG 75/5/1 17 February 1997 p.3 para 13

68 See LEG 75/5/1 at p.4 para 16

in claiming from the respondents, and Sweden on the other hand had experienced major bunker spills between 1986 and 1995 where the total cost ranged from US$ 120,000 to US$ 2,000,000 out of which many were left unidentified. In the United Kingdom, the Marine Pollution Control Unit[69] has carried out clean-up operations costing US$ 1,410,000 from thirteen non-tanker spills that occurred since 1993. These figures exclude similar and parallel efforts undertook by local authorities and ports which in turn amount to a higher total value actually utilized by United Kingdom.[70] In view of the said, the respective States' experiences in handling bunker oil spills proved the necessity and urgency of addressing the need for an international regime on the subject at the IMO.

The Underlying Doctrines

Concerning the need for a competent regime on bunker oil pollution damage, and especially focusing on the main features of the proposed instrument, one may discover two main theories that run through its veins, namely the polluter pays principle[71] and strict liability[72] concept. At the initial stage of drafting, the proponents were fully aware of the fact that these doctrines shall operate as fundamental theories of the whole regime, as was evident in former instruments in the field of marine pollution liability and compensation. Therefore, it would be highly necessary to outline these underlying doctrines used in the Bunkers Convention, before landing on the main discussion concerning the nature and applicability of the said.

---

69 The Coastguard Agency's Marine Pollution Control Unit (MPCU) was formed in 1967 following the *Torrey Canyon* incident, to provide a command and control structure for decision making and response following a shipping incident that causes, or threatens to cause, pollution in UK waters. This MPCU was then restructured during the merger between Marine Safety Agency and The Coastguard Agency in 1998 to become the Counter Pollution and Response (CPR) Branch of the Maritime Coastguard Agency. See http://www.mcga.gov.uk/c4mca/mcga07-home/emergencyresponse/mcga-pollutionresponse/mcga-dops_cp_environmental-counter-pollution.htm

70 See LEG 75/5/1 at p.5 para 24–25.

71 See *Supra* n. 25. See also Organization for Economic Co-operation and Development, OECD Home Page, www.oecd.org

72 Strict liability is a concept of law where a tortfeasor becomes liable for damage and loss caused by his act or omission regardless of culpability or fault.

Polluter Pays Principle

The relevant principle doesn't seem to be new in the field of environmental law and policy. The basis for liability and compensation was well founded to be a derivate of the relevant area in both domestic and international forums for a long period of time, although it was later accepted universally as a customary rule.[73] In essence, it says nothing else than the polluter should pay for the consequential harm created by his act or omission. Having been legally derived from the work of the OECD[74], in its recommendations in 1974 and later through its policy on environment, this principle has acquired some significant recognition from the Rio Declaration commonly known as the Agenda 21[75] and the EC Treaty[76]. According to the OECD mandate, the "polluter should bear the cost of measures to reduce pollution decided upon by public authorities to ensure that the environment is in an acceptable state".[77] Although this principle belong to the domain of the public law, it deals with cost of environmental protection measures valued in terms of economic means rather than bearing a theory on *ex post facto* environmental harm.[78] In effect, the OECD's Recommendation in 1989 suggests that "in matters of accidental pollution, the polluter pays principle implies that the operator of a hazardous installation should bear the cost of reasonable measures to prevent and control accidental pollution from that installation, which are introduced by public authorities in member countries in conformity with domestic law prior to the occurrence of an accident in order to protect human health or the environment.

Even though, the Rio Declaration upheld this principle, it at the same time subjected it to several qualifications stating further that "National authorities could endeavour to promote the internalisation of environmental costs and the use of economic instruments, taking into account the approach that the polluter should, in principle, bear the cost of pollution, with due regard to the public interest and without distorting international trade and investment".[79] The European Commis-

73 Polluter Pays Principle is also known as Extended Polluter Responsibility (EPR) as described by the Government of Sweden as early as 1975. Nowadays, various taxes in the form of 'ecotax' seem to underpin environmental policy as a means of a mere fine rather than a mode of deterrence.

74 See *Supra* n. 25

75 Found in Principle 16 of the Rio Declaration on Environment and Development, adopted under the auspices of the United Nations Conference on Environment and Development (UNCED), in Rio-de-Janeiro, Brazil, in June 1992.

76 Commonly known as *Maastricht Treaty*, formally known as *Treaty on European Union* (TEU), signed February 7, 1992 and entered into force on November 1, 1993 available at http://eur-lex.europa.eu/en/treaties/dat/11992M/htm/11992M.html#0001000001

77 See *Supra* n. 25

78 See L. Bergkamp, *Liability and Environment*, Kluwer 2001, p.15

79 *Supra* n. 75, Principle 16.

sion on the other hand has endorsed a Green Paper on remedying environmental damage using this principle[80] to justify civil liability arising out of such incidents that result in causing pollution damage.[81]

These theories have well been asserted into regimes dealing with liability and compensation for pollution damage caused, in general contexts. Thereby, the polluter pays principle forms the most basic element of maritime liability regimes including the CLC, HNSC and the Bunkers Convention. In first place, it is important to identify the polluter and the real cost of the damage in order to impose liability and to allocate the appropriate costs. According to Bergkamp[82], the success of the imposition of this principle depends on the way the cost is internalised, and the objectives of the liability systems should be to deter, spread the associated risk and lower the polluter's activity.[83] However, in commercial maritime, such objectives must work in compliance with the objectives of protecting the interests of the trade, which is indeed an essential element of a respected legal regime.[84] In order to serve both these aspects, international legal drafters have taken use of the concept of strict liability subject to certain limitations based on economic terms.

Imposition of strict liability

Prior to the adoption of any of the international marine pollution liability and compensation instruments by the IMO way back in 1960s, the related matters had to be judged according to the domestic laws of the States concerned or under any prevailing international instruments that were existence.[85] At international level, such in-

80 Communication from the Commission to the Council and Parliament and the Economic and Social Committee, "Green Paper on Remedying Environmental Damage", Brussels, 14 May 1993. COM (93) 47 Final

81 See M. Hinteregger, Environmental Liability and Ecological Damage, Part I: International and Supranational Systems of Environmental Liability in Europe, Cambridge University Press, 978-0-521-88997-1 available at http://assets.cambridge.org/97805218/89971/excerpt/9780521889971_excerpt.pdf

82 *Supra* n. 78

83 *Id*, at 118.

84 See Prof. S. Jayakumar, Deputy Prime Minister, Keynote Address at the Singapore Conference on Freedom of Seas, Passage Rights and the 1982 Law of the Sea Convention (January 9, 2008) available at http://app.mfa.gov.sg/pr/read_content.asp?View,10946. See also Peter Ehlers, *Origin and Compensation of Marine Pollution – A Survey*: Pollution of the Sea – Prevention and Compensation, Hamburg Studies on Maritime Affairs, Part III, 2007, Springer Berlin Heidelberg

85 See T. O'Riordan, J. Cameron, *Interpreting the Precautionary Principle*, Earthscan Publications Ltd., London, 1994; David Freestone, Ellen Hey, *The Precautionary Principle and International Law: The Challenge of Implementation*, Kluwer, 1996;

struments were unidentifiable as admissible for several reasons although few could be taken into account with respect to limitation of tort-feasor's liabilities. These instruments were of little or no value comparing to the amount of liability that had to be bourn by such party and compensation to be paid to the victims. In such circumstance, most of the States used their own legislative tools to provide solutions to this problem, basically depending on the remedies of their own legal systems. As a result, the common law and civil law remedies were taken into consideration as the foundation to resolve such disputes by those respective States, which were able to utilize them to the extent they had been recognized by such legal system.[86]

The prevailed tools

As the two main legal systems that exist in modern legal society, civil law and common law remedies gained much of the recognition needed to solve marine pollution liability disputes. Under the English Common Law[87], a victim of a tort has three main causes of action for tortuous or quasi-tortuous liability of such nature, namely trespass[88], negligence[89] and nuisance[90]. In bringing actions un-

---

A. E. Boyle, *Globalizing Environmental Liability: The Interplay of National and International Law*, Journal of Environmental Law, 2005, 17(1): 3–26, Oxford

86 See M. Mason, *Civil Liability for Pollution Damage: Examining the Evolving Scope for Environmental Compensation in the International Regime*, Marine Policy, Vol. 27, Issue 1, Jan. 2003, p 1–12, Elsevier Science Ltd.,; A. Bianchi, *The Harmonization of Laws on Liability for Environmental Damage in Europe: An Italian Perspective*, Journal of Environmental Law, 1994, 6(1): 21–42, Oxford; J. Brunnee, H. M. Kindred, P. M. Saunders, *International Law, Chiefly as Interpreted and Applied in Canada*, 7th Ed, 2006, Emond Montgomery Publications Ltd., Ch.13, pp 1001–1008.

87 English Common Law is the legal system of England and Wales and the basis of common law legal system used in Commonwealth Countries and the United States. See Van Caenegem R. C., *The Birth of the English Common Law*, Cambridge Univ. Press, 1973 (Rep. 1988); Potter, Harold, *An Historical Introduction to English Law and its Institutions*, London, Sweet and Maxwell, 1932 (4th Ed. 1958); Maitland, Frederic, William, *English Law and Renaissance*, Cambridge Univ. Press, 1901 (Rep. The Law Book Exchange Ltd., 2000)

88 See John Kruse, *Bialiff's Law: A Lawful Trespass: v 1: Development of the Common Law to the Present Day*, Wildy, Simmonds and Hill Publishing, 2009; P. Birts, *Remedies for Trespass*, Sweet & Maxwell, 1990

89 See C. Walton, S. E. Wood and R. Cooper, *Charlesworth and Percy on Negligence: Mainwork and Supplement*, Sweet & Maxwell, 2009; A. Beever, *Rediscovering The Law of Negligence*, Hart, 2009; G. Chakrabati, *Law of Negligence*, R. Cambray & Co. Ltd., 2007

90 See D. Elvin, J. Karas, *Unlawful Interference with Land: Nuisance, Trespass and Covenants and Statutes*, Sweet & Maxwell, 2002; R. McCracken, G. Johns, J. Pereira, *Statutory Nuisance*, Bloomsbury Publishing PLC, 2008

der trespass, one should prove that the tortuous action was committed with the intent to unjustifiably interfere with the victims' property[91] in a direct manner.[92] In marine pollution cases it will not be so possible for the victim to establish the proper intent and direct interference components, as the true facts would prove to be negative to this end. Much English jurisprudence have shown such negative results in applying trespass as a fruitful cause of action to marine pollution cases.[93] Negligence on the other hand does not provide a suitable solution either, due to its requirements as to duty of care, breach of such duty and consequential loss occurred, which falls outside a provable competency.[94] In marine pollution cases there exists of a burden on the shipowners that they have a duty not to carry out any action which they might reasonably consider that may pollute the coastline.[95] In carrying out such duty of care, the ship owner is well informed of the foreseeability of damage that could occur. In the absence of a proof to the end that a breach of the duty on the part of the owner including his employee(s) responsible existed, it would be difficult for one to prove negligence on the part of the respondent. However, if it is satisfied that there existed a certain breach of duty, how could one prove that the damage which is linked to such breach has affected his private property, since environment does not fall into the subject of private possessions? In this event, negligence cannot be used as a proper cause of action in marine pollution damage litigation. However, the court in this respect would be able to apply the maxim, *res ipsa loquitur*[96], under which the rebutting the presumption of negligence shifts to the respondent. Here, the respondent will have to take his advantage of proving his difficulty of preventing the damage on matters outside his control or let it go to the victim to prove that the respondent

---

91 W. V. H. Rogers, *Winfield and Jolowicz on Tort*, Sweet & Maxwell, 12th Ed, 1984, at p. 360

92 Abecassis and Jarashow, *Oil Pollution from Ships,* Steven & Sons, 2 Ed, 1985, p. 358

93 See *Southport Corporation v. Esso Petroleum Co. Ltd.* (1956) 2 W.L.R. 93, *Fowler v. Lanning* (1959) 2 W.L.R. 249, *Letang v. Couper* (1964) 2 All E.R. 929

94 See Oil Pollution Claims in Australia: The Legal Remedies available at http://www.au.findlaw.com/article/4410.htm

95 Wu Chao, *Pollution from the Carriage of Oil by Sea: Liability and compensation*, Kluwer Law International, 1996, p. 16

96 It is a common law theory on the use of circumstantial evidence in tort of negligence, where it means in *Latin* 'the thing speaks for itself'. The use of this maxim allows the claimant to use circumstantial evidence to meet the burden of proof in negligence actions with respect to the first two elements i.e. duty of care and breach of such duty and proving that the respondent was negligent. However, the claimant would still be under the burden of proving that the respondent's negligence was the cause-in-fact of the harm and proximate cause. This is applied as a matter of law where there is sufficient evidence to allow a Trier-of-fact in arriving at a conclusion that the respondent is liable

did not take necessary measures to prevent the damage being occurred. Therefore, this would facilitate the victim to bring his case to a certain favourable extent in order to claim damages under the tort of negligence.[97]

The third option in such cases is nuisance, which is two folded as to public nuisance and private nuisance. In private perspective, it is the illegal interference with one's right to enjoy his property(s)[98]. As far as marine pollution is concerned, there would be no grounds to substantiate any property rights neither in the sea nor on the coast. However, marine pollution incidents could well be grounded under the tort of public nuisance, due to its requirements which could be satisfied under prevailing circumstances of such a case. In English Common Law, public nuisance is defined as "any nuisance which materially affected the reasonable comfort and convenience of life of a class of Her Majesty's subjects."[99] This has a broad coverage under which pollution to coasts and territorially acclaimed sea would certainly fall within such scope. But however, this would be able only if it could be proved that it affected a certain portion of the public and not to any specific person or persons on private basis. In a marine pollution incident, the damages are normally numerous and of a particular nature as Wu Chao explains[100], and this together with its general character of violating the rights of general public as described above forms the basis for the argument that such an incident falls within the ambit of a public nuisance.

As in the case of common law, fault liability[101] is in existence in civil law system[102]. This paved way for the States which have civil law system to apply it as the basis against pollution damages. Not only when the ship owner himself has made a fault, but also for the faults of his servants under vicarious liability[103].

97 See *Byrne v. Boadle Court of Exchequer*(1863) 2 H. and C. 722, 159 Eng. Rep. 299; *Fontaine v. British Columbia (Official Administrator)* (1998) 1 S.C.R. 424 (Canada); *Sanfield Building Contractors Ltd v. Li Kai Cheong* (2003) (Hong Kong); *Scott v London & Catherine Dock Co* (1865) 3 H&C 596 (Scotland)

98

99 The modern definition is found in 31–40 Archbold (2005) where it states that "[A] person is guilty of a public nuisance (also known as common nuisance), who (a) does an act not warranted by law, or (b) omits to discharge a legal duty, if the effect of the act or omission is to endanger the life, health, property, morals, or comfort of the public, or to obstruct the public in the exercise or enjoyment of rights common to all Her Majesty's subjects.

100 *Supra* n. 95, at p. 19

101 See J. Coleman, *Theories of Tort Law*, Ch. 2, *The Difference between Fault and Strict Liability*, available at http://plato.stanford.edu/entries/tort-theories/#2

102 See M. Lukas, *Fault Liability, Tort Law of the European Community*, (2008) Springer; C. von Bar, *The Common European Law of Torts*, vol. 2, Oxford, 2005.

103 It is a form of strict, secondary liability that arises under the common law doctrine of agency, *respondeat superior*, where the responsibility of the superior for the acts

Undoubtedly such existing provisions gave rise to various pollution based liabilities in maritime pollution liability laws in various civil law countries which were engaged traditionally in shipping and other marine activities. Although much of the said legislations were not of marine pollution aspects in specific terms, they could be taken into account as binding rules on adjudication of liability and compensation in conjunction with the prevailed civil law remedies. This trend was not only confined to civil law countries, but also to common law countries in general, which intended to create specific legal rules on marine pollution liabilities and consequently compensate the victims to a reasonable extent. In such circumstance, courts of law in these different jurisdictions tend to rely on both legislative instruments as well as the principles of its own legal system. English law, in particular enacted the Merchant Shipping Act 1894[104] as a remarkable step under the common law system and as a civil law country; France for example has included provisions dealing with related issues in its Commercial Code 1807. Apart from that, basically most of the States involved in maritime activities and more particularly coastal States had indeed included such provisions in their respective civil codes. States which have a mixed law system normally tend to use the more favourable rules governing liability and compensation for marine pollution in order to satisfy the fulfillment of basic rights of both parties and more importantly for reasonably compensating the victims. More noticeably it is quite clear from an analysis that absolute liability has not been enforced in any of the States or more commonly in most of these States, which have subjected such liability to various limitations depending on ones' determination as to damage that could occur within its territorial limits and the compensation that would have to be bourn by the shipowner under reasonable terms, paying attention to the interests of the trade.[105]

During the early stages of law making, ascertainment of liability on the part of the violator based on conventional provisions and providing effective compensation as a result to the victims, were the key issues that were in place. At the time of drafting a proper convention to meet these ends, what became later the CLC

---

of their subordinate, or in a broader sense, the responsibility of any third party that had the 'right, ability or duty to control' the activities of a violator. See C. Y. C. Chu, Y. Qian, *Vicarious Liability under the Negligence Rule*, Int'l Review of Law and Economics, vol. 15, Issue 3, Sep. 1995, pp. 305–322.

104 Full text is available at http://www.archive.org/stream/merchantshipping00grearich#page/n3/mode/2up

105 See M. G. Faure, J. Hu, *Prevention and Compensation of Marine Pollution Damage: Recent Developments in Europe, China and the US*, Kluwer, 2006; A. Khee-Jin Tan, *Vessel Source Marine Pollution: the Law and Politics of International Regulation*, Cambridge Univ. Press, 2006; Edgar Gold, *Liability and Compensation for Ship-Source Marine Pollution: The International System*, available at http://www.fni.no/YBICED/99_02_gold.pdf;

69[106], the primary issue was to base liability under a certain acceptable form to all participating States. As far as the victims are concerned, the best suitable option was to apply the common law approach of presumption of fault on the part of the shipowner or to make himself liable under the strict liability regime.[107] In such approach the legislators also had the option of introducing certain limitations based on accepted international conventions and national laws, coupling it with compulsory insurance that have to be taken by the responsible entity.[108] Hamburg Rules relating to Carriage of Goods by Sea[109] and the Convention Relating to the Carriage of Passengers and Other Luggage by Sea[110] adopted the system of presumed fault in the aftermath of CLC, but the latter adopted a strict liability regime although there was much persuasion from Comite Maritime International (CMI)[111] to adopt the system based on presumed liability.[112] But however, the adoption of strict liability for oil pollution damage resulted due to the grave and imminent outcomes which were dangerously affecting the well-being of the environment and persons who had to suffer from ultimate danger and other causes directly linked to oil pollution from ship-based sources. It was evident at these initial law making processes that legislators tended to opt for the imposition of strict liability as a tool of providing favourable outcomes to the victims. In the meantime it was subjected to certain limitations in order to afford some sought of comfort for the liable party for several reasons that have been discussed before, without extending its scope to absolute liability. [113]

The evolution of the concept

The roots of the strict liability concept dates back to 1865 where it lies with a much controversial judgment of the English legal system pronounced in *Rylands*

106 *Supra* n.3 and n. 36

107 See W Chao, *Liability and Compensation for Oil Pollution Damage: Some Current Threats to International Convention System*, Spill Science and Technology Bulletin, vol. 7, Issues 1–2, Jun 2002, pp 105–112; Oil Spill and Compensation available at http://www.unescobkk.org/fileadmin/user_upload/westpac/documents/News/Marine_oil_spill_Apr_09/Invited_lecturers/20_Apr/04_Liability___Compensation.pdf

108 See Colin de la Rue and P. Murray, *Introduction – Compensation for Pollution from Ships' Bunkers* in *Oil Pollution from Ships – Current Legal Issues*, Shanghai International Maritime Forum, 2005, Ince & Co., available at http://www.docstoc.com/docs/12380468/POLLUTION-FROM-SHIPS-BUNKERS

109 http://www.admiraltylaw.com/statutes/hamburg.html

110 http://www.admiraltylawguide.com/conven/passengers1974.html

111 http://www.comitemaritime.org/

112 http://www.comitemaritime.org/histo/his.html

113 See *Supra* p. 23

v. Fletcher.[114] Inadequacy of proper rules in attributing liability under then existing tort law paved way for the creation of this new concept based on the objectives of compensating the victims in events of damage in the absence of fault on the part of the respondent. Application of this judgment was initially restricted to land based incidents as the events of the founding case were confined to the use of the land.[115] In essence, the *obiter dictum* stipulates[116] that "anyone who in the course of non-natural use of his land accumulates thereon for his own purposes anything likely to do mischief it if escapes, is answerable for all direct damage thereby caused".[117] Although, this original restrictive theory wasn't expanded to include anything outside the scope of land use and the substance which cause damage, the development of the law in many respect opened up the gates in spreading the concept into many areas of damage, where the previous remedies could not be used in the absence of fault on the respondent. As a result, this rule had been used in many instances in answering liability issues involved in acts which likely to do mischief, and English case law proves the use of this rule in a variety of subject matters including oil.[118]

Prior to the imposition of strict liability in the CLC, the law makers had to face many difficulties in justifying the use of *Rylands v. Fletcher* rule. Among numerous things, they had to overcome several fundamental issues such as, applying this rule to publicly owned properties such as the sea, which was somewhat contrary to the original reference that to private properties such as lands. Hence, the carriage of oil in tankers being not non-natural use[119] and problems associated with ownership of oil as cargo and the ship itself belonging to different entities[120] created further complications. Despite all these, the international legislators opted for the imposition of strict liability in CLC 69[121] which was followed by Fund 71, CLC 92[122], Fund 92, HNS 96[123] and the newest addition, the Bunkers

---

114 (1865) 3 H. & C. 774; 159 ER 737 (Court of Exchequer); (1866) LR 1 Ex. 265 (Court of Exchequer Chamber); (1868) LR 3 HL 330 (House of Lords)

115 See J. Murphy, *The Merits of Rylands v. Fletcher*, Oxford Journal of Legal Studies, vol. 24, no. 4, 2004, 643–669

116

117 See J. Murphy, H. Street, *Street on Torts*, Oxford, 13th ed, 2007, p 471; A. R. B. Amerasinghe, *Rylands v. Fletcher in the Law of Ceylon*, International and Comparative Law Quarterly (1962), vol. 11, pp 937–975

118 *Mulholland & Tedd Ltd., v. Baker* (1939) 3 All ER 253

119 See J. Murphy, H. Street, *Street on Torts*, Oxford, 13th ed, 2007, p 474–479

120 See UNCTAD, *Review of Maritime Transport* (2008) available at http://www.unctad.org/en/docs/rmt2008_en.pdf

121 Art. III.1

122 Art. 4.1

123 Art. 7.1

Convention[124]. At national level, many States had already incorporated strict liability for oil pollution damage by ships by virtue of ratification of those regimes, with the exception of the United States which has indeed imposed this regime in its OPA 90 in addition to number of state legislations dealing with environmental hazards.[125]

Although there have been many criticisms for the imposition of strict liability for damages arising out of the carriage of oil by sea[126], there could be found many justifications to this end. The American Law Institute – Restatement of the Law (Second) upheld the imposition of this concept in circumstances of carrying abnormally dangerous activities that result in harm to persons or properties.[127] It has further clarified the meaning of 'abnormally dangerous' character of an activity including *inter alia* the existence of high degree of risk of some harm to person, land or chattels of others; likelihood of a great harm; inability to eliminate risk by means of reasonable care; extent to which the activity is not a common usage; inappropriateness of the activity with regard to the place it is carried on; and extent to which its value to the community is outweighed by its dangerous attributes.[128]

The modern use of strict liability especially in maritime liability legislations has taken a new path in applying the classical doctrine in a way suitable outcomes could be achieved.[129] There seems to be many reasons, namely its compatibility with polluter pays principle in environmental legislations, substituting the use of negligence, nuisance or trespass as a remedy which had several defects in the application to oil pollution at sea especially with regard to matters concerning proof, duty of care and vicarious liability issues and more importantly low cost and speedy recovery of damages by the victim. Although the defences that are accorded remain in the existing legislations accordingly to the doctrine of strict liability, the respondents' undertaking of the absolute liability has been simplified

---

124 Art. 3.1

125 See U. S. Environmental Protection Agency, Regulatory Information for the Petroleum Sector, available at http://www.epa.gov/regulations/bizsector/petroleum.html

126 See M. D. Goener-Oezbek, *The Carriage of Dangerous Goods by Sea*, Hamburg Studies on Maritime Affairs 12, Springer, 2008; G. Gauci, *Limitation of Liability in maritime Law: an anachronism?*, Marine Policy, vol. 19, Issue 1, Jan. 1995, pp 65–74

127 At para 519

128 See L. F. E. Goldie, *Liability for Damage and the Progressive Development of International Law*, Int'l & Comparative Law Quarterly (1965), Cambridge University Press, 14:1189–1264; P. Giliker, *Codifying Tort Law: Lessons from the Proposals for Reform of the French Civil Code*, Int'l & Comparative Law Quarterly (2008), Cambridge University Press, 57:561–582. See also *Philosophy and the Law of Torts* (G. J. Postema ed. 2002) Cambridge, 2002

129 See Modern Maritime Law and Risk Management, 2nd ed., 2009, ISBN 978 1 84311 823 7

by introducing limitations as to payable compensation.[130] This remarkable move has been done with the intention of facilitating the accession of insurance for risk and as a matter of safeguarding commercial interests.[131] Furthermore, the use of the classical doctrine in *Rylands v. Fletcher* has been restricted by introducing the rule that foreseeability[132] of the type of damage as a necessary prerequisite in order to address causation[133].

In such theory, the liable party becomes strictly liable for his acts or omissions that resulted in damage been caused, without the opportunity of laying it on issues of fault, accident or unforeseeability of the relevant incident. Especially, in a case of such scenario where the shipowner undertakes to carryout somewhat dangerous business that might end up in harm being caused to public and private interests at large, adopting a strict liability regime would bring justifiable ends. Even though one might argue that shipping is not a dangerous business or sailing a ship does not mean introducing dangerous consequences to the marine environment, effects such as oil pollution damage need to be addressed in a more effective and comprehensive manner, taking into consideration of the disastrous events that have taken place in the maritime environment[134]. It is important here to discuss why such regime was chosen by the legislators instead of using other common law approaches which were available to them at such material time.

---

130 See G. Gauci, *Limitation of Liability in maritime Law: an anachronism?*, Marine Policy, vol. 19, Issue 1, Jan. 1995, pp 65–74; Sandra Spears, *Probing the Limits of Maritime Regimes*, llyodslist.com available at http://folk.uio.no/erikro/WWW/corrgr/presseklipp/probing.pdf

131 See Ling Zhu, *Compulsory Insurance and Compensation for Bunker Oil Pollution Damage*, Hamburg Studies on Maritime Affairs 5, Springer, 2007

132 See W. P. Stastky, Ch. 2, Foreseeability in Tort Law in *Essentials of Torts*, 2nd ed, West Legal Studies, Thomson Larning, 2001, pp 13–22; *Law in Environmental Decision-Making: National, European and International Perspectives*, 127–129 (T. Jewel, J. Steele eds.,1998), Clarendon Press Oxford

133 See R. W. Right, *Causation in Tort Law*, Dec. 1985, 73 Calif. L. Rev. 1735; W. M. Landes, R. A. Posner, *Causation in Tort Law: An Economic Approach*, Journal of Legal Studies, vol. XII, Jan. 1983, The University of Chicago; S. Shavell, *An Analysis of Causation and the Scope of Liability in the Law of Torts*, Journal of Legal Studies, 1980, The University of Chicago

134 See J. Lu, H. Lim, S. C. Liew, M. Bao and L. K. Kwoh, *Oil Pollution Statistics in Southeast Asian Waters compiled from ERS SAR Imagery*, Centre for Remote Imaging, Sensing and Processing, National University of Singapore, available at http://esapub.esrin.esa.it/eoq/eoq61/oil.pdf; G. F. Birch, *Marine Pollution in Australia, with special emphasis on central New South Wales estuaries and adjacent continental margin*, Int'l Journal of Environment and Pollution, vol. 13, Nos. 1–6/2000, pp 573–607; D. R. Vieites, S. Nieto-Roman, A. Palanca, X. Ferrer and M. Vences, *European Atlantic: the hottest oil spill hotspot Worldwide*, Naturwissenchaften, Springer, vol. 91, No. 11/Nov. 2004, pp 535–538

There are numerous advantages in applying strict liability for oil pollution damage, referring to the impact it causes to the environment at large. Firstly, it can be used effectively to reduce such environmental damage in compatible with the polluter pays principle under which the ship owner becomes liable without having the opportunity of laying the onus on the plaintiff to prove, as in the case of negligence and under certain exceptional circumstances as to issues relating to vicarious liability. Secondly, it enables the victim to recover compensation in a speedier and under less expensive litigation costs. Thirdly, the ship owner could protect himself under such circumstances by obtaining liability insurances[135] based on the assumption of risks. Lastly, this principle can cope effectively with environmental hazards by way of minimizing environmental damages and justifiably compensating the victims of all natures in a more reasonable manner through environmental liability legislation, at both national[136] and international level.[137]

Although, most delegations favoured such imposition referring to the "polluter pays principle"[138] that is widely used in pollution based instruments and the requirement under the Rio Declaration,[139] this move was not entirely accepted initially by some delegations and participating organizations. Initial critics were submitted by the CMI citing several conceptual shortcomings that could topple the proper application of the whole instrument. It pointed out such issues in comparison with the CLC regime which fits into a different set up than that of the bunkers regimes. It was of the view that a reference of the shipowner as the 'polluter' in the case of damage caused by bunkers where such casualties frequently occur due to the fault of the persons other than the owner of the ship, would be misleading and may encourage an unduly emotive approach to the principle of polluter should pay. On the other hand, the absence of a second tier fund[140] as in the case of CLC regime and consequent inability to share the financial burden

135 See K. S. Abraham, *Environmental Liability and the Limits of Insurance*, Columbia L. Rev., 1988, p 942

136 See A. Alberini, D. Austin, *On and Off the Liability Bandwagon: Explaining State Adoptions of Strict Liability in Hazardous Waste Programs*, Journal of Regulatory Economics, Springer, vol. 15, No. 1/Jan. 1999, pp 41–64

137 See M. Jacobson, Director, IOPC Fund, *Forward to the First Edition of* Wu Chao, *Pollution from the Carriage of Oil by Sea: Liability and Compensation*, Kluwer, 1996, at xiii; R. Wolfrum, C. Langenfeld, P. Minnerop, *Environmental Liability in International Law: Toward a Coherent Conception*, Max Planck Institute for Comparative Public and Law and International Law, Research Paper for the Federal Environmental Agency, Nov. 2004 available at http://www.umweltdaten.de/publikationen/fpdf-k/k2942.pdf

138 *Supra* p. 17–19

139 LEG 77/WP.3, 22 April 1998, p. 24

140 e.g. ITOPF

with the oil industry would put the shipowner in an uneasy situation[141] when he is ought to meet liability in stricter terms.[142] But however, such reluctances did not scatter the motivation of most delegations who favoured the strict liability regime in upholding the idea that compensation should not be deprived in the face of any difficulties as to proof of fault. In addition, a number of delegations justified the use of the concept as the basis for liability by way of citing number of domestic legislations which have already enacted this concept[143] and furthermore opted in balancing the statutory imposition of strict liability with certain limitation regimes as gap filler to the above argument.[144]

The Instrument

The draft free-standing convention[145] presented by the sponsoring delegation[146] and the final text of the Bunkers Convention were modeled on more similar principles that had been used throughout the former regimes dealing with liability and compensation within the framework of the IMO[147], namely, the CLC 1969/1992 and HNS Convention 1996. The main difference one may find is the relevant subject, the convention has in mind to address[148]. As a starting point in analyzing the convention and substantiating the main topics at discussion throughout this paper, it is important to discover the subject matter. The entire convention deals with pollution damage that caused by bunker oil originating from a ship. It can be understood firmly by the words that have been used in the Preamble to the Convention citing:

---

141 See Oil Companies International Marine Forum: OCIMF's comments following the EP Transport Committee meeting on 28th and 29th May 2001, p. 1–2, available at http://www.ocimf.com/view_document.cfm?id=322

142 LEG 74/4/2, 9 August 1996, p.2

143 LEG 76/WP.3, 13 October 1997, p.14

144 *Id* at p 11

145 See P. Griggs, *International Convention on Civil Liability for Bunker Oil Pollution Damage 2001*, p. 2, http://www.bmla.org.uk/documents/imo-bunker-convention.doc

146 See Peter Morgan, *News from Intergovernmental and International Organizations: News from IMO: IMO Legal Committee – 80th Session*, CMI News Letter, No. 4–1999, p 7, available at http://www.comitemaritime.org/news/pdfiles/nl99_4.pdf

147 IMO: Legal Committee, 80th Session, 11–15 October 1999, http://www.imo.org/Newsroom/mainframe.asp?topic_id=280&doc_id=752

148 i.e. liability and compensation for bunker oil pollution damage caused by oil carried in the bunkers of the ship

"Considering that complementary measures are necessary to ensure the payment of adequate, prompt and effective compensation for damage caused by pollution resulting from the escape or discharge of bunker oil from ships".[149]

Therefore, it is clear that the foremost attention be paid to two very important terms used in the convention with regard to the subject matter, namely, "Bunker Oil" and "Ship".

"Bunker Oil"

This indeed the main subject matter for which an international regime for liability and compensation was inevitable in cases of pollution damage. In former international regimes relating oil pollution and liability issues, this term was used only as a part of oil that could contribute towards pollution damage rather than the main subject matter.[150] In such regimes, the term "oil"[151] was given much of the concern as a whole, citing bunker oil only as one of the potential components of the entire issue. This was due to the main fact that the bunker oil was used in its term only as the main, or more precisely the sole oil contributor to the operation of the ship.[152] It was decided in the Legal Committee at its 80th Session to provide the definition of oil as meaning "Bunker oil"[153] rather than citing it as "oil", which was presented in its original form at the initial proceedings of the work, drafted similarly to former conventions.[154] According to Article 1(5);

"Bunker oil" means any hydrocarbon mineral oil, including lubricating oil, used or intended to be used for the operation or propulsion of the ship, and any residues of such oil.

As mentioned above, the earlier proposal was to leave the term as oil and not as bunker oil, but however, due to definitional problems that could arise in contradiction with other similar regimes as to oil pollution, liability and compensation,

149 Para 6 of *Preamble*

150 Pollution damage caused by bunker oil in tankers are covered under CLC 1969 by its art. II read in conjunction with Arts. I.5, I.6 and 1.7 and Protocol 1992 to CLC 1969 by its Art. 3 read in conjunction wit Arts. 2.5 and 2.6

151 The terms as defined in CLC 1969 Art. I.5 includes "persistent oil such as crude oil, crude oil, heavy diesel oil, lubricating oil and whale oil…" while in Protocol 1992 Art. 2.5 includes "any persistent hydrocarbon mineral oil such as crude oil, fuel oil, heavy diesel oil and lubricating oil…"

152 See Bunkering Operations, http://www.pka.gov.my/Bunkering%20Checklist%20Appendix%20II.pdf

153 *Supra* n. 2

154 This change was made due to the suggestions made by the Finnish delegation at LEG 80, and the term 'Bunker' was subsequently used within square brackets.

the decision was more successful this time, overcoming the burden of interpreting what has been meant by this particular regime, separating the definitions provided by the said other former regimes for the same term of 'oil'. More clarifications as to this effect were given by the Legal Committee during its 81st Session, where it stated that the text was updated to ensure that the subject matter is properly addressed throughout the entire text as it is the oil including lubricants which is used for the operation or propulsion of the ship and any residues of both bunker and lubricating oils.[155]

Although, this definition seems to be more straight forward and easy to understand in a general context, it was not the case with regard to the former regimes on oil pollution, liability and compensation, which paved way for the proposed text at the inception of its work for the Legal Committee. At its initial meetings much emphasis were made in order to draft Bunkers Convention taking into account of the success and somewhat similar provisions of former regimes such as the MARPOL 73/78, CLC 69/92 and the HNS Convention. But however, understanding the more basic requirements for which the present convention in focus is inevitable, the Legal Committee paid much attention towards creating rules restricting to the sole application of the desired objectives. Therefore, certain terms had to be eliminated from the present regime with regard to the term "Bunker oil" than that of the earlier ones, and confine strictly to the subject matter. Obviously, it must have been the reason to include the term "Bunker oil" rather than leaving it as "oil" on common terms. Nevertheless, it is important to analyze the existing previous regimes and further compare what categories of oil fall within the present convention and what fall outside.

The CLC 69 defines 'oil' with reference to specific categories that are carried in tankers as cargo or oil in bunkers, namely crude oil[156], fuel oil[157], lubricating oil[158] and whale oil[159].[160] But here, the important difference that could be contrasted from the Bunkers Convention is the requirement of the oil under the CLC 69 confining it to be 'persistent oil'[161]. This requirement was further extended under the Fund Convention 71 defining oil as persistent hydrocarbon mineral[162] oil.[163]

155 LEG 81/4, 21 January 2000
156 See *What is Crude Oil*, http://www.opec.org/library/FAQs/CrudeOil/q1.htm; *Petroleum*, http://en.wikipedia.org/wiki/Petroleum
157 See *Supra* n. 2
158 See *Lubricating Oil for Small Scale Live Steam Locomotives*, http://www.southern-steamtrains.com/notes/lubeoil.htm
159 See *Whale Oil*, http://en.wikipedia.org/wiki/Whale_oil
160 Art. 1(5)
161 http://www.itopf.com/marine-spills/fate/persistence-of-oil/
162 See http://en.wikipedia.org/wiki/Hydrocarbon
163 Art. 1(2).

Much of the attention was focused on this issue as to what types of oil fall within the meaning of persistent oils, and further, scientific illustrations with regard to the nature of persistent hydrocarbon mineral oils.

In a more scientific language, Crude oil is a complex of hydrocarbons with 4 to 26 or more carbon atoms[164] in one molecule[165]. In addition, non-hydrocarbons such as Sulphur and Vanadium compounds can also be found in crude oil representing to around 25 percent of the oil. Therefore, hydrocarbons are the largest contributors that are present in chemical components of crude oil. In the refining process of crude oil, many different types of oil varying from less weighing gasoline to heavily weighing tar can be found as results. Bunker fuel is a product of the intermediate process, which becomes the residue of Naphtha used in the petrochemical industry. Water soluble components of crude oil and other refined outputs consist of many different compounds which are toxic to plants and species of the marine environment. Scientific experiments have proved that aromatic substances are more toxic than middle-molecular substances and high-molecular substances such as tar. In view of such results, it is a proven fact that diesel[166] fuel with its high aromatic content is more damaging than bunker fuel, which indeed has a low aromatic content.[167] Nevertheless, the toxicity of fuel used in ship bunkers, namely the bunker oil, poses severe threats to the livelihood of marine environment[168] including plants and species[169] at first sight, and thereafter poisons the organisms on the beach, once this oil comes in contact with beaches although its

164 http://www.edinformatics.com/math_science/c_atom.htm

165 http://www.historyoftheuniverse.com/molecule.html

166 http://en.wikipedia.org/wiki/Diesel_fuel

167 See J. G. Speight, *The Chemistry and Technology of Petroleum*, 4th ed, Taylor & Francis Group, CRC Press, 2007, pp. 757–758

168 See G. J. Mulkins-Phillips, J. E. Stewart, *Effect of Environmental Parameters on Bacterial Degradation of Bunker C Oil, Crude Oils, and Hydrocarbons*, Appl Microbiol, 1974 Dec.; 28(6): 915–922, corrected by Appl Microbiol, 1975 April; 29(4): 569; P. K. Shin, *Effects of a spill of bunker oil on the marine biological communities in Hong Kong*, Environment Int'l (USA), vol. 14:6, pp 545–552; Finnish Environment Institute, *The Impacts of Oil on the Marine Environment*, Doc. 9/14/2009, available at http://www.ymparisto.fi/print.asp?contentid=334186&lan=en&clan=en

169 See C. J. Strobel, A. H. Brenowitz, *Effects of Bunker Coil on Juvenile Horseshoe Crabs (Limulus polyphemus)*, Estuaries and Costs Journal, vol. No. 2/ June 1981, Springer; J. A. Schwartz, B. M. Aldridge, J. L. Stott and F. C. Mohr, *Immunophenotypic and functional effects bunker C fuel oil on the immune system of American mink (Mustela vison)*, Veterinary Immunology and Immunopathology, vol. 101, Issues 3–4, Oct. 2004, pp. 179–190; A. H. Lasday, Texaco, Inc,; E. W. Martens, CheyronResearch Co., Fate and Effects of Oil on Marine Life: Progress Report on Research Sponsored by the American Petroleum Institute, Off-Shore Technology Conference, 3–6 May 1976, Huston, Texas, Paper No. 2449-MS, available at http://www.onepetro.org/mslib/servlet/onepetropreview?id=OTC-2449-MS&soc=OTC

toxicity decrease due to the length of the period that it takes to reach coastlines. Depending on the place of the tragedy, the impact of the effects pose towards the marine environment as a whole varies, but however, in any case, the major part of the damage remains on one side of the marine environment, either the beach or living and non living species of it.

Crude oils fall into the category of 'persistent oils' and for a long period of time it was considered to be of persistent nature, until 1993 in Cohasset and Panuke fields off Nova Scotia where a new grade of crude oil was produced and found to be of non-persistent nature. The term "persistent" became a major point of argument during and after the CLC 69 and Fund Convention 71 were drafted, for which no one was able to provide a proper Interpretation. The absence of a real definition in both these conventions made it somewhat difficult for interested parties to understand this term and properly execute applications for claims against pollution damages. It resulted in drafting a practical guide under the Fund Convention 71 providing instead a definition for 'non-persistent oil'.[170] The said defines "non-persistent oil" as "oil which, at the time of the shipment consists predominantly of non-residual fractions and of which, more than 50 percent by volume distils as a temperature of 340 degrees centigrade when tested by the ASTM Method-D 86/78 or any subsequent revision thereof".[171] This shows that the persistent nature depends on the volume of distillation of the oil at a given temperature and the component of which needs to be of non residual fractions. Wu Chao correctly points out that the term 'persistent' refers to "oil which considered damaging by virtue either of the proportion which remains in the marine environment after evaporation has ceased or of the time which it takes to dissipate naturally".[172]

Under the above method of classifying persistent oils, several categories of oils have found to be of persistent nature, such as crude oil, marine diesel oil[173], No. 4 Fuel Oil[174], residual oils (e.g. marine fuel oil, Bunker 'C'), No. 5 Fuel Oil[175], most lubricating oils, Bitumen[176] and Whale oil.

---

170 The practical guide was prepared in 1981 under the Fund Convention 1971

171 *A Non Technical Guide to the Nature and Definition of Persistent Oil*, p.10

172 Wu Chao, "*Pollution from the Carriage of Oil by Sea: Liability and compensation*", Kluwer Law International, 1996, p. 86

173 See United States' Environmental Protection Agency (EPA), *In-Use Marine Diesel Fuel*, Air and Radiation, EPA420-R-99-027, available at http://www.epa.gov/oms/regs/nonroad/marine/ci/fr/dfuelrpt.pdf

174 See Hess Material Data Sheet, *No. 4 Fuel Oil*, MSDS No. 15054, available at http://www.hess.com/ehs/msds/15054No4FuelOil.pdf

175 See Hess Material Data Sheet, *No. 5 Fuel Oil*, MSDS No. 0332, available at http://www.hess.com/ehs/msds/0332No5FuelOil.pdf

176 See K. Kris Hirst, *Bitumen: A Smelly but Useful Material of Interest*, http://archaeology.about.com/od/bcthroughbl/qt/bitumen.htm

The CLC 92 has inserted the same definition, but with the omission of Whale Oil from CLC 69, and further, broadening the term as 'any persistent hydrocarbon mineral oil' taken from that of the Fund Convention 71, which in turn paved way for such omission of Whale Oil that do not fall within the meaning of 'hydrocarbon mineral oils'. A more different approach is evident in the OPA 1990, which has provided a much broader definition to the term of 'oil', ensuring that no over-laps could occur between that and that of the CERCLA.[177] Section 2701[23] of the OPA 1990 provides this broader definition, stating that any kind or any form of oil, without limiting to the list of oils it has stated within this definition, and further separating types of oils which had already been addressed under a different regime, namely the CERCLA, can be applied for claims under such categories of oil involving an incident.[178] The United States legislature has had not paid any attention to the term 'persistent' nor 'hydrocarbon mineral oil', thereby prevented itself from using these terms, thus giving a description of the types of oil that fall within the definition of 'oil'. Therefore in international context it is led to confusion, more than such prevails in certain domestic laws; mainly the United States, which is not a party to CLC 92 and its supplementary arrangement, namely the Fund Convention 92 as in force at present and the Protocol 2000[179].

As long as Bunker oil is concerned, the use or non use of the term 'persistent' does not exists of due to the nature of bunker oil being persistent in character. Unlike in the situation of tankers, where different kinds of oil been traded worldwide[180], ships in particular consume diesel oil and/or bunker oil depending on its dynamics. This issue was noted by the Legal Committee at its 82nd Session, in dealing with the minimum gross tonnage issue and ascertaining the types of ships that would fall into the category of conventional ships under the proposed

177 See, United States Code Congressional and Administrative News (101st Congress, 2nd Session 1990) Volume5, p. 780.

178 Section 2701[23], 33 U.S.C. 40 defines 'Oil' as meaning oil of any kind or in any form, including, but not limited to, petroleum, fuel oil, sludge, oil refuse, and oil mixed wastes other than dredged spoil, but does not include petroleum, including crude oil or any fraction thereof, which is specifically listed or designated as a hazardous substance under subparagraph (A) through (F) of section 101(14) of the Comprehensive Environmental Response, Compensation, and Liability Act (42 USC 9601) and which is subject to the provisions of that Act.

179 Protocol of 2000 to the International Convention on the Establishment of an International Fund for Compensation for Oil Pollution Damage, 1971, Leg/Conf.11/6, 27 Sep. 2000, available at https://www.ccaimo.mar.mil.br/secimo/convencoes/Prot2000.pdf

180 See M. Stopford, *Maritime Economics*, Routledge, 1997 (Rep. 1999, 2000) pp 406–412

Bunkers Convention.[181] A survey conducted by Hong Kong and China, proved that most of the ships under the gross tonnage of 2,000 use only diesel as fuel due to its cost effective character, and engines above this level use persistent oils as fuel, which could easily accommodate the installation of fuel treatment plants[182], that cannot be installed in small ships. Nevertheless, the main idea behind the definition of "Bunker oil" in the present Convention is not to limit the use of the scope of the term to any specific kinds of fuel used in the bunkers of a ship, but to include all and any kinds of fuel that may be used or intended to be used for the operation and propulsion of the ship in general. In this context, the term "Bunker oil" would certainly have a wider scope of use than leaving it as meaning "persistent hydrocarbon mineral oils". The ultimate objective of this Convention is to include all types of hydrocarbon mineral oils, lubricating oils and their residues, without taking the burden on the use of specific types of oils that would exclude some categories of oils used by small ships and thus lead to fall outside the scope of this Convention. Therefore, it is quite clear that the use of the term "Bunker oil" in this Convention does not mean Bunker oil according to scientific identification as "Bunker C" or any similar or other substance, but fuel that is been used by the ship for its operation or propulsion in a more general context.

Considering the nature of damages, which possibly occur due to bunker oil spills been a major issue aftermath of several marine disasters in the recent decades[183], a more positive approach towards imposing liability burdens on the liable party and subsequently demanding proper compensation was inevitable for a long period of time. During the initial stages of drafting, the International Union for Conservation of Nature and Natural Resources (IUCN)[184] pointed out the importance of welcoming a convention for the protection of marine environment from ship-source pollution, as a measure to fill the gap that exists in this area of concern. It has stressed the need for a new convention in such context due to the fact that all ships carry fuel oil, which pose tremendous difficulties in the clean up process and the ultimate damage it causes, than the oil transported as cargo in tankers.[185] Understanding such requirements in a international set-up has led to adopt rules pertaining to liability and compensation for bunker oil spills, and further dedicate the application of the new convention solely to damages occur from such bunker oil. The recognition of the importance of such requirements and subsequent consideration paid with respect to ensuring adequate, prompt and effective compensation for damages arising out of 'Bunker oil' spills from ships is evident from the

181 LEG 82/3, 30 June 2000

182 See http://nudeal.com/selling-lead/ywc-type-bilge-water-treatment-plant-used-ship

183 See *Supra* pp 1–2

184 See IUCN Home Page http://www.iucn.org

185 LEG 79/6/3, Importance of the new convention.

preamble to the Bunkers Convention[186]. Any application for compensation under this convention need to prove that the damage was caused as a result of bunker oil linked to the causation discussed in the Bunkers Convention 2001.[187]

"Ship"

The preliminary element around which the entire regime operates would be the main issue on focus in finding the basics of the subject matter in a discussion. As far as these basics are concerned, maritime activities wouldn't have fall into such discussions if ships were not in existence. Dating back to several centuries, ships were considered an important mode of transportation for different purposes[188] from social reasons to commercial activities, and have developed from floating wooden platforms to modern day high technological machines with enormous capabilities.[189] From the inception of historical rules of customary laws up to modern contemporary binding legal principles dealing with maritime activities, the term 'ship' has not only been a word to define but also a debatable subject as a whole. Although, defining of or searching for the acute meaning had lost its significance through a continuing development of the subject matter, one could still argue on its real definition pertaining to different applications it is been used.[190] These arguments were evident during international law making processes at various junctures of maritime law, where the delegates of different origins put forwarded their opinions[191] based on both international competence and domestically available legal definitions. The Bunkers Convention is no exception of such an ongoing conflict of interests.

Accordingly, the Legal Committee adopted a much simple and basic interpretation for the term "ship"[192] instead of making restrictions based on different aspects, although there were quite an amount of definitions put forwarded by respective delegations. Within its meaning;

186 "[R]ECALLING ALSO article 235 of that Convention, which provides that, with the objective of assuring prompt and adequate compensation in respect of all damages caused by pollution of the marine environment,…

187 Discussed in detail in Ch. 5, *infra* pp. 158–172

188 See *Supra* n. 180, Ch. 14, pp 567–612

189 See R. J. Dupuy, D. Vignes, Hague Academy of International Law, *A Handbook on the new law of the sea*, vol. 2, Kluwer

190 See B. Grant, *What is a "ship": R v. Goodwin in the Court of Appeal*, [2006] 2 Web JCLI, available at http://webjcli.ncl.ac.uk/2006/issue2/grant2.html ; R. Shaw, *What is a ship in maritime law*, JIML 11(2005), 4:247

191 See *Supra* p 30

192 See *Infra* n. 193

"Ship means any seagoing vessel and seaborne craft, of any type whatsoever".[193]

With respect to the above definition, it is not possible to leave it without any objections on the basis that it provides a simple and basic meaning. In general, it is clear that the ship in this case shall be of a sea going character and not anything which is made solely for the use within internal waters[194]. The complications seem to exist once a careful observation is carried out with respect to the other terms used within this definition, such as the meaning of 'seagoing vessel' and 'seaborne craft'[195]. Throughout the historical development of the definition of ship, it has proven drastic changes in different applications depending on the purpose and desired outcomes. Drawing the attention to some of these main developments would provide certain justifications in easing the present task. The MARPOL definition is somewhat fundamental providing a list of marine vehicles that are in existence with different modes of dynamics. Accordingly, a 'ship' is defined as any vessel of any type whatsoever operating in the marine environment and includes hydrofoil boats, air-cushion vehicles, submersibles, floating craft and fixed and floating platforms.[196] The earliest difficulty of separating the term vessel from that of ship constituted much confusion among interested parties of both legal and technological expertise. In classical theories of law, much domestic legislation defined these terms separately in trying to minimize ambiguous and over lapping situations. The Merchant Shipping Act 1894 is quite an example, where it defined 'vessel' and 'ship' separately in its section 742, as a 'Vessel' includes any ship or boat, or any description of vessel used in navigation, and a 'Ship' includes every description of vessel used in navigation not propelled by oars. It is quite evident from this that the English Law has not been successful in overcoming the said ambiguous and over lapping situations, *prima facie*, with complicated issues still been in existence. The attempt made by the CLC 69 was successful to a certain extent in its interpretation as, "any seagoing vessel and seaborne craft of any type whatsoever..."[197] This covers a much wider area of application without causing much confusion as to different notions. Being the grand-father model definition that was used and still been in existence in subsequent conventions, such as the CLC 92[198], HNS[199] and at present in the Bunkers Convention, this has proven much success, overcoming complicated si-

193 Art. 1(1).
194 As interpreted by UNCLOS, Art. 8 it means "...waters on the landward side of the baseline of the territorial sea form part of the internal waters of the State."
195 See *Infra* pp 40–43
196 Art. 2(4)
197 *Supra* n.193
198 Art. 2.1
199 Art. 1.1

tuations that existed in the past. Consequently, The Merchant Shipping Act 1995 has changed its classical interpretation absorbing some of the internationally agreed derivatives and the amendments, by virtue of section 2(1) (a) of Schedule 4 of the Merchant Shipping (Registration, etc) Act 1993 which reads as, a "ship includes every description of vessel used in navigation."[200] This change has taken the effects of interpreting a ship within the modern international usage of the terms as to sea-going and seaborne craft of whatsoever. Depending on the purpose of the conventional objective realizations, the whole definition vary from one convention to the other, that would be discussed later, but the main types of ships have been confined to sea-going vessels and seaborne crafts of any type whatsoever. As Abecassis rightly argues, a vessel which is exclusively river-going would not fall within the category of ships the 1969 Convention was intended to be applied.[201] This idea followed in the subsequent conventions as stated before, but however, the HNS and the Bunkers Convention remain as the only international conventions that do not impose limitations as to other requirements so mentioned in CLC 69[202] and 92[203]. Obviously, one has to pay attention to the desired outcomes of those regimes that were evident especially with respect to their conventional goals. The approach that has been adopted in the OPA 1990 is somewhat different but pays attention to the term 'vessel' as the main subject matter of concern, and furthermore provides a much wider definition to its effect. According to it, "vessel" means every description of watercraft or other artificial contrivance used, or capable of being used, as a means of transportation on water, other than a public vessel.[204] This indeed would include oar-propelled vessels and also non self-propelled vessels.

As far as the Bunkers Convention is concerned, the requirement of the vessel being a sea-going one or a seaborne craft needs to be fulfilled in order to institute any action under the regime. Being the definition proposed by draft Article 1(1) in LEG 78/5/2, this falls identical to the meaning provided by the HNS. The main absence in both conventions is that there is no definition or clarification to the terms "sea-going" and "sea-borne". Although it was discussed in the Legal Committee during its 79th Session, whether it means a ship that is (a) at sea; (b)

200 Sec. 313 of the Merchant Shipping Act now provides the definition of 'ship'.
201 D.W. Abecassis, *supra* n. 92, p 356
202 "...actually carrying oil in bulk as cargo"
203 "...constructed or adopted for the carriage of oil in bulk as cargo, provided that a ship is capable of carrying oil and other cargoes shall be regarded as a ship only when it is actually carrying oil in bulk as cargo and during any voyage following such carriage unless it is proved that it has no residues of such carriage of oil in bulk aboard"
204 Sec. 1001(37)

designed for operation on the high seas[205]; or (c) beginning or completing a voyage by sea;[206] much attention has not been paid entirely to neither of these aspects. The main question at that juncture was not with regard to the gravity of the harmful effects of oil pollution that occur within the territory or territorial sea[207], but where a ship moves from one location to another within the territorial sea at the time an incident occurred be still considered as 'sea-going'. Unlike 'tankers' which not only use oil for its operation and propulsion, but also as a means of transporting oils, vessels and crafts falling under the meaning of the Bunkers Convention must have the ability to operate and propel using bunker oil, and at the same time to posses the ability of sea-going or of seaborne character in order to qualify for its application. This clearly represents vessels and crafts of only such category within the meaning of the convention and further consideration of anything which falls outside, such as yachts, sailing boats and other types of boats used for pleasure, fishing and other purposes in the absence of any dynamics to use bunker oil to operate or propel, are of no significance.

The main opposition that could pose some confusion or ambiguity within this meaning would be on Mobile Offshore Units[208] (MOUs).[209] This was in debate during the drafting of CLC 69 and 92, and clear clarifications as to their inclusion within the meaning of ship were given upon extensive debate. These MOUs have many different characteristic features and operating systems depending on the purposes they've been used. With regard to Floating Storage Units (FSUs) and Floating Production, Storage and Offloading Units (FPSOs), which do not have the ability to propel or steer on its own, does not lack the criteria of being a sea-going vessel or seaborne craft at first sight. But however, the situation under the Bunkers Convention could only be isolated in such circumstance that they fail to fulfill the requirement of using bunker oil for its operation or propulsion. During the 79th Session of the Legal Committee, the International Association of drilling Contractors forwarded its comments on the situation with regard to Mobile Offshore Drilling Units (MODUs)[210], and further attention was paid by the Committee whether such units that are not buoyant during substantial period of operations could still fall within the meaning of 'ship' under the Bunkers Con-

205 See Part VII, Art. 86 of UNCLOS

206 LEG 79/6/2, 12 February 1999

207 See LOSC 1982, Part II, Articles 2–4

208 A Mobile Offshore Unit is an offshore unit that is not intended to be placed permanently or long term on the field, but is designed to be moved from one location to another whether or not it has any means of propulsion or of lowering legs to the seafloor

209 See S. Flogeland, *DNV Classification of mobile Offshore Units*, http://www.dnv.ru/Binaries/dnv_classification_mobile_offshore_units_rev2_tcm83-10777.pdf

210 See http://www.mms.gov/ooc/Assets/KatrinaAndRita/BackgrounderMODU.pdf

vention. Although the insertion of these units within the compulsory insurance requirement was avoided by the CLC regimes, the Committee's view was that these would still need to maintain compulsory insurance as required by Article 7 of the Bunkers Convention, referring to their operations in the Exclusive Economic Zone[211] (EEZ) and subsequently becoming a subject under the control of the coastal State.[212] Therefore it is quite clear from such requirements that MODUs fall within the meaning of 'ship' under this Convention, and further, any such unit would fall into the meaning when they are actually engaged in floating on sea waters, to the extent such units use bunker oil for their operation or propulsion irrespective of been stationed for certain period without operating as vessels or crafts. In addition, the definition would cover wrecks[213] of vessels if at the time of the incident, the wreck was actually carrying bunker oil which was used or intended to be used for the operation or propulsion of that ship from which the wreck broke out. The main objective being the inclusion of all types of vessels and crafts within this meaning as a whole, and to provide a wide coverage, while exceptions seem to be extremely less than that were observed in the CLC regimes. Nevertheless, the situation with regard to flying-boats[214] would still be a question that needs much consideration as seen in the English case of *Polpen v. Commercial Union*,[215] where it was held that the main purpose of a flying boat been flying rather than navigation[216], and therefore do not fall within the meaning of a 'ship' in a marine insurance contract.[217] This question reverts to the earlier discussion, whether the definition requires a sole sea-going character or quasi sea-going character as in the case of stationed crafts and units. Hence, it further complicates the situation in ascertaining whether the main objective of the ship need to be navigation or one of the purposes been navigation is sufficient. The definition of 'ship' does not provide a proper answer to this situation, but however, one may consider that while such boat is on the surface of the sea, it could be termed as a floating unit that may fall into the definition itself, and therefore

211 See Art. 55 of LOSC 1982; D. J. Attard, *The Exclusive Economic Zone in Inetrnational Law*, Oxford University Press, 1986; B. Kwiatkowska, *The 200 mile Exclusive Economic Zone in the new Law of the Sea*, Springer, Dordrecht, 1989

212 *Supra* n. 205

213 A shipwreck is the remains of a ship that has wrecked, having either been sunk or beached

214 See E. Alkema, *Seaplane a "Vessel" for Purpose of Salvage*, Michigan Law Review, vol. 52, No. 8 (Jun., 1954), pp 1229–1230

215 [1943] 1 K.B. 161

216 See M. Ashford, *A Jet Ski; a vessel, boat or ship? R v. Goodwin (2006)1 Lloyd's Rep. 432*, https://maritimejournal.murdoch.edu.au/index.php/maritimejournal/article/viewFile/22/43

217 See S. Hodges, *Law of marine insurance*, Cavendish, 1996, p 317

constitutes the requirements needed under this Convention subject to the type of its operational fuel falls within the category provided under the meaning of 'bunker oil'.

Although all the requirements are fulfilled, governmental ships operating on non-commercial service[218], warships[219], naval auxiliary or other ships owned or operated by a state[220], and used for the time being for such purposes do not fall within the scope of application of the Bunkers Convention,[221] which would be discussed later. This situation remained the same in all previous private liability conventions[222] respecting the status of a sovereign state and principles of 'sovereign immunity'[223].

---

218 See A. Aust, *Handbook of International Law*, Cambridge, 2005, pp 317–318

219 Id; see also Tim Hillier, *Sourcebook on Public International Law*, Routledge – Cavendish, 1998, p 422; A. Watts, *The International Law Commission: 1948–1998*, 1999

220 Warship is defined in Ar. 29 of LOSC as a ship belonging to armed forces of a State; bearing the external markings distinguishing the character and nationality of such ships; under the command of an officer duly commissioned by the government of that State and whose name appears in the appropriate service list of officers; manned by a crew which is under regular armed forces discipline. Naval auxiliaries are vessels other than warships, that are owned by or under the exclusive control of the armed forces, and because they are state owned or operated and used for the time being only on government non-commercial service, auxiliaries enjoy sovereign immunity, the same as warships

221 See Art. 4(2)

222 See Art. XI for CLC 69; Art. XI of CLC 69 remained unchanged in Protocol 92; Art. 4.4 for HNSC 96

223 See C. M. Vazquez, *Sovereign Immunity, Due Process, and the Alden Trilogy*, Yale Law Journal, vol. 109, No. 1927, June 2000; R. S. Neyland, *Sovereign Immunity and the Management of United States Naval Shipwrecks*, http://www.history.navy.mil/branches/org12-7h.htm ; E. Lauterpacht, *International Law – Sovereign Immunity – Proof of Interest*, The Cambridge Law Journal (1955), 13: 19–23

Chapter 2: Scope of Application

The foremost and most vital issue that need to be addressed formally in a discussion relating to a convention or its implication in a given situation rests with the scope of application of the particular subject. As far as legal instruments are concerned, scope of application plays an important role in all spheres with respect to enforcement of laws, administration of justice and adjudication, in both domestic and international legislations. Its role in international maritime legislations pertaining to liability and compensation is no exception, but in fact forms highly sensitive issues. Proper functioning of such instrument greatly depends on its scope of application, and its veins run through the whole regime touching most of the important sections that govern its successful implementation. The scope of application is multi faceted, inter-dependant and involves most vital aspects of the instrument. With regard to the present work in dealing with the Bunkers Convention, the next juncture would be to analyze its scope of application for the purposes of realizing the application, implementation and desired objectives under the Convention.

(a) Geographical Scope: *jurisdiction ratione loci*

Article 2 of the Convention, which is originally based on Article II of the CLC '92 and follows its same pattern, expressly declares that;

"This Convention shall apply exclusively:
(a) to *pollution damage* caused:
  (i) in the *territory*, including the *territorial sea*, of a *State Party*; and
  (ii) in the *exclusive economic zone* of a *State Party*, established in accordance with *international law*, or if a *State Party* has not established such a zone, in an area beyond and adjacent to the *territorial sea* of that State determined by that State in accordance with *international law* and extending not more than *200 nautical miles* from baselines from which the breath of its *territorial sea* is measured;
(b) to *preventive measures, wherever taken*, to *prevent or minimize* such damage. (emphasis added)

The above provision which emphasizes the scope of application of the Convention has within its meaning, several important branches that run through different aspects of the topic in discussion. Initially, it would be meaningful to take out the geographical scope with a view of ascertaining the physical operation of the Convention itself. The application of the geographical scope could be separated into three main areas that have been mentioned as falling within, namely, the "territory", "territorial sea"[224], and the "exclusive economic zone"[225].

"territory"

In absence of a clear definition of what it means, the general interpretation of territorial sovereignty of States[226] could be used to fill the gap as seen in entirely all former maritime liability regimes. Acceptance through international customary law[227] as to the recognition of ones' territorial integrity and sovereignty in international context has eased any difficulties in understanding the present phrase. However, the public international law of the sea derivatives as to internal waters[228] regimes has penetrated by virtue within private law perspectives enclosing them in ones' territorial jurisdiction as to its territory. Furthermore, states with archipelagic waters[229] would eventually gain territorial sovereignty to such waters as respected under international customary law.[230] In the entire context, coastal states enjoy sovereign rights within its territory without any external ex-

---

224 See Arts. 2–4 of LOSC

225 See *Supra* n. 211 at 42

226 See J. E. Thomson, *State Sovereignty in International Relations: Bridging the Gap Between Theory and Empirical Research*, International Studies Quarterly (1995), Blackwell Publishers, 39, 213–233; T. J. Biersteker, *State, Sovereignty and Territory* in *Handbook of International Relations* (W. Carlsnaes, B. A. Simmons, T. Risse, 2002 ed. 2006), SAGE Publications; M. Anderson, *Frontiers: territory and state formation in the modern world*, Blackwell, 1996 ed. 1997

227 See A. E. Roberts, *Traditional and Modern Approaches to Customary International Law: A Reconciliation*, The American Journal of International Law, vol.95, 757, available at http://www.asil.org/ajil/roberts.pdf; Malcolm Shaw, *International Law*, 6 ed., 2008, Cambridge

228 Art. 8, LOSC

229 Art. 49, LOSC

230 See *Supra* n. 189, Part I; see also M. Shaw, *International Law*, 5th ed., Cambridge, pp 502–504; J. O'Brian, *International Law*, Ch. 13, 401–403, Cavendish, 2001 rep. 2002; Prof. B. H. Oxman, *The Development of the Modern Law of the Sea* in *Law of the Sea*, UN/UNITAR Fellowship Programme in International Law, The Hague, 2008, http://www.law.miami.edu/news/images/999images/Oxman%20UNITAR%20Outline.pdf

ceptions, including *inter alia*, the land, internal waters as mentioned above including and comprising of bays, ports and estuaries and waters enclosed by straight baselines[231].

"territorial sea"

Likewise the former liability regimes, the Bunkers Convention too is silent with regard to the definition and breath of the territorial sea. It clearly proves the intention of the Legal Committee to copy the same text with respect to the scope of application directly from the CLC and include it in its draft[232], though not at the initial stage, but followed right throughout the entire negotiation period till the final stage. Although, international law was developed rapidly since the adoption of the CLC 69, and the later protocols which amended it made significant improvements to that effect, and none of these succeeding instruments including the present Convention has made significant changes to include either a proper definition or a numerical value to its breath. On the other hand it would be proper to note that such absence would have been made to gain the participation and interests of states which do hold a 'persistent objector'[233] to such extent, and in turn with the objective of gaining maximum ratifications.

The classical doctrine has tend to deteriorate gradually upon the signing of the Montego Bay Convention of 1982[234], which imposed a 12 nautical mile breath for the territorial sea that has been accepted by Contracting States, [235] and subsequently enacting their national legislations to that effect, implying it to have been derived from international customary law[236]. Presently, all coastal States

231 See R.R. Churchill, A.V. Lowe, *The Law of the Sea*, 3 ed., 1999, p 60

232 The scope of application under the initial draft was based similarly as the HNS but later amended in the same pattern as the CLC 92 due to the views expressed by the delegations and especially by the International Chamber of Shipping, that the proposed Bunkers Convention should be modeled as the CLC regime in order to cover the same types of claims, and as a matter of bringing it in line with the need of filling the gap between compensation regime for tankers and other types of ships for which the proposed convention is aimed at adopting. See LEG 74/4/4 dated 13 September 1996

233 See *supra* Churchill, Rowe, n. 231, p 78

234 See http://www.un.org/Depts/los/convention_agreements/convention_overview_convention.htm

235 Both the 1930 Hague Conference and 1958 Geneva Convention on the Territorial Sea and the Contiguous Zone failed to reach an international agreement on the breath of the territorial sea

236 See United Nations, Division for Ocean Affairs and the Law of the Sea, *Digest of international cases on the law of the sea, Part 54*, United Nations, 2007, pp 96–110

are entrusted with the right of enlarging their national jurisdiction with regard to sovereignty of the territory to territorial sea, and therefore one could justify the inclusion of both these terms in one paragraph in the Bunkers Convention[237], as an intelligent move.

"exclusive economic zone"

Considering the gravity and magnitude of damage that eventually caused to the marine environment and related objects resulting from oil spillages, and especially the various threats they pose, many delegations at pre-CLC 69 negotiations were of the view that the application of that Convention should be extended.[238] Failing to obtain sufficient support in favour of such view due to the momentum that developed around the issue of the freedom of the high seas[239] and opposing views concerning economic and commercial reasons[240], pushed back any persuasions that were in existence. Succeeding developments in the public law sector especially after the adoption of the LOSC through dramatic works of the UNCLOS III[241] turned pages of history in 1984[242], by way of unveiling a Protocol to the CLC 69.[243]

Although it came out to be unsuccessful due to certain conventional requirements regarding ratification and lack of support by the United States in particular, the 1984 Protocol was the first in line to include the "exclusive economic zone

237 See Art. 2 (a) (i)

238 O.R. 1969, LEG/CONF/C.2/SR.8, p 660

239 Even though the International Convention Relating to Intervention on the High Seas in Cases of Oil Pollution Casualties 1969 granted coastal states certain rights with regard to preventive measures, it was highly felt that a liability and compensation regime should observe the freedom of the high seas

240 For example, unnecessary preventive measures that might be taken by states, which would demand higher costs and in certain instances that would lead states in using dangerous substances for neutralization, and the inflation of insurance costs putting more burdens on the shipping industry

241 The United Nations' Conference on the Law of the Sea Sessions were held from 3rd December 1973 to 24th September 1982 and was responsible in drafting the Law of the Sea Convention, which was adopted in 1982 and came into force in 1994

242 The initial work for the consideration of draft protocol to revise the International Convention on Civil Liability for Oil Pollution Damage, 1969 and the International Convention on the Establishment of an International Fund for Compensation for Oil Pollution Damage, 1971, was started in 1981 and continued till 1984

243 The International Conference on Liability and Compensation for Damage in Connection with the Carriage of Certain Substances by Sea held in London from 30th April to 25th May 1984 adopted the 1984 Protocols to amend the CLC 69 and FUND 71

(EEZ)" within the scope of application of international maritime legislations concerning liability and compensation. Despite of the numerous opposing views as to the inclusion of the EEZ or rather the extension of the scope towards such, the Working Group at the Legal Committee was able to obtain international agreement as to the intended extension. One of the most important issues was the realization of a 'demonstrated necessity' in view of drastic consequences upon oil spillages coupled with prompt and adequate compensation that need to be paid. Much of the objections were raised on the points of increased risks for volatile insurance market and proven low-rated effects on the marine environment as a whole due to the naturally deteriorating characteristic behaviour of oil.[244] But however, considering the whole issue in the light of extended national sovereignty rights of states by virtue of the incorporation of the LOSC, and as a matter of harmonization, and especially concerning the adverse effects of oil pollution on fish and fishing industry to the extent that fishermen should still need to enjoy their rights under domestic and new international legal order[245], namely the LOSC, the overall extension was accepted as inevitable. Aftermath of the *Exxon Valdez* disaster off the coast of the United States[246], and the enactment of the OPA in view of imposing liability on the respondents, paved way for the IMO to adopt the much successful 1992 Protocol[247] with the same scope of application as of 1984 version, though under less conditions as to its acceptance by State Parties.[248]

As mentioned earlier, this development has penetrated thereafter into later instruments dealing with marine liability issues, namely the HNS and Bunkers Convention. However, the deficiencies as to definitions of the terms in question at present have not been fully overcome, leaving it to be filled using international public law concerning the sea.[249] Therefore, one may have to refer to the definition provided in the LOSC as gap filler in private liability issues. According to the LOSC, the "exclusive economic zone" is an area beyond and adjacent to the territorial sea[250], which shall not extend beyond 200 nautical miles from the baselines[251] from which the breath of the territorial sea is measured[252]. Un-

244 O.R. 1984, LEG/CONF.6/C.2/SR.5

245 *Id.*, at 14

246 Occurred in Prince William Sound, Alaska, on 24th March 1989; see P. Saundry, *Exxon Valdez Oil Spill*, http://www.eoearth.org/article/exxon_valdez_oil_spill

247 Entered into force on 30th May 1996

248 The 1984 Protocol required ratification by 6 states with not less than one million units of gross tonnage, whereas the 1992 Protocol eased that requirement to 4 states, together with certain other changes as to compensation criteria.

249 See *Supra* p 47–49

250 See Art. 55

251 See Art. 5

252 Art. 57

like the regime of territorial sea, the exclusive economic zone is *sui generis* in character, and operates as a separate functional zone of the coastal state in which such state could exercise certain rights and obligations[253] vested by the LOSC. The lawful establishment of this zone needs to fulfill the requirements laid down by the LOSC, and further in accordance with the international law, namely the said LOSC, the customary laws and other sources of international law accepted by such states as binding. This indeed been mentioned in Article 2(a) (ii) of the Bunkers Convention as requiring such established zone to have been *established in accordance with international law* as a necessary requirement to be named it in such term. Presently not all states have declared their respective EEZ, due to various reasons varying from non-acceptance of the concept, opposing conflicts to confusions regarding the Continental Shelf[254] of such states. This scenario has well been covered in Article 2 with much certainty as to the proper identification of the scope of application under the Bunkers Convention. The meaning of the area and the breath mentioned in the first part of Article 2 (a) (ii) is well included in the latter part, taking a combined text of Articles 55 and 57 of the LOSC, in order to fill the gap that would exists in cases of a non-declaration of the EEZ.

According to Article 2 (a), the application of the Convention depends on the location in which the pollution damage occurred. As described above, the said location could be further categorized into three main areas, namely, the territory, territorial sea or in the exclusive economic zone. It is quite clear from these straight forward wordings that the place in which the oil actually escaped does not play a significant role, as the most important aspect being the place where the damage was sustained.[255] These wordings are the same comparing to the CLC regime[256] and the purposes for which that regime was established under liability rules. Accordingly, the only requirement is to prove that the pollution damage occurred on or within one or more of the above said locations.

Part (b) of the scope of application deals with the preventive measures that would be, taken to prevent or minimize the damage.[257] The interesting view point associated with this part is the inclusion of the words "wherever taken". It is quite evident that there is no restriction as to any specific place in which such measures

253 See Art. 56

254 See Part VI, Art. 76 of LOSC; see also P. A. Symonds, O. Eldhom, J. Muscle, G. F. Moore, *Characteristics of Continental Margins*, Ch. 4, in *Continental Shelf Limits: the scientific and legal interface*, (P. J. Cook, C. Carleton), Oxford, 2000, pp 25–60; R. Y. Jenings, *The Limits of Continental Shelf Jurisdiction: Some Possible Implications of the North Sea Case Judgment*, Int'l and Comparative Law Quarterly, (1969), 18:819–832, Cambridge

255 See Art. 1.8

256 As in Art. 1.8 of CLC 69

257 See Art. 2 (b)

should be applied in order to qualify under the Conventional meaning. Although one might judge this aspect as an intelligent inclusion, it is wiser to say that it was a timelier move upon the adoption of the 1984 Protocol through which the shortcomings of the CLC 69 regime were adjusted.[258] It was indeed an intelligent and positive move to wipe out the ambiguous meaning left unsettled before such amendment was considered during the preparation for the 1984 Protocol.[259] But however, the delegation at those meetings faced with severe obstacles of arriving at a compromise, without leaving any possibility for States to use unnecessary preventive measures that would heat up compensation claims brought against responsible authorities, [260] and more specifically in avoiding any State in using environmentally harmful substances in the combat operations[261]. Hence, such actions that could well be taken in high seas[262] under the proposed wordings would certainly fall into many disputes concerning the principle of freedom of the high seas[263]. In turn, it would be possible to argue that such a right is available under the 1969 Intervention Convention[264] since it came into force much before these amendments were brought in[265], but however, it was merely a convention dealing with prevention rather than a private liability and compensation regime.[266] Nevertheless, under the present context, all preventive measures taken with respect to bunker oil pollution incidents fall under the Bunkers Convention, irrespective of the place in which such measures were taken.[267] Therefore, the scope of application with respect to preventive measures extends to greater areas without any specific limitations as to the maximum stretch, to the extent that such actions do not fall in contravention with legal and legitimate rights of other neighbouring coastal states.

258 Article 2 of the CLC 69 reads "...and to preventive measures taken to prevent or minimize such damage"
259 *Supra* n. 244, p 3
260 *Id.*, at 17
261 *Id.*, at 13
262 See Part VII, Arts. 86–90, LOSC
263 See *Id*, Art. 87
264 Art. 1.1
265 See full text at http://sedac.ciesin.org/entri/texts/intervention.high.seas.casualties.1969.html
266 See P. Sands, *Principles of International Environmental Law*, Cambridge, 2003, pp 449–450; IMO, MEPC, *Manual on oil pollution*, Ch. 2, IMO, (1983 2nd ed., 1997); R. Price, A. Haberbeck, *The maritime laws of the Arabian Gulf Cooperations Council States*, vol. 1, Graham & Trotman, 1986, pp 174–176
267 *Supra* n. 255 at 51

(b) Pollution Damage and Preventive Measures: *jurisdiction ratione materiae*

Having discussed the scope of application with regard to geographical terms, a careful examination into material scope is inevitable with the view of understanding the relevant Convention as a whole. The material scope of application is an important aspect in analyzing the operation of the instrument not only as a matter of identifying the essential criteria, but as a central point on which the Convention is aimed to lay on. The objective of the present Convention is to focus on the civil liability aspect of pollution damage caused by bunker oil spills, and therefore, the issue of "pollution damage"[268] gathers much of the attention as a whole. But however, both "pollution damage" and "preventive measures"[269] are centralized as inseparable as could be seen in the foregoing discussion, and both these aspects are inter-connected needing careful consideration as to their definitions and inter-play throughout the Convention. Both these aspects form the subject matter which gathers much of the momentum throughout the realization of conventional goals.[270] For proper understanding it is advisable to take these two issues separately in order to avoid possible confusions as to the application of these aspects, and further to disallow any situations of ambiguity.

"Pollution Damage"

Before landing on the issue of "pollution damage" as described under the Convention, it would be much beneficial to overcome any confusion that might arise on the issue as a whole. As a convention dealing with the private liability aspect of bunker oil spills from ships, as the main reason for which this Convention was drafted, such attribution of liability need to be connected to an actual damage that had been occurred. It is well understood that in the absence of such damage, the objectives of the Convention cannot be achieved. Therefore, the most important feature is the existence or the presence of damage, in particular of the relevant objective under the present Convention, the "pollution damage". This was the

268 See Art. 1.9

269 See Art. 1.7

270 See *Supra* M. Mason at n. 86; Wu Chao, *Liability and Compensation for Bunker Pollution*, 33 J. Mar. L. & Com. 553 (2002); W. Oosterveen, *Some Recent Developments Regarding Liability for Damage Resulting from Oil Pollution – from the Perspective of and EU Member State*, Environmental Law Review, vol. 6, issue 4, 2005, pp 223–239; M. Mason, *Transnational compensation for oil pollution damage: examining changing spatialities of environmental liability*, at http://eprints.lse.ac.uk/570/1/RPESA-no69(2002).pdf

central issue in all such similar regimes that presided over the present Convention such as the CLC regime and the HNS.[271] In turn, the Bunkers Convention was adopted as gap filler to the prevailing international regimes dealing with oil pollution compensation system, and this fact is clear from its preamble which states that;

"Considering that complementary measures are necessary to ensure the payment of adequate, prompt and effective compensation for *damage caused by pollution* resulting from the escape or discharge of bunker oil from ships" (emphasis added).[272]

During the initial period of drafting this Convention, the sponsoring delegations laid on the belief that it should complement the existing regimes such as the CLC 92 and HNSC. In their view, the new convention should be aimed to cover claims for all types of damages which bunker spills cause to third parties and be able to meet all such risks posed by ships' bunkers.[273] Therefore, the draft text submitted by the sponsoring delegation was based primarily on issues relating to the CLC and HNSC regimes. The notable exclusion from that of the CLC was the absence of the term "pollution damage", instead, referring to the term "damage" under the relevant provision dealing with the scope of application.[274] This was strongly opposed by the drafters citing in chief, "the main reason for the need of a bunkers convention as a matter of covering damages caused by ships which do not fall under the meaning of the CLC regime", and therefore emphasizing on the issue of following the CLC precedent on types of claims so covered. On the other hand, it was the view that, enlarging the definition following the HNSC pattern would create unnecessary inconsistencies with the much similar CLC regime.[275] Therefore, it was decided by the Legal Committee to follow the same trend as the previous liability regimes designed to deal with similar issues relating to oil pollution damage.[276] Following the said trend, it has not only influenced the objective criteria, but also the wordings of the definition of 'pollution damage", which in turn has followed the same as described under the CLC 92 with the addition of the term bunker oil within the meaning of the kind of oil involved in the subject matter.[277] According to its Article 1(9);

271 Although the HNS is expected to deal with number of other types of damages than the pollution damage unlike the CLC, it is used to combat against pollution risks associated with the carriage of hazardous and noxious substances.

272 Paragraph 6 of the Preamble to the Bunkers Convention 2001.

273 LEG 73 agenda item 12, 12 October 1995

274 *Id,Draft* Art. 2, Annex, p.3

275 *Supra* n. 62, p. 12

276 *Supra* n 181

277 Art. 1.9 (a)

"Pollution damage" means:

(a) loss or damage caused outside the ship by contamination resulting from the escape or discharge of bunker oil from ship, wherever such escape or discharge may occur, provided that compensation for impairment of the environment other than loss of profit from such impairment shall be limited to costs of reasonable measures of reinstatement actually undertaken or to be undertaken; and
(b) the costs of preventive measures and further loss or damage caused by preventive measures.

Components of Damage

From its outset one may identify three classes of loss and damage for which compensation is attainable, namely;

i) the damage caused;
ii) the costs of preventive measures; and,
iii) further damage caused by such preventive measures.

Therefore, the principal aim is to compensate these three main types of damage with reference to pollution incidents. However, the present instrument has not provided a definition for the term "pollution", although it refers to pollution incidents out of which compensation is payable for loss or damage relating to such pollution. None of the previous liability regimes dealing with similar topics have committed towards defining what is meant by "pollution", but have made similar reference, leading the matter to confusions arising at the starting point. Many authors who have dealt with the previous Conventions have made efforts to define this term using other authorities such as dictionaries and public law instruments such as the UNCLOS, the very distinct nature of the use of action oriented term restricts one to rests on the used term in this private liability law instrument rather than taking such definition of only a part of the whole term, for granted. Nevertheless, in the absence of an important issue connected to the main idea cannot be left aside in understanding the entire subject. Therefore, it might be necessary at this juncture to simply define what it means as "pollution" and what types of damage can be categorized as "pollution damage" in such context.

In the absence of a straight forward definition under the CLC regimes or related domestic legislations such as the Merchant Shipping Act or the OPA 90[278], it is possible to turn to certain persuasive definitions provided by different state laws of the United States and other jurisdictions.[279] For example, the legislation of States such as Louisiana[280] and Florida[281] dealing with pollution make reference to the presence of certain substances as polluting marine environment rather than citing external introduction of such substances by the physical involvement of human beings. As the liability law deals with the subject of responsibility of the authoritative object so responsible, an action oriented definition is of high value. In such a context, it would be more reasonable to refer the definitions that deal with physical involvement of the man, as can be found under the UNCLOS 82. Accordingly;

"Pollution of the marine environment means, the introduction by man, directly or indirectly, of substances or energy into the marine environment, including estuaries, which results or is likely to result in such deleterious effects as harm to living resources and marine life, hazards to human health, hindrance to marine activities including fishing and other legitimate uses of the sea, impairment of quality for use of the sea water and reduction of amenities". [282]

This definition would somewhat ease tensions pertaining to the understanding of the term "pollution damage" as a whole used in the Bunkers Convention in developing a criteria for what falls within. The abovementioned definition would help develop a list of categories in this respect assuming to fall within the meaning of "pollution damage" aftermath an oil spill.[283] These effects would spread on many

278 However, the pre-existed Act which was repealed by the enforcement of the OPA 90 did contain a definition for pollution under its Title 43, Chapter 36 dealing with the Outer Continental Shelf Resource Management, in its paragraph 1811

279 See B. G. Rabe, *Federalism and Entrepreneurship: Explaining American and Canadian Innovation in Pollution Prevention and Regulatory Integration*, Policy Studies Journal, vol. 27, 1999; P. Clarke, *Environmental law toolkit – NSW: a community guide to environmental law*, 5th ed., The Federation Press; M. W. D. White, *Australian Marine Pollution Laws*, 2nd ed., The Federation Press, 2007; N. Emmott, N. Haigh, *Integrated Pollution Prevention and Control: UK and EC Approaches and Possible Next Steps*, Journal of Environmental Law, 1996, 8(2): 301–311, Oxford; R. N. Reeve, *Introduction to environmental analysis*, University of Sunderland, 2002; A. C. Gross, N. E. Scott, *Comparative Environmental Legislation and Action*, Int'l and Comparative Law Quarterly (1980), 29: 619–663, Cambridge

280 Oil Spill Prevention and Response Act, R.S. 30, Paragraph 2454 (24)

281 Pollutant Discharge Prevention and Removal, Florida Chapter 376, paragraph 376.031 (17)

282 Art. 1(4).

283 See R. V. Randle, *Oil Pollution Deskbook*, Environmental Law Institute, Washington, 1991, pp 164–172; US Fish & Wild Life Services, *Effects of Oil Spills on Wildlife*

aspects such as health hazards, possessions of claimants including *inter alia* movable and immovable properties, fishing boats, recreational boats, beaches, piers and coastal properties[284] on one hand and the marine environment which does not fall as a subject of private ownership on the other. Therefore, any damage occurs as a result of pollution, after an oil spill would fall within the meaning of "pollution damage" in a much greater perspective.[285] The International Union for Conservation of Nature and Natural Resources (IUCN) forwarded their proposal seeking a more user-friendly definition of "pollution damage" that would address the elements of the definition in a more separated and expressed manner. Based on the same old definition taken out of the CLC regime, it categorized types of damages which would fall within the meaning of the proposed Convention through direct and positive approach for the convenience of understanding.[286]

As mentioned earlier, the described damage under the Convention is three folded. The first deals with the actual damage caused[287] and the rest is associated with preventive measures[288]. The latter would be discussed later in this Chapter when dealing with that subject matter as a separate issue. Therefore, it is of much importance to discuss the issue of damage at this juncture in order to proceed with a clear mind.

The important characteristics

Externality

Paragraph (a) expressly declares that the *'loss or damage caused must be outside the ship'* in order to be entitled for compensation. It is well understood that internal damages are not compensated under this Convention including damages

and Habitat: Alaska Region, Dec. 2004, available at http://alaska.fws.gov/media/unalaska/Oil%20Spill%20Fact%20Sheet.pdf; S. R. Pezeshki, M. W. Hester, Q. Lin, J. A. Nyman, *The effects of oil spill and clean up on dominant US Gulf coast marsh macrophytes: r review*, Environmental Pollution, vol. 108, Issue 2, pp 129–139; C. A. Fleischer, *Liability for Oil Pollution Damage Resulting from Offshore Operations*, available at http://www.cenneth.com/sisl/pdf/20-4.pdf; M. D. French, R. J. Jennings, W. Nicole, S. Sankar, E. D. Schmidt, *Estimation of potential impacts and natural resource damages of oil*, Journal of Hazardous Materials, Elsevier – Amsterdam, 2004, vol. 107, pp 11–25

284 See Edgar Gold, *Gard Handbook on Marine Pollution*, 2nd ed., page 78

285 See F. Maes, *Marine resource damage assessment: liability and compensation for environmental damage*, Springer, 2005

286 LEG 79/6/3, 18 March 1999

287 Paragraph (a)

288 Paragraph (b)

caused to cargo irrespective of its nature whether carried on board the vessel or in its tanks in case of oil carried as cargo. Both the CLC and HNS regimes bare the same principle with respect to this aspect and as an instrument that followed the same trend and wordings as the CLC 92 in particular, the Bunkers Convention is intended to apply similarly.[289] However, taking into consideration that only "pollution damage" will be compensated under this regime, the question of internal damages would not fall within the ambit of claimable. Although the IOPC Fund has in certain instances allowed such internal damages to be compensated within its operation as an oil pollution compensation fund watching the interests of the shipping trade as well[290], a stricter rule is expected under the present Convention, which however lacks such fund to that end, in compensating actual victims outside the vessel itself. Therefore, the main issue would be to ascertain whether the victim claiming compensation under this Convention could satisfy its requirement of being a party who claims for pollution damages occurred solely outside the vessel including its contents.

Contamination

The second most important characteristic is that the *damage must result from contamination*. This is one of the most controversial issues involved in the definition, as damages which could occur for reasons other than due to contamination do not fall within this compensation regime.[291] Once again, the successive repetition of the CLC 92 interpretation has proved to have carried along the same contents to the new regime. The influence of the *Torrey Canyon* disaster which was centralized on the issue of contamination as the main source of "pollution damage" tend to have moved forward into the present regime, rather than coping with new developments of the relevant area in matters arising out of other sources such as collisions, enlargement in sunken wrecks, explosion or fire. This concept was later accepted, upon being settled at the Tokyo Conference convened to adopt CLC 69[292], by the IOPC Fund in the case of *Kasuga Maru No.1* recognizing the

289 See Fowler, Rodriguez, Kingsmill, Flint, Gray, Chalos, *The International Convention on Civil Liability for Bunker Oil Pollution Damage, 2001*, available at http://www.frc-law.com/files/pub_international-convention-civil-liability.pdf

290 See IOPC Fund Annual Report 1990, at pages 46, 52 dealing with an incident that occurred in Japan

291 See CMI Guidelines on Oil Pollution Damage, Resolution 1994, http://web.uct.ac.za/depts/shiplaw/cmi/cmioil.htm

292 The term "contamination" that was omitted from the draft Convention tabled at the 1969 Diplomatic Conference was later revived during that Conference under a proposal made by the United Kingdom, O.R., 1969, LEG/CONF/C.2/WP.9, p.566, but

importance of defining the relevant sections of the Convention based on the clear language that has been used.[293] On one hand it would have been difficult to obtain general agreement at the negotiations, if all types of causes were allowed to be included within the liability and compensation regimes for oil spills, due to the complexities that would arise in the entire field of maritime law. [294] Therefore, limiting it to contamination as one of the chief reasons for damage would suffice to fill the gap that exists in this area of private international law. Although, some delegations spoke in favour of including fire and explosion as a means of compensable damage in terms of the bunkers Convention, the majority was in favour of consistency with the CLC regime.[295] On the other hand, the margin of appreciation tends to be somewhat narrow in terms of identification as to the reason for the existence of damage in question. This would be the hardest point to overcome as it would need clear observation data which are sometimes difficult to obtain in practical terms. In this respect it is clear that a contamination followed by a fire or explosion would fall within the allowed criteria and situations where the damage occurred initially due to contamination and further victimized through fire or explosion. In such terms, a claimant will only be successful in maintaining his action for compensation under the Convention if the damage resulted from contamination and not due to any other reason such as fire or explosion. A justification to this end can be found in the jurisprudence of the IOPC Fund in the case of *Aegean Sea* incident[296] where it was settled by the Executive Committee that the

---

the French delegation opposing such inclusion pointed out discrepancies that exist in the *definition* in the absence of damages occur due to fire and explosion, O.R., 1969, LEG/CONF/4, p.446

293 IOPC Fund Annual Report 1989, p.41

294 See P. Simon, La réparation civile des dommages causes en mer par les hydrocarbures (Civil compensation for damage at sea caused by oil), unpublished thesis, 1976, p.23

295 LEG 75/11, 29 April 1997, p. 10

296 On Dec. 3, 1992, the Greek OBO Carrier laden with 80,000 tons of North Sea Brent Crude Oil ran ground during heavy weather while approaching the port of La Coruna on the Galician Coast, North West Spain. See D. Pastor, J. Sanchez, C. Porte, J. Albaiges, *The Aegean Oil Spill in the Galicia Coast (NW Spain). I. Distribution and Fate of the Crude Oil and Combustion Products in Subtidal Sediments*, Marine Pollution Bulletin, vol.42, Issue 10, Oct. 2001, pp 895–904; M. Sole, C. Porte, X. Biosca, C. L. Mitchelmore, J. K. Chipman, D. R. Livingston, J. Albaiges, *Effects of the "Aegean Sea" oil spill on bio transformation enzymes, oxidative stress and DNA-adducts in digestive gland of the mussel (Mytilus edulus L.)*, Comparative Biochemistry and Physiology Part C: Pharmacology, Toxicology and Endocrinology, vol. 113, Issue 2, Feb. 1996, pp 257–265; J. L. Gomez Gesteira, J. C. Dauvin, *Impact of the Aegean Sea Oil Spill on the Subtidal fine sand macrobenthic community of the Ares-Betanzos Ria (Northwest Spain)*, http://otvm.uvigo.es/investigacion/informes/documentos/archivos/aegeansea_gomezgesteira.pdf

damage caused by oily smoke resulting from fire, was due to contamination.[297] In cases such as the *Betelgeus*[298] case and *Castillo de Bellver*[299], there had been situations where damage occurred due to rain drops contaminated with oily smoke aftermath explosion and fire, and still falling within the meaning of the 1969 Convention. These are clear examples of the enlargement of risks associated with the normal operations and commercial activities conducted by the shipping trade.

The approach introduced in the HNS Convention deviates from that of general oil pollution compensation stream. In the application of the specific instrument for the intended purposes, the HNSC instead define "damage" referring to a list of consequences in a detailed manner.[300] Rather than using an ambiguous language as the CLC regime, it deals with the issue of contamination only in respect of losses or damages to the environment.[301] Unlike in the case of the CLC and Bunkers regimes, other types of damages are not linked to contamination, thus allowing claims arising out of fire or explosion to be compensated as well.[302] Obviously, the different purposes for which these regimes are applied and the hazardous nature of the substances so included, compel the HNSC to deal in such a way allowing non-environmental aspects to be compensated to a wider spectrum. Hence, the use of the term as "damage" serves the idea of covering all possible consequences without limiting them to ones of polluted nature.

The domestic law counterparts of oil spill damages take different approaches varying from one State to the other. Neither the Merchant Shipping Act 1995 nor the OPA 1990 expressly provides a clear definition to "pollution damage" as stated before. The Merchant Shipping Act 1995 bears the same provision[303] as in the Merchant Shipping (Oil Pollution) Act 1971[304] stating in addition, 'damage includes loss'.[305] Although the said United Kingdom legislation requires such damage or loss to have occurred due to contamination as in the case of similar

297 FUND/EXC. 38.9, 11 February 1994, at section 3.3.3

298 In Bantry Bay, Ireland, January 1979, tanker Betelgeuse caught fire and exploded at the oil jetty, sinking and causing deaths of more than 50 men while unloading 70,000 tons of Arabian heavy crude oil 40,000 Arabian light crude oil and the letter oil cargo was on board when the vessel exploded

299 Off Cape Town, South Africa, 1983, the vessel caught fire while carrying 252,000 tons of light crude oil (Murban and Upper Zakum). See A. G. S. Moldan, et al, *Some Aspects of the Castillo de Bellver oil spill*, Marine Pollution Bulletin (1985), vol. 16, Issue 3, pp 97–102; J. Wardley-Smith, *The Castillo de Bellver*, Oil and Petrochemical Pollution (1983), vol. 4, Issue 1, pp 291–293

300 See Art. 6

301 See Art. 6(c)

302 See Arts. 6(a), (b) and (d)

303 Sec. 170 and Sec. 170 of Schedule 4

304 See Sec. 1(1)

305 Sec. 20(1) of the 1995 Act

international regimes, the OPA 1990 offers no such requirement, rather referring to listed damages and costs resulting out of the incident.[306]

As a result of contamination, there could exists of many damages, leaving the victims in grievances. Depending on the quality of oil and its deteriorating effects, "pollution damage" could extend from environmental harm, human hazards to private property loses and damages in both physical and economic perspectives. These aspects would later be discussed in detail during the discussion on types of damages suffered by claimants.[307]

Escape or discharge

The third and final qualification with respect to the definition of "pollution damage" is that *the contamination must result from the escape or discharge of bunker oil*[308] *from ship*. In this context, it is clear that only escapes or discharges of bunker oil which result in contamination are attributed towards liability. The mode of such escape or discharge is not expressly mentioned in the definition, thus creating controversies as to its actual implementation. Both CLC 69 and 92 wordings bore the same implications leading it to highly controversial debates among conventional lawyers and legal experts. Even during the debates at Diplomatic Conferences held prior to the adoption of the 1969 Convention, this issue gathered much momentum leaving opinions remain divided. One section of experts relied on the fact that, escape or discharge shall have arisen out of accident and the other section relied on both operational and accidental aspects. In practice, accidental discharges are much severe in result whereas operational discharges cause only minor damages for which remedies are available under the laws of the respective jurisdiction in terms of fines. Professor Lucchini goes further to state that pollution is not pre-meditated and is unforeseeable, so what must be taken into consideration is only accidental discharges.[309] Much of the controversies er-

306 See Sec. 1002

307 *Infra* pp 163–168

308 Although, the word "oil" appeared in the original draft without any reference to "bunker oil", there was overwhelming support of the delegations to insert the word "bunker oil" due to the subject of the Convention on which it is intended to apply. Thereafter, square brackets were inserted around the term "bunker", but later decided by the Committee to delete the brackets in order to comply with the rest of the provisions of the Convention. See LEG 81/WP. 2, 30 March 2000, p. 11

309 Lucchini, "La pollution des mers par les hydrocarbures : les Conventions de Bruxelles de novembre 1969 ou les fissures du droit international classique" ("Marine oil pollution: the 1969 Brussels Conventions or the cracks in conventional international law"), JDI, 1970, p.815

upted after the removal of the term "deliberate discharge" that was inserted in the draft text presented at Tokyo and London[310] Conferences from the final text.[311] Based on this, Ganten, the ex-Director of the IOPC Fund, states that "each escape or discharge of oil is covered, irrespective of whether it is the result of an accident or whether it is intentional discharge".[312] Abecassis holds the same view in allowing both accidental and intentional discharges.[313] As evident from the above discussion, the opinions are so divided on this issue making it difficult in arriving at a true conclusion. Therefore, one may take up this issue in conjunction with the situation and location such escape or discharge is associated with, namely with respect to the interpretation of the term "incident".

According to Article 1(8) of the Bunkers Convention, an;

> "Incident" means any occurrence or series of occurrences having the same origin, which causes pollution damage or creates a grave and imminent threat of causing such damage.

As a repetition of both CLC and HNSC regimes, this term gathers much of the attention in interpreting the physical component as described under the term "pollution damage". This was quite evident when the German delegation at the 1971 Diplomatic Conference took it for granted in stating that tank washing and other normal operations of the vessels fall outside the meaning of "escape or discharge" used in the text.[314] Though, such statement does not bear any undertaking as to the entirety of the participants, and neither of possessing any binding nature, the attempt is indeed notable. Pollution incidents due to oil spillages have accounted for many minor and major incidents that have taken place in many jurisdictions. These have occurred not only as a result of major maritime casualties[315] such as accidents, but also as a result of mishaps and intentional discharges varying from bunkering to other oil transfer operations such as off-shore ship-to-ship oil transfers.[316] All these situations fall into the meaning of "incident", which might lead to minor or major pollution damages depending on the remoteness or seriousness of such occurrence. These concerns penetrated into the tough decision making process at the Committee level convened to discuss the adoption

310 O.R., 1969, LEG/CONF/4, September 1969, p.458

311 The preparatory draft text read as "… The owner shall in any case be liable for any pollution damage caused by oil deliberately discharged from his ship."

312 Ganten, "The International Oil Pollution Compensation Fund", Environmental. Policy and Law, vol. 12, February 1984

313 D. W. Abecassis, *Supra*, n. 92, p 198

314 O.R , 1971, LEG/CONF.2/C.1/SR.11, p.389

315 See N. Hooke, *Maritime Casualties, 1963–1996*, 2nd ed., 1997, LLP Limited, London

316 See Colin de La Rue and Charles B. Anderson, "Shipping and the Environment", LLP-London/Hong Kong 1998, Ch. 9, at p.375

of the Bunkers Convention. It was suggested by many delegations that damage resulting from bunkering operations should also be covered, but were uncertain whether this should cover both shore-to-ship and ship-to-ship or only the latter.[317] As these are very sensitive issues which had to be dealt with careful studies, especially in avoiding contradiction and conflicts with national laws pertaining to port-State controls[318], the issue did not receive much back-up in later discussions of the Committee. Turning to such domestic laws, the United States' Bunkers Liability Regime is centered on the concept of a general no-discharge policy[319], under which it imposes liability by way of civil penalties on the spill itself[320] and *inter alia*, allowing civil suits to penalize discharges of oil, caused due to gross negligence or willful misconduct[321].

However, in the absence of any clarification to this end, whether the "escape or discharge" is meant to be operational or accidental, it could be concluded that, referring to the clear language used in the Convention, any "escape or discharge' would fall within this meaning that has led to the occurrence of the final result, namely "pollution damage". Although such stricter method is often conclusive in theory, different jurisdictions would interpret this meaning based on its own domestic tools of interpretation. Therefore, this confusion would create chaos in future determinations on liability and compensational schemes for oil spills from ships.

"Preventive measures"

The definition of "Pollution damage" does not limit itself to losses and damages cause naturally as a result of oil spillages, but extends further to actions taken aftermath of the incident and subsequent occurrences that may exist upon such measures. This is expressly indicated in the paragraph (b) of the definition of "Pollution damage" stating that *the costs of preventive measures and further loss or damage caused by such preventive measures* should be covered within its meaning. Therefore, as shown above in the discussion on "pollution damage" as

317 See LEG 76/WP.3, 13 October 1997, p. 13/14

318 See IMO, *Port State Control*, IMO, (1995 ed., 2001); IMO, *Procedures for Port State Control*, IMO, 2001; S. Knapp, M. van de Velden, *Visualization of differences across port state control regimes by means of correspondence analysis*, Econometric Institute Report, Erasmus University Rotterdam, EI 2007-32, Revision, 2009-07-29; D. Anderson, *The Effect of Port State Control on Substandard Shipping*, Maritime Studies, Issue 125, July/Aug 2002, pp 20–25

319 Title 33 U.S. Code, Sec. 1321 (b) (1)

320 *Id.*, (b) (6) and (7) (A)

321 *Id.*, (b) (7) (D)

a whole issue, the two other classes of "Pollution damage" are associated with the second part[322] of the definition dealing with "preventive measures".[323] In turn, the Convention defines what it means by the term "Preventive measures" in Article 1(7). According to that;

> "Preventive measures" means any reasonable measures taken by any person after an incident has occurred to prevent or minimize pollution damage.

From its outset, the above expression is clear, plain and straight forward in meaning. But an inside observation will bring about several confusions regarding positive application. Some of these could be put forward in the following manner; what are reasonable measures[324]? Does any person who takes such measures will have *locus standi*[325] in a proper legal proceeding? Will measures taken to avoid an oil spill just after a collision or from a sunken vessel be eligible to qualify? Are pre-spill measures covered within the meaning?

These are indeed highly sensitive issues that need careful reasoning under respected laws. Firstly, it should be borne in mind that there can exist of various types of "preventive measures" that could be taken to prevent or minimize pollution damage.[326] These vary from technical methods to somewhat political, commercial and economic tools of propaganda and advertisement. Although, the actual combat should start at aiming to disseminate the spilled oil and avoid the expanding of damage, providing compensation for "preventive measures" of other nature are sometimes allowed legally. Secondly, the national courts are empowered under the Convention to determine the reasonableness of such efforts taken as "preventive measures" in events of disaster.[327] Though certain other ju-

---

322 Paragraph (b)

323 See *Supra* at pp 57–58

324 See T. J. Wagner, *Recoverable Damages under the Oil Pollution Act 1990*, S. U. S. F. Mar. L. J. 283 (1992–1993); O. Thebaud, D. Bailly, J. Hay, J. Perez, *The cost of oil pollution at sea: an analysis of the process of damage valuation and compensation following oil spills*, http://otvm.uvigo.es/investigacion/informes/documentos/archivos/Prestige_Hayetal.pdf; J. Hay, O. Thebaud, J. P. Agundes, *Preventing pollution through the compensation for damage? An appraisal of the European experience in the field of marine oil spills*, http://www.uibcongres.org/imgdb/archivo_dpo4084.pdf; C. A. Kontovas, H. N. Psaraftis, *Marine Environment Risk Assessment: A Survey on the Disutility Cost of Oil Spills*, http://www.martrans.org/docs/ws2009/Kontovas%20Psaraftis%20-%20Disutility%20Cost%20of%20oil%20spills.pdf;

325 See D. Shelton, A. Kiss, *Judicial Handbook on Environmental Law*, UNEP, 2005, available at http://www.unep.org/Law/PDF/JUDICIAL_HBOOK_ENV_LAW.pdf

326 See M. Hasebe, *The International Liability and Compensation Regime*, http://www.cep.unep.org/events-and-meetings/11th-igm/oil-spill-compensation-funds.pdf

327 See http://www.upv.edu.ph/oilspill/compensation.php; H. Xue, *Transboundary damage in international law*, Cambridge Studies in International and Comparative

risprudence such as the opinions of the IOPC for example, could play persuasive roles, the determination factor rests with such national judicial systems.[328] Nevertheless, the development of the idea of the scope of application with regard to "Preventive measures" may provide enhanced accessions to disputes relating to such interpretation.

The notion of 'reasonableness'

As a starting point, it would be prudent to take up the issue of 'reasonableness' involved in the measures so taken. Though it is mentioned in the definition, the Convention is silent on what it means by the term '*any reasonable measures*', and do not even provide clarification as to the term '*reasonable*'. This is not a deficiency that is peculiar to the Bunkers Convention, as it was also evident in the CLC 69[329] and 92[330], Fund Convention 71[331] and 92 and HNSC[332], by the same effect that appeared in those as well. However, the formula adopted by the Seventh Intersessional Working Group attending to Fund Convention 1971[333] provides certain satisfactory base in terms of persuasive nature.[334] It has stressed the importance of evaluating "reasonableness" in terms of objective criteria based on the use of best possible technical measures upon effective advice. Obviously, this would employ considerable technical expertise upon scientific research and related practical experiences. In addition, constant review of the effectiveness of the measures would in turn leave the idea of reasonableness under scrutiny. In such terms, the costs incurred should be assessed on reasonable grounds which correspond to measures taken in actual proximity leaving out remote overheads.[335] These would relate to effectiveness of the measures taking into account of handling the issue on technical appraisals which benefit the end result, rather

---

Law, Cambridge, 2003, pp 56–72; Dr. I. C. White, *Facilitating the speedy payment of oil spill compensation claims under the CLC and FUND Conventions*, http://www.itopf.com/_assets/documents/tianjin.pdf; M. Jacobson, *The International Oil Pollution Compensation Funds and the International Regime of Compensation for Oil Pollution Damage*, http://www.icao.int/icao/en/anb/atm/SARafrica/docs/Saly_Conf/en/IOPC_Funds_3-2_en.pdf

328 See Art. 9

329 See Art. 1 (7)

330 Provisions of CLC 69 with regard to the definition of 'preventive measures' was not replaced by the Protocol 92 and has been retained unaltered

331 Art. 1 (2)

332 *Supra*, n. 147

333 FUND/WGR.7/21, para. 7.2.8

334 For further details see CMI Guidelines on Oil Pollution, Part III, 1994

335 See FUND/WGR.7/21, Annex 1, para. 2.3

than creating negative after-effects. These technical appraisals would evolve in using the best techniques in addressing the public of the nature and effects of the spill in terms of media based forms, rather than welcoming unreliable sources in the method of informing them.[336] Although such technical appraisals are considered to be the best sources in arriving at decisions at IOPC Fund and related authorities dealing with compensation, such as P&I Clubs[337], national courts of law of respective States, would not be bound on such, leaving the matter open to be decided on laws pertaining to 'reasonableness' in such jurisdiction.[338] This would purely be the case with respect to Bunkers Convention as matters are solely dependant on national compensational structure, except in circumstances involving ships which must possess compulsory insurance.[339] Likewise, the issue of 'reasonableness' lies on measures taken, it is also associated with costs. As reasonable measures should not correspond to unreasonable costs, courts would not allow exaggerated costs set unreasonably. This theory that is respected in law of tort according to Colin Rue and Charles Anderson[340] citing *The Pactolus*[341], has also been the view point of the IOPC Fund.[342] As the main issue involved in this compensation regime being "pollution damage", and that it is associated with the *costs and further loss or damage caused by preventive measures,*[343] it would be the costs that matters in any litigation constituted under the Convention. Therefore, it implies that the costs of preventive measures reasonably taken, gathers much of the attention in litigations, on which, courts would lay high emphasis. In this case, quantification of actual costs would fall into high scrutiny which would in turn be weighed in terms of 'reasonableness'. The absence of any implication as to reasonableness of costs incurred in the Convention has further been overshadowed in paragraph (a) of Article 1(9) dealing with the definition of "pollution damage" which reiterates costs *for impairment of the environment other than loss of profit from such impairment, to reasonable measures of reinstatement actually undertaken or to be undertaken*, leaving again the matter on reasonableness of the measures taken or to be taken.

---

336 For more details see, Appendix 13/1: ITOPF, Preventive Measures and Property Damage, para. 2.1.1-2.1.5.

337 See Home Page, International Group of P & I Clubs, http://www.igpandi.org/Home/Sitemap

338 See A. Aarnio, *The rational as reasonable: treatise on legal justification*, Kluwer, 1987, pp 90–98; E. Pattaro, H. Rottleuthner, R. A. Shiner, A. Pec, *A treatise of legal philosophy and general jurisprudence*, vol.1, Springer, 2005, pp 74–77; D. J. Gifford, *Understanding the Australian legal system*, Cavendish, 1997, pp 5–12

339 see Art. 7

340 *Supra*, n. 316, p 402

341 (1856) Swab.173

342 See FUND/WGR.7/21, para. 7.2.7

343 Art. 1(9) (b)

As mentioned earlier, the term 'reasonable' does not limit only to measures taken by using technical forms, rather extending its application to other forms of preventive measures, such as public awareness through promotional activities. These, which fall under the category of 'pure non-pecuniary costs' that lacks proper justifiable methods of assessment should be assessed in the light of accepted criteria.[344] The IOPC Fund in several occasions arrived at varying decisions depending on the acceptability of such claims on grounds of reasonableness.[345] In response, the Seventh Intersessional Working Group[346] adopted a criteria that would evaluate 'reasonableness' of the measures and in turn consider as preventive measures taken according to the definition.[347]

The right to act

The next important aspect with regard to claims on costs of preventive measures lies on the issue of right of action by the claimant, where the principle of '*locus standi*' come into play. In any legal system, this requirement is considered foremost in any legal proceeding. Tort law is no exception to this rule, which requires the fulfillment of such or any satisfactory legal relationship connected to the claimant against whom such prejudice has been caused. Although there appeared to be voluntary groups and interested parties who join in taking up measures such as clean-up in events of oil spills as a matter of minimizing pollution damage,[348] common law jurisdictions have faced severe obstacles in allowing compensation in the absence of any legal duty upon which such parties act. As it is necessary in common law jurisdictions to prove a legal duty for such parties to perform in wake of civil wrongs, the international compensational system needed suitable legal maxims to apply in response. The CLC 69 overcame this situation by introducing *costs of preventive measures and further loss or damage caused by preventive measures* within the meaning of "pollution damage" and further defining "preventive measures" as "*...taken by any person*". Therefore, the *locus*

344 See *infra* p.161–162

345 For example in the *Haven* case costs of publicity incurred towards counteracting reputation damage was found to be unacceptable on the ground of lack of proof to that end. FUND/EXC 35.10, 8 June 1993, para 3.2.18

346 Para 2.4.4., of Annex I to the Report of that Working Group, FUND/WGR.7/21, 20 June 1994

347 This was a result of the *Braer* case which involved a huge promotional campaign

348 See IPIECA / ITOPF, *Oil Spill Compensation: a guide to international conventions on liability and compensation for oil pollution damage*, Feb. 2007, pp 10–11, available at http://www.ipieca.org/activities/oilspill/downloads/publications/reports/english/Compensation.pdf

standi issue extends to any person taking "preventive measures" irrespective of having any proprietary rights or interests on the affected subjects. Exclusion of such provision would have created chaos in this whole system, in particular, where there does not exist such rights or interests concerning environmental damage. In practice, the owner of the ship normally carries out the "preventive measures" in preventing and minimizing pollution damage, as he is the most competent person to handle the contains with the knowledge of their types and behaviours. In return, the shipowner is entitled under the CLC regime to initiate claims against the fund seeking reimbursement.[349] Although, such a provision is not available in the Bunkers Convention due to the absence of a Fund that has to be constituted in the jurisdiction having competency to deal with such matter, it can be understood that the Bunkers Convention does not disallow such claims, considering the pure language used in the context of "preventive measures". Practically the situation here is quite different. Although the shipowner could institute such a claim against the insurer when there exist of insurance, this opportunity does not exist in the absence of any insurance policy, as only a certain category of ships are required to hold compulsory insurance under the Convention.[350] This suggests that it would be beneficial for all shipowners to hold insurance with respect to their ships irrespective of its type or category. But however, non-requirement of insurance for certain category would put the Convention under criticism as such ship-owners could easily evade their responsibilities under national laws with respect to the taking of the said "preventive measures" as a duty. The Convention has indeed remedied such situations by including the words '... by any person' meaning public authorities which, could take-over "preventive measures" in such absence on the part of the shipowner and institute its claim against him for reimbursement. The legislations of the United States' compel ship-owners or other responsible parties, to response in preventing damage and further spreading of oil from their ships, and they are under a legal obligation, failing which, public authorities take over such actions and claim for the expenses.[351]

The appropriate instance

The third important feature in the definition, deals with the issue of exact stage at which, "preventive measures" could be taken according to the Convention. Although it is expressly stated that such measures are compensable '... after

349 CLC, Art. 5 (8)

350 See Art. 7

351 See W. S. Frye, *Oil Spill Prevention Measures: tackling the tough SPCC issues*, http://www.epa.gov/OEM/docs/oil/fss/fss04/frye_04.pdf

an incident has occurred', it seems to spread into a wider area of application once a careful observation is laid on the interpretation of the term "incident" which means to include the phrase '… creates a grave and imminent threat of causing such damage". This ambiguous situation creates '*prima facie*' difficulties as to a proper understanding. But a closer look will eradicate such difficulties. As "pollution damage" relates to costs of "preventive measures and further loss or damage caused by such "preventive measures" and that the "preventive measures" relate to measures taken after an incident has occurred is further connected to the term "incident" meaning any incident creating grave and imminent threat of causing such damage. Therefore, measures taken to prevent oil been spilt are covered under the meaning of the Convention in general. These measures which are well known as "pre-spill" measures were not covered under the CLC 69, but the 1984 Protocol inserted the new additions in the definition of "incident", and later came into force through the CLC 92. These changes came into existence as gap filler to the international regime on oil spill damages from ships in response to threat removal measures that existed in former TOVALOP[352] and CRISTAL[353] voluntary schemes.[354] Therefore, it is clear that any "preventive measures" taken in response to a situation which creates a grave and imminent threat of causing pollution damage, qualify for compensation notwithstanding the fact that the actual occurrence of discharge or escape of oil from the ship faced with a casualty, thus existed.

Further loss or damage by Preventive Measures

Finally, the third category of "pollution damage" involved with oil spill incidents relate to '*further loss or damage caused by preventive measures*'. The interesting point involved here is that it does not deal with the aspect of 'contamination'. Although the primary loss or damage is connected to that issue, consequential loss or damage caused by preventive measures does not fall on such a requirement. It

352 Tanker Owners Voluntary Agreement Concerning Liability for Oil Pollution (TOVALOP), signed on 7 January 1969

353 Contract Regarding a Supplement to Tanker Liability for Oil Pollution (CRISTAL) signed on 1 December 1971

354 See R. J. Dupuy, D. Vignes, *A Handbook on the new law of the sea*, vol. 2, Kluwer, pp 1158–1168; A. M. Sheppard, *Modern maritime law and risk management*, Routledge-Cavendish, (2001, 2nd ed., 2007)pp 977–978; S. Bloodworth, *Death on the High Seas: the demise of TOVALOP and CRISTAL*, Florida States University Journal of Land Use and Environmental Law, 1998, available at http://www.law.fsu.edu/journals/landuse/Vol132/Bloo.htm; S. Baughen, *Shipping Law*, *supra* n. 354, pp 338

does not mean that contamination will not be considered as a source of damage at all, but it will not be the only issue. However, it is important to note the extent to which such loss or damage is admissible, namely, any measure aimed *to prevent or minimize pollution damage*.[355] Therefore, it is quite evident that any measure which was used to prevent other types of damages, other than for the required purpose, according to the Convention, would not fulfill the aspect of obtaining compensation. A clear determination of the intention in carrying out the measures would pave way for allowing or disallowing compensation in such situations.[356] There have been many instances where different entities making provisions in their respective claims seeking compensation for loss or damage occurred due to preventive measures, but not everyone is successful in such attempts, due to the non-application of the specific action or remoteness of such, with regard to the exact purpose for which it should have been carried out. Among those recoverable, there may prevail cases with regard to human health problems directly resulting from such situations including personal injuries, damage to clean-up vehicles, damages and repair costs of roads that had been used by heavy machinery and clean-up vehicles in such unordinary process[357], and other such loss or damage occurred during salvage and clean-up operation as a matter of taking

355 See J. Basedow, U. Magnus, *Prevention of Marine Pollution*, Pollution of the Sea – Prevention and Compensation, Hamburg Studies on Maritime Affairs, vol. 10, part II, Springer, pp 41–61; N. P. Ventikos, I. D. Genikomsidou, E. D. Grigoriou, *Oil Pollution in Greek Seas: Critical Review of Confrontation and Waste Management*, Submission of Greece to the 9th Meeting of the Focal Points of the Regional Marine Pollution Emergency Response Centre for the Mediterranean Sea (REMPEC), Malta, 21–24 April 2009, available at http://www.rempec.org/admin/upload/publications/FPM09WG30INF9(E)Consolidated.pdf; K. W. Ketkar, *The Oil Pollution Act of 1990: A Decade Later*, Spill Science & Technology Bulletin, vol. 7, Issues 1–2, June 2002, pp 45–52; J. Hay, O. Thebaud, J. P. Agundez, *Preventing pollution through the compensation for damage? An appraisal of the European experience in the field of marine oil spills*, No. 2007, available at http://www.uibcongres.org/imgdb//archivo_dpo4084.pdf; M. Turan, *Turkey's Oil Spill Response Policy: Influences and Implementation*, 2009, available at http://www.un.org/depts/los/nippon/unnff_programme_home/fellows_pages/fellows_papers/turan_0809_turkey.pdf

356 See IOPC Fund Annual Report, 1989 on *Kasuga Maru* No.1 incident, p. 35

357 For example, the *Tanio* incident, France 1980, where the vessel sailing from Wilhelmshaven to Civitavecchia with 26,000 tons of heavy fuel oil and 9,000 tons of bunker fuel oil was caught with a violent storm that spilt the tanker into two. See ITOPF, http://www.itopf.com/information-services/data-and-statistics/case-histories/tlist.html#TANIO; R. H. Ganten, *The Tanio Spill: A case history illustrating the work of the International Oil Pollution Compensation Fund*, Proceedings of the 1985 Oil Spill Conference, 135–139, API Publication No. 4385, American Petroleum Institute, Washington DC, USA; C. Brac, *Operation Tanio*, La Nouvelle Revue Maritime, July 1981, 363, pp 10–25

"preventive measures" including pure economic loss such as loss of business[358] or interruption to normal working schedules of people in the vicinity of the area under concern. The causation plays a vital role in determining whether such claims have actual base in maintaining the case seeking compensation. As mentioned earlier, remoteness or absence of the said causal link between the preventive measures and the harm,[359] or gross negligence or contributory negligence on the part of the claimant[360] would weaken such claims.

Exclusions

The application of the Convention is restricted to a certain extent by Article 4, which is expressly titled as 'exclusions'. Therefore, one must pay attention to the purview of those excepted situations, before proceeding further on the Bunkers Convention with regard to any incident that cause pollution damage. Basically, this provision deals with two types of situations where such exception applies. Accordingly, it could be categorized as oil pollution incidents that fall within the ambit of the CLC regime[361] and to ships which enjoy the benefits of state immunity[362]. In order to obtain a clear understanding, it would be prudent to take these issues separately.

Situation under the CLC

Though, it is not the intention to open up a comprehensive discussion on the application of the CLC regime at this juncture, as most of its provisions are dealt in respective stages of this work, one may still finds it necessary to discuss its relevance to the present situation. Article 4.1 expressly excludes the application

---

358 See *Aegean Sea* incident, *Supra* n. 296. See also Review of Decisions 1979–93, FUND/WGR.7/3 para. 3.1.3

359 *Supra*, pp 59–65

360 *Supra*, pp 20–24

361 " [1] This Convention shall not apply to pollution damage as defined in the Civil Liability Convention, whether or not compensation is payable in respect of it under that Convention"

362 "[2] Except as provided in paragraph 3, the previsions of this Convention shall not apply to warships, naval auxiliary or other ships owned or operated by a State and used, for the time being, only on Government non-commercial service
[3] A State Party may decide to apply this Convention to its warships or other ships described in paragraph 2, in which case it shall notify the Secretary-General thereof specifying the terms and conditions of such application"

of the present regime to the extent it falls within the meaning and application of the CLC. It specifically rules out any pollution damage that cause within such conventional meaning, irrespective of being able to succeed or fail in recovering damages and compensation for the alleged incident. Evidently, the CLC provide the possibility for the recovery of compensation for pollution damage that occur from bunker oil used in ships that fall within the provided meaning of that regime. Even though, the definition of 'pollution damage' does not provide a straight forward wording to this effect[363], the term 'oil' as defined[364] under the specific Convention holds positive justifications. Accordingly, the CLC regime allows the application of its provisions with respect to pollution damage caused by specified categories of oil "whether carried on board a ship as cargo or in the bunkers of such a ship".

As the CLC applies to pollution damage caused by ships that are commonly known as 'tankers' carrying oil as cargo, consequently, these ships fall into the category of 'CLC ships'. Any bunker oil spill associated with such types of ships would certainly fall inside that conventional meaning. Therefore, such CLC ships are excluded from the scope of application of the Bunkers Convention. Resultantly, any pollution damage that is caused by bunker oil carried in the bunkers of such ship has to be addressed only under the provided criteria of the CLC regime. Throughout the drafting of the Bunkers Convention, it was the intention of the delegations to avoid any over-lapping situations of having one incident been addressed in more than one regime of similar nature. On the other hand, the HNSC has overcome the same effect by its corresponding provisions in Article 4.3(a). However, the Bunkers Convention does not exclude the application of any pollution damage caused by any bunker oil used by HNS conventional ships, as it is clear from the relevant provisions dealing with exclusions in both these regimes.[365]

Apart from avoiding over-lapping situations, there are several other reasons for justification of its actual exclusion. One strong conviction is that such situation would have paved way for complexities in the judicial process requiring the shipowner to constitute more than one limitation fund, and or institute multiple limitation proceedings based on those separate regimes. Furthermore, much severe technical problems would have occurred in separating types of damages based on different types of oil, namely damage that occur due to cargo oil and that of bunker oil. Therefore, the provided exclusion is certainly a convincing approach.

363 See Art. 1.6; "Civil Liability Convention" means the International Convention on Civil Liability for Oil Pollution Damage, 1992, as amended

364 See Art. 1.5

365 See A. E. Boyle, *Globalising Environmental Liability: The Interplay of National and International Law*, Journal of Environmental Law, 2005, 17(1):3–26

Likewise, other instruments that are applied in civil navigation, all categories of state ships[366] fall outside the scope of this Convention. Article 4.2 expressly excludes ships used in military services such as warships and naval auxiliaries, and other ships that are used in Governmental services to the extent that they are engaged on non-commercial service operations.[367] However, the margin of appreciation has been narrowed by the use of the words *only on Governmental non-commercial service for the time being.* Therefore, ships used for quasi-Governmental purposes comprising of some commercial character would thereby fall within the meaning of the Convention. In such a situation, the onus of proof would rests on the victim in alleging that such ship was partially used for commercial purposes at the time of the bunker oil spill which caused pollution damage. In circumstances where State ships have been used for commercial purposes[368] outside their normal operations at governmental service levels, such ships would in no way fall outside the conventional meaning. In events of pollution damage that has been caused by bunker oil used by such ship, the State Party in question would no more be allowed to apply the rule of state immunity, in order to escape liability[369] for such occurrence. Resultantly, such State Party would have to undergo the normal procedure of the law laid down by the Bunkers Convention.

Having been derived from the corresponding provisions of the CLC[370] and HNSC[371] regimes, the Bunkers Convention uses the same wordings in excluding ships owned and operated by States, from its ambit. However, States Parties are vested with the option of applying this Convention to warships and such other ships that were mentioned above, at their discretion, by virtue of the proceeding paragraph.[372] In such circumstance, the State Party which so decides shall notify the Secretary-General, meaning the Secretary-General of the IMO[373]. Such State

366 See C. R. Dunlop, *Immunity of State Ships*, Journal of Comparative Legislation and International Law, Third Series, vol. 6, No. 4 (1924), Cambridge, pp 271–277

367 See *Supra* pp 43–44

368 See M. H. Nordquist, S. N. Nandan, S. Rosenne, *United Nations Convention on the Law of the Sea, 1982: a commentary*, Kluwer, 2002, pp 237–239; S. K. Verma, *An Introduction to Public International Law*, pp 165–166

369 See Art. 4.4

370 See Art. XI.1

371 See Art. 4.4; "With respect to ships owned by a State Party and used for commercial purposes, each State shall be subject to suit in the jurisdictions set forth in article 9 and shall waive all defences based on its status as a sovereign State"

372 See Art. 4.3; "A State Party may decide to apply this Convention to its warships or other ships described in paragraph 2, in which case it shall notify the Secretary-General thereof specifying the terms and conditions of such application"

373 See Art. 1.13; "Secretary-General means the Secretary-General of the Organization"

may further specify any underlying terms and conditions that it intends to impose for such application.

Chapter 3: Status Quo – Liable Agrieved Parties

This Chapter focuses on the broad issue of the liable party to whom liability should be attributed in terms of the Bunkers Convention, and the potential victims who would ultimately fall into grievances upon an incident of pollution damage caused due to oil spills from ships. The question of determining the "liable party" in particular has gained enormous attention throughout the development of this area of the law, dating back to the era of CLC 69.[374] It has not only gathered high controversies during all such law making processes, but also being one of the most sensitive issues which needed careful consideration and drafting. The development of the law in general has posed severe questions that need positive solutions, especially with regard to the identification of the exact party so liable, in overcoming poor situations of drafting. Therefore, much of the attention was paid during the drafting of the Bunkers Convention in addressing this issue as a whole with the view of overcoming such shortfalls. Attribution of liability to the person so liable for the wrong doing, being the key issue in any litigation, received tremendous backup from the participated delegations in arriving at a workable and acceptable solution.

Personal Scope of Application: jurisdiction ratione personae

The Liable Party

The issue of the liable party gains the central point of importance in any legal proceeding of this nature. In international maritime jurisdiction for liability and compensation it plays an important role in identifying the exact person liable for such wrong doing. Most importantly it becomes the central issue to be resolved before applying strict liability for pollution damage and consequently determining against whom the action be instituted. Following the same pattern as adopted

374 See A. Mandaraka-Sheppard, *Modern Admiralty Law: with risk management aspects*, Cavendish, 2001, pp 944–946; G. Doeker, T. Gehring, *Private or International Liability for Transnational Environmental Damage – The Precedent of Conventional Liability Regimes*, JEL 1990, Oxford, 2(1):1–16

under the CLC 69[375] & 92[376] and the HNS Convention[377], the Bunkers Convention lays the liability on the "shipowner". Article 3(1) states *inter alia*;

"[...] the *shipowner* at the time of an incident shall be liable for pollution damage caused by any bunker oil on board or originating from the ship..." (emphasis added).

Therefore, it is important to investigate the intention of the drafters in including the term "shipowner" within the ambit of liable party under the Convention with a view of ascertaining exactly, the personalities that fall under the given definition. The Legal Committee had to deal with complex issues before the final draft was presented for adoption concerning the interpretation of this said term. The following includes an analysis taken from the *travaux préparatoires* of such heated debates that took place prior to the arrival at an acceptable compromise.

"Ship Owner"

Originally, there were four options for the definition of 'shipowner' during the initial work at the Legal Committee.[378] These options were separately considered by the delegations that represented, and many views were in favour of inserting the best definition that could cover all types of ownerships[379] and consequently attaining the best solution.

Option 1:

"Shipowner" means the owner [, charterer, manager and operator] of the ship.

Option 2:

"Shipowner" means the owner, charterer, manager and operator of the ship; and "Responsible party" means the owner of the ship, or any other organization or

375 See Art. III.1
376 See Art. III.1
377 See Art. 7.1
378 See LEG 78/5/2
379 See K. Mitroussi, *The role of organizational characteristics of ship owning firms in the use of third party ship management*, Marine Policy, vol. 28, issue 4, July 2004, pp 325–333; G. V. Scammell, *Shipowning in England c. 1450 – 1550*, Transaction of the Royal Historical Society (1962), 5th Series, Cambridge, 12:105–122; J. D. Tracy, *The Rise of Merchant Empires: long- distance trade in early modern world 1350 – 1750*, Cambridge (19990 rep. 1993)

person who or which has assumed responsibility for compliance with the insurance requirements of this Convention.

Option 3:

"Shipowner" means the owner of the ship, or any other organization or person who or which has assumed responsibility for the operation of the ship.

Option 4:

"Shipowner" means the person or persons registered as the shipowner or, in the absence of registration, the person or persons owning the ship. However, in case of a ship owned by a State and operated by a company which in that State is registered as the ship's operator, "shipowner" shall mean such company.[380]

Viewing these four options from the outset it is clear that the intention of the delegations was to insert the most appropriate way of identifying the liable party having effective control of the ship. Option 1 was straight forward and much easier in channeling liability to one person rather than making confusions with an extensive use of any associated terms of identification. On the other hand, option 2 would have covered more liable parties, but of the views of the delegations which supported that idea, it needed further drafting including in its definition, the 'registered owner'.[381] Obviously, a full coverage of liable parties would enhance benefits for the victims, but this needs careful consideration in order to list exactly, the persons so liable. The main question that arose in respect of this option was whether the ship owner could arrange insurance for all liable parties. The absence of a positive solution with respect to such an issue would result in creating disputes concerning multiple insurances.[382] In a volatile insurance market for shipping industry[383] this could mean potential rejections of many insurance policies by respective under-writers in respect of the same ship. Therefore, the matter could gather more negative outcomes in such a context. In view of these concerns, the Committee took the decision to attend to the liability principle at first sight due to its importance of identifying the most appropriate person as the

380 LEG 78/11, 2 November 1998

381 Option 2 used the text based on the ISM Code. See LEG 79/6/1, 12 February 1999, Para 8

382 LEG 79/6/1, 12 February 1999, para 106

383 See S. S. Huebner, *Marine Insurance*, Biblio Bazaar, 2008; P. Bennet, *Mutual risk: P&I insurance clubs and maritime safety and environmental performance*, Marine Policy, vol. 25, issue 1, Jan 2001, pp 13–21

responsible party, and then to pay attention to the insurance aspect.[384] This was indeed quite clever and timely motivation. Although there were much support for option 3, some of the delegations were quite uncertain and questioned the use of the term 'charterer'[385] used in it and in option 1 and 2 as well, since it could also mean a 'voyage charterer'[386] who would not have control over the operation of the ship in normal circumstances.[387] In their view, the liable person must be as close as possible to the operator of the ship, and one delegation noted further that the owner of the ship might not always be the shipowner.[388] The Option 4 was based on the CLC 69 & 92 definition of "owner", which gathered much of the support from participated delegations. Formally, the four options were then reduced to two after the conclusion of the 78th Session, and left option 1 and 4 for further consideration at the next session.

The central issue of adopting a concrete definition for 'shipowner' remained unresolved before the delegations attended the 79th Session of the Legal Committee.[389] Much of the concentration was centered on the 1st and the 4th options that were presented during the preceded sessions. The words "bareboat and demise"[390] matched the definition more than leaving the word 'charterer'[391] open, which could possibly lead to complex situations such as meaning 'term charterer'[392] or 'time charterer'[393] who do not have controlling powers over the management of the

384 *Supra* n. 382, para 112

385 Charterer is a party which enters into a contract for the use of an entire vessel for a certain voyage or a period of time

386 In voyage charter party, the shipowner undertakes to transport specified cargo under an agreed payable freight for the voyage between designated ports

387 *Supra*, n. 382, para 108

388 *Id,* para 109

389 See LEG 79, 12 February 1999

390 These two terms are used in a common context to mean charter-parties that lease ships. The ship itself is hired empty, where it is manned and operated by such charterers for their account with the owner, upon taking possession and control of such ship. They are ultimately become responsible for all navigational, managerial and commercial activities carried out by them during such use. See Dr. A. Odeke, *Bareboat charter (ship) registration*, Kluwer, 1998, pp 43–58; I. Carr, P. Stone, *International Trade Law*, 3rd ed., Cavendish, 2005, pp 165–168

391 See The Charterers P&I Club Newsletter, *The Charterer*, July 2009, available at http://www.else.co.uk/Documents/The%20Charterer%20July%202009.pdf

392 Chartered for a certain period. See M. Dockray, K. R. Thomas, *Cases & materials on the carriage of goods by sea*, 3rd ed., Cavendish, 2004, p 5

393 Chartered for a certain period of time. See M. Dockray, K. R. Thomas, *Cases & materials on the carriage of goods by sea*, 3rd ed., Cavendish, 2004, p 3–4; Robert Gay, H. Dickinson, *Time charterparty – whether implied term that charterers would not undertake additional voyage – whether promissory estoppel against charterers*, JIML 15[2009] 3: 213; T. Nunes, Charterer's Liabilities under the Ship Time

vessel likewise 'voyage charterer'.[394] Although they have a slight legal difference in meaning, they were recognized in terms of long term charter interests they possess. But however, these words were included in square brackets for further understanding as to the specificity of the term 'charterer'.[395] In respect of option 1, the Legal Committee was able to secure much support due to its consistency with the term 'shipowner' defined in the LLMC Convention,[396] which indeed was influential in setting up limitation levels in the Bunkers Convention. It was the idea of the delegations that this definition would suffice the imposition of liability on a group of persons through an effective responsibility approach rather than citing one person, and a further insertion of the words 'joint and several' on each party so responsible under the liability clause[397], which would enable the victims in obtaining compensation faster and without confusions as to the exact person so liable or as to each others' component of liability. The intervention of the US delegation, drawing examples from its national legislation,[398] made a huge impact on accepting a concept having a wider coverage for 'responsible party'.[399] At the very beginning, it was felt by the Committee that the liability should not attach to the 'registered owner' merely because he is so registered as the owner of the ship, but to impose on the party so responsible for day-to-day operations of the ship, such as the charterer, manager or operator. Unlike in situations of tankers, the proposed Convention deals with bunkers, where all ships are equipped with on board, and damage caused by such bunkers is directly linked to the operation of the ship.[400] Nevertheless, it would be much beneficial on the part of the victims, to impose liability on the 'registered owner' in order to ensure speedy settlements of

---

Charter, Houston Journal of International Law, vol.26, 2004; IMT *Shipping and Chartering GmbH v. Chansung Shipping Co. Ltd., (The Zenovia)* [2009] EWHC 739 (Comm)

394 See C. J. Giaschi, *Carriage of Goods, Bills of Lading and Charterparties*, U. B. C. Maritime Law 332, available at http://www.admiraltylaw.com/UBC%20Law332/carriage_of_goods_outline.pdf;

395 *Supra*, n. 389

396 Art. 1.2 of the LLMC defines 'shipowner' as "the term shipowner shall mean the owner, charterer, manager and operator of a seagoing ship"

397 Article 3(1bis) in the draft presented at LEG 79

398 The US motivation was primarily based on its OPA 1990, and as a matter of public law in combating oil spill prevention, spill response while emphasizing on compensational aspects. They relied on the fact that when more parties are held responsible, it has become workable, practical and simple process in their jurisdiction. See LEG 77/WP.3, 22 April 1998, p.23

399 *Supra*, n. 378, p. 14

400 The International Chamber of Shipping pursued this idea by way of note presented to the Legal Committee favouring the inclusion of the concept 'responsible party'. See LEG 74/4//4, 13 September 1996

claims and convenience in identifying the liable party.[401] It was further considered that the inclusion of 'registered owner' in both options would guarantee an exact identifiable person rather than leaving it as with reference to the term 'responsible party', which was in existence in the initial drafts during the preceded sessions and subsequently deleted upon the acceptance of the term 'registered owner' in favour of the actual responsibility concept. Furthermore, it was the idea that such 'registered owner' would be required to maintain insurance and other financial security as needed by draft Article 7,[402] consisting one of the three main elements on which this liability regime has been developed. Therefore, the inter-play between the terms 'shipowner' and the requirement of compulsory insurance from the 'registered owner' would avoid any misunderstandings and further develop a criteria for the recognition of one prime person for identification as to channeling of liability compatible with direct actions brought against the insurer.[403] This would further enhance the LLMC model definition for 'shipowner', evading the situation where each of the listed persons required having separate insurances.[404]

Referring to Option 1, the real need for the inclusion of the "demise charterer" in the definition and furthermore remove the square brackets that were inserted within, was highlighted by some delegations at the 81st Session due to conceptual and practical difficulties arising out of actual operations of the ship, as the "bareboat charterer" does not participate in the crewing of the ship.[405] On the other hand, most of the other delegations were in favour of dropping out the term "demise charterer" from the definition, pointing out the modern developments of the area under which, "demise charterer" fell within the meaning of the concept of "bareboat charterer".[406] The latter persuasion led the Committee to delete the term "demise charterer" from the remaining definition in Option 1.[407] As an alternative to channel responsibility to one person as a matter of enabling the victims to institute direct actions and evade confusions as to the identification of

401 *Supra*, n. 51, p 2

402 *Id*

403 LEG 80/4/1, 13 August 1999

404 *Supra*, n. 378, p. 2

405 For terms and conditions in various bareboat charterparties, see Miami Charterers – Bareboat Charter Agreement, http://www.miamicharters.com/2002/vessels/Fastrack%20Bareboat%20Contract.doc; Winds of Ireland, http://www.windsofireland.net/files/contract.pdf. See also E. T. Laryea, *Paperless trade: opportunities, challenges and solutions*, Global Trade and Finance Series, Kluwer, 2002, pp 57–59; Judgment – *BP Exploration Operating Co. Ltd. v. Chevron Transport (Scotland)*, [2001] UKHL 50

406 See Dr. A. Odeke, *supra* n. 390, p 57; Dr. V. J. G. Power, *EU competition law on sharing information between competitors*, JIML 15 [2009] 3: 224

407 LEG 81/WP.2, 30 March 2000, paras. 45–47

the person so liable, it was the idea of the Committee to re-arrange the definition with much clearer terms. Being the pioneers of such, the International Group of P&I Clubs indicated the practical difficulties which are associated with the use of several responsible parties in one definition, making them jointly and severally liable, in situations where some of them are not known to the Club due to the absence of mutual agreements between those persons and the clubs.[408] In turn it is much easier to include a definition for the term "registered owner" who is liable in such circumstances allowing the victims to institute direct actions against their respective insurers, and leave the term "shipowner" as agreed under Option 1 in which case the right of recourse is reserved for such responsible party to institute actions in other forums with regard to their *status quo*,[409] arising out of contractual obligations under insurance arrangements. Acceptance of this provision led to the deletion of Option 4, which was based on the definition of "shipowner" in the CLC 92, and subsequently introducing instead, the term "Registered owner" in facilitating the liability aspect and that of the compulsory insurance at once.[410] At the end of this Session, the Committee was able to arrive at a compromise defining both the terms "Shipowner" and "Registered owner" in allowing a proper identification of the liable party and further to facilitate direct actions against the insurer or financial security provider in cases of absence of a clear ownership.

The ultimate compromise

Due to the aforementioned reasons and as an ultimate compromise, the final draft includes definitions as to both "shipowner" and "Registered owner" in identifying the proper responsible and liable party to oil spills from the bunkers of ships, ensuring the payment of adequate, prompt and effective compensation.[411] Accordingly;

"Shipowner" means the owner, including the *registered owner,* bareboat charterer, manager and operator of the ship.[412] (emphasis added); and

"Registered owner" means the *person* or *persons* registered as the owner of the ship, or in the absence of registration, the *person* or *persons* owning the ship. However, in the case of a ship owned by a State and operated by a company which in that State is registered as the ship's operator, the *"registered owner"* shall mean such company.[413] (emphasis added).

408 *Supra*, n.403, p. 2, at para. 8
409 *Supra*, n. 407, para. 27
410 *Id.*, at para. 49
411 See Preamble to the Bunkers Convention
412 Art. 1.3
413 Art. 1.4

In further clarification of the term "*person*", the Bunkers Convention uses the same definition that is available under the CLC 92[414] (as in-existence in CLC 69[415]) and HNSC[416], to mean;

"… any individual or partnership or any public or private body, whether corporate or not, including a State or any of its constituent subdivisions".[417]

Legal implication on responsible corporate entity

It is well understood that natural persons possess the competency of appearing in any law suit made by or against such person.[418] But the situation with regard to a corporate body varies from one jurisdiction to another.[419] One strong objection is that the Convention has nowhere defined what is meant by the term 'corporate', whether it refers to the interpretation given under a respective national law in cases of litigation or based on the private international soft-law interpretation, such as the International Safety Management (ISM) Code 2002[420], which defines a "company" as "the owner of the ship or any other organization or person such as the manager, or the bareboat charterer, who has assumed the responsibility for the operation of the ship from the shipowner and who, on assuming such responsibility, has agreed to take over all duties and responsibility imposed by the Code"[421].

414 Art. I.2
415 Art. I.2
416 Art. 1.2.
417 See Art. 1.2
418 See J. Teichman, *The Definition of Person*, Philosophy, Cambridge, vol. 60, No. 232 (Apr., 1985), pp 175–185; L. J. Kirmayer, *Psychotherapy and the Cultural Concept of the Person*, Transcultural Psychiatry, vol. 44, No. 2, 232–257 (007); S. Deneulin, M. Nebel, N. Sagovsky, *Capabilities and Rights*, in Transorming Unjust Structures The Capability Approach, Springer, 2006, pp 17–26
419 See J. Dewey, *The Historic Background of Corporate Legal Personality*, The Yale Law Journal, vol. 35, No. 6 (Apr., 1926), pp 655–673; K. Iwai, *Persons, things and Corporations: The Corporate Personality Controversy and Comparative Corporate Governance*, The American Journal of Comparative Law, vol. 47, No. 4 (Autumn, 1999), pp 583–632; P. I. Blumberg, *The Multinational Challenge to Corporation Law: The Search for a New Corporate Personality*, Oxford, 1993; K. E. Goodpaster, *The concept of corporate responsibility*, Journal of Business Ethics, Springer, vol. 2, No. 1 (Feb., 1993), pp 1–22; G. Berens, C. van Riel, *Corporate Associations in Academic Literature: Three Main Streams of thought in the Reputation Measurement Literature*, Corporate Reputation Review, Palgrave McMimillan, vol. 7, No. 2 (July 2004), pp 161–178(18)
420 See full text, http://www.imo.org/HumanElement/mainframe.asp?topic_id=287. See also R. Asariotis, *Implementation of the ISPS Code: an overview of recent developments*, JIML 11[2005] 4: 266
421 Part A, I General, 1.1.2

Although, the above interpretation has a certain persuasive value[422] it does not provide a standard basis under the Bunkers Convention. Therefore, the absence of a clearer meaning to this end would not only create chaos in the Convention itself, but also lead to uncertainties with regard to several other issues, such as requirements for legal recognition in a given State as to its validity and other issues pertaining to the nationality or flagship of a Ship operated by such company.[423] Apart from registration, the most complex issue arises with regard to, applicable law to a 'corporate' body in case of liability and right of action. Before attending to the merits of the case, a national court would always require to identify the legal implications on the recognition of the parties, such as a corporate body in that particular jurisdiction. Absence of uniform rules in private international law concerning, mutual recognition of legal personality of companies, would lead to many conflict of law issues. The failure in 1956[424] in arriving at a uniform set of rules concerning this issue has enlarged the controversies, leaving it to be settled under respective national laws. The situation remains at high risk due to different States recognizing 'incorporation theory'[425] or 'real seat theory'[426] under their respective company law. Based on such laws in Europe[427] and US, Drury forwards the point that a company, being a creature of the legal system which created it, can have no existence outside the ambit of that law.[428] The classical doctrine of recognizing corporate entities within one's own legal system has tended to dete-

422 See A. M. Sheppard, *Modern maritime law and risk management*, Ch. 8, 2nd ed., Routledge – Cavendish, 2007; I. C. Varotsi, D. A. Pentsov, *The International Safety Management Code* in *Maritime Work Law Fundamentals: Responsible Shipowners, Reliable Seafarers*, Springer, 2008, pp 640–664; *A.* Rodriguez, M. C. Hubbard, Fowler, Kingsmill, Flint, Gray, Chalos, LLP, *The International Safety Management (ISM) Code: A New Level of Uniformity*, available at http://www.frc-law.com/files/pub_ISM.pdf

423 See R. Schaffer, F. Agusti, B. Earle, *International Business Law and its Environment*, South-Western Cengage, (2005 ed., 2009) p. 60

424 1956 Draft Treaty of the Hague Conference on Mutual Recognition of the Legal Personality of Companies.

425 Used mostly in common law countries and several other countries such as The Netherlands and Switzerland, where the company is governed by the law according to which it is duly formed. Once such company has met formation requirements, then that company is recognized anywhere

426 Applied in countries such as Germany, France and Italy. The 'real seat theory' requires a company to have a registered office for mere controlling policies

427 See P. Pelle, *Companies crossing borders within Europe*, Utrecht Law Review, Igitur, vol. 4, Issue 1, (March), 2008, available at http://www.utrechtlawreview.org/publish/articles/000056/article.pdf; *EC* Law Aspects of Hybrid Entities, Part II, available at http://www.ibfd.org/portal/pdf/Fibbe_Sampleexcerpt.pdf; P. P. Craig, G. de Burca, *EU Law: Text, cases, and materials*, 4th ed., Oxford, 2008, pp 807–819

428 Drury (1998) at p. 176

riorate in wake of the modern legal theories, where status of a company is defined by the *lex societatis*,[429] rather than *lex fori*. Accordingly, the incorporation theory overruns the real seat theory in several respects.[430] In present day commercial structure where multinational companies emerging at a greater speed and non existence of a registered office of a company duly established, the 'incorporation theory' becomes more suitable.[431] Hence, the recognition of such company as a duly incorporated entity in a State where incorporation is adhered makes it acceptable to all States adhering to incorporation theory. The differing situation of law relating to recognition of companies tends to become a more serious issue in borderline cases. As Rammeloo points out, both civil and common law courts make use of the substantive-law oriented device of good faith in resolving such conflicts of laws.[432] Anyway, a court attending to a liability of a corporate body or the right of such a company to sue any liable party with respect to the proper law, as mentioned and defined under the interpretation of 'person' in the Bunkers Convention will first have to solve the matter of due recognition, before dealing with the action, in order to ascertain whether such corporate body is recognized as a legal personality in that jurisdiction having competency to deal with the matter.[433] In failing such requirement, the corporate body in question will have to be represented by natural persons representing such company inside the corporate

429 See S. Ebert, *The law applicable to groups of companies involving European companies (Societas Euopea)*, Company Lawyer, vol. 24, no. 4, 2004, pp 108–113; M. Johnson, *Roll on the 14th Directive – Case law fails to solve the problems of corporate mobility within the EU – again*, Hertfordshire Law Journal, vol. 2, no. 2, 2004, pp 9–18; S. Ebert, *European Union: Company Law – Freedom of Establishment*, International Company Commercial Law Review, vol. 14, no. 5, 2003, pp 51–52

430 See http://www.itutorials.co.uk/freedom_of_establishment.php. See also S. Chertok, T. Coburn, *Jurisdictional Competition in the European Community*, Expresso Preprint Series, 2005, paper 870; P. McEleavy, E. Micheler, *Current Developments – Private International Law*, International and Comparative Law Quarterly, vol. 52, no, 2, 2003

431 See H. Xanthaki, *Centros: Is this really the end for the theory of Siege Reel*, Company Law, vol. 22, no. 1, 2001, pp 2–8; T. Koller, *The English Limited Company ready to invade Germany*', International Company and Commercial Law Review, vol. 15, no. 11, 2004, pp 334–343. See also *Centros*, [1999] E. C. R. I-1494; *Uberseering BV v.. Nordic Construction Company Baumanagement GmbH (NCC)*, [2002] E. C. R. I-0991

432 Stephan Rammeloo, "Corporations in Private International Law, A European Perspective", Oxford, 2001, p. 241

433 See K. Baelz, T. Baldwin, *The End of the Real Seat Theory (Sitztheorie): the European Court of Justice Decision in Uberseering of 5 November 2002 and its impact on German and European Company Law*, German Law Journal, available at http://www.germanlawjournal.de/article.php?id=214

veil.[434] In conclusion, it is worthwhile stating that, the absence of uniform rules pertaining to recognition of legal personality of international companies has left the issue totally dependent on the national laws governing this area of the law.

Turning over to the main discussion, one may find somewhat broad and detailed definitions as to the liable party that falls within the context of a potential defendant in law suits brought by the victims of oil spill pollution damages aftermath such incidents. One could list different types of such persons against whom successful actions could be instituted and maintained under the Bunkers Convention. These categories of different responsible parties are examined below, but however, are not intended to be limited to these discussed categories alone, as the complete list of potential defendants could be exhaustive.

(a) The owner / registered owner

The Bunkers Convention imposes liability on the "shipowner", alleging the owner at the time of the incident. It includes the registered owner of the ship, or in the absence of such ownership by registration, the person or persons owning the ship. The burdens on the registered owner seem to be much greater than other persons listed in the definition of shipowner. Especially with regard to the compulsory insurance requirement laid down in Article 7, it is the registered owner who is primarily and solely under the obligation of satisfying such insurance and financial security requirements.[435] However, there is a broad coverage under this definition where number of different entities could fall into the meaning as described. It covers both natural and juridical personalities whether registered or not. Furthermore, State owned ships operated by companies registered as respective operators in such States also fall into the category of registered ships within the meaning of the Convention. Although it provides a reasonable test to identify

434 See J. Rush, M. Ottley, *Business Law*, Thomson, 2006, pp 200–201 *D.* Kelly, A. E. M. Holmes, R. Heyward, *Business Law*, 5th ed., Cavendish, 2005, pp 313–314; S. Goulding, *Company Law*, 2nd ed., Cavendish, 1999, pp 66–80; M. Ottley, *Briefcase on company law*, 2nd ed., Cavendish, 2002, pp 2–8. See also *Hashem v. Shayif & Anor* [20008] EWHC 2380 (Fam)

435 Art. 7.1: "The registered owner of a ship having a gross tonnage greater than 1000 registered in a State Party shall be required to maintain insurance or other financial security, such as the guarantee of a bank or similar financial institution, to cover the liability of the registered owner for pollution damage in an amount equal to the limits of liability under the applicable national or international limitation regime, but in all cases, not exceeding an amount calculated in accordance with the Convention on Limitation of Liability for Maritime Claims, 1976, as amended."

the owner of a ship[436] based on the definition, it will always not be straight forward in certain circumstances.

Application of 'alter ego'

There have been overwhelming situations where ships are registered by way of companies behind the corporate veil,[437] but posing complex issues as to their true existence in different legal dimensions with respect to validity and operation. In order to establish liability, it is important to identify the exactly liable natural or legal person. Therefore, ownership upon mere façade would pose severe questions as to the existence of such company as an owner of a ship. Moreover, operational misconducts of the actual operators who hide behind such corporate veils would have to be taken into consideration in resolving complex issues. Especially in circumstance where the corporate liability is limited,[438] piercing the corporate veil would enable to compensate extensive pollution damages as a matter of remediation and restoration.[439] It results in lifting the corporate veil[440] by the respective courts[441] in identifying natural personalities behind such companies who are sometimes personally responsible in situations of corporate bankruptcy[442], but the actual practice seemed to be very limited. Obviously, a court would lift the veil in circumstances of a clear *prima facie* fraud leading to the discovery of a invalid formation or existence, fraudulent conveyance or as the case may require the court to do so in identifying the actual responsible person against whom a

436 See Lloyd's Register, Fairplay, http://www.lrfairplay.com/IMO/imo.html. See also IMO Ship identification number scheme, http://www.imo.org/Facilitation/mainframe.asp?topic_id=388.
437 See OEDC, Maritime Transport Committee, *Ownership and Control of Ships*, March 2003, available at http://www.oecd.org/dataoecd/53/9/17846120.pdf
438 See A. M. Sheppard, *supra* n. 374, p 122–124; R. Price, A. Haberbeck, *The maritime laws of the Arabian Gulf Corporations Council States*, vol. 1, Graham & Trotman, 1986, pp 197–200
439 See Bergkamp, *supra*, n. 82, p 307
440 See J. Payne, *Lifting the Corporate Veil: A Reassessment of the Fraud Exception*, The Cambridge Law Journal (1997), 56:284–290; K. Roepke, *Gluabigerschutzregime im europaischen Wettbewerb der Gesselschaftsrecht* in *Schriften zum europaischen unde internationalen privat-, Bank- und wirtschaftsrecht*, De Gruyter , 2007, p 104
441 See C. C. Nicholls, *Corporate Law*, Ch. 7, Edmond Montgomery, 2005, pp 188–214; L. S. Seally, S. Worthington, *Cases and materials in company law*, 8th ed., Oxford, 2008, pp 52–72
442 See F. B. Cross, R. L. Miller, *The Legal Environment of Business: Text and Cases*, 7th ed., South Western, Cengage (2007 ed., 2009), p 450

true legal action could be sustained, but it would not always be done for mere adjudication. This generates high controversies in situations where directors or officers of such corporate entity are the real operators of a ship. Many States tend to depend on the concept of 'alter ego'[443] in such circumstances, especially where the responsibility is directed towards the real commanding officer of a corporate entity.

There indeed had been number of maritime cases in which such practice as to the piercing of corporate veil and applying this concept had been carried out in number of jurisdictions. Especially in the United Kingdom, since the enactment of the Merchant Shipping Act 1894, the issue obtained tremendous attention. In the case of *Lennard's Carrying Co.*,[444] the court held that the liable party is someone who acts in his official capacity for which the company becomes liable in such a way that his action is the very action of the company itself. At present the United Kingdom, by virtue of the Merchant Shipping Act 1995 allocates liability only on one person who is registered as the 'registered owner';[445] that is the corporate entity in such a case. On the other hand, it is not clear from the OPA 1990; to what extent a US court would allow to impose liability on natural persons acting in violation of their official capacities in carrying out the required functions as operators, upon piercing the corporate veil.[446] In Germany, Belgium and The Netherlands, it would be possible only in circumstances of abuse of limited liability of companies, by the real actors. Naturally, a corporate officer is individually liable for torts committed personally, and therefore is not allowed to hide behind the corporate veil. Although such situations would attribute vicarious liability on the corporate body,[447] the secondary liability of that individual will remain as a clear case.

---

443 See M. A. Sheppard, *supra* n. 374, pp 919–920; M. Spadaccini, *Forming an LLC: In Any State*, Entrepreneur Media Inc., 2007, pp 87–88; M. Spadaccini, *The Operations Manual for LLCs*, Entrepreneur Media Inc., 2008, pp 237–241

444 [1915] A.C 705

445 1995 c.21

446 See W.T. Miller, , "A Summary of Intragroup Liability in Light of the Oil Pollution Act of 1990", document contained in documentation of conference 'Oil Pollution – Claims, Liability and environmental Concerns', IBC 23 October 1991, p. 4

447 See G. Stessens, *Corporate Criminal Liability: A Comparative Perspective*, Int'l and Comparative Law Quarterly (1994), 43:493–520; N. Campbell, J. Armour, *Demystifying the Civil Liability of Corporate Agents*, The Cambridge Law Journal (2003), 62:290–304; R. B. Grantham, *The Limited Liability of Company Directors*, Univ. of Queensland TC Beirne School of Law Research Paper No. 07–03, Maritime and Commercial Law Quarterly, 2007; R. R. W. Brooks, *Liability and Organizational Choice*, Journal of Law and Economics, vol. 45, No. 1, (Apr., 2002), pp 91–125; G. Gunasekara, A. Sims, *Statutory Trends and the Genetic Modification' of the Common Law: Company Law as a Paradigm*, Statue Law Review (2005), 26(2), 82–102. See also Pakistan National Shipping Corp., Case [2003] 1 All ER 173 at 180

The physical ownership than that of a mere documentary one is highly contrasted in situations where a ship is under a mortgage[448] or financial facility. [449] In many instances there appears to be ships which are possessed by an owner who does not have a legal ownership in the absence of a proper registration.[450] Obviously, the registered owner in such circumstance would be the mortgagee or financial facility provider who would retain legal ownership of the ship, which is under such facility. The interesting question here arises as to who is liable under the Convention, or more specifically, against whom an action for compensation shall be instituted? The definition of 'shipowner' is silent on this issue and does not provide a straight forward answer. Depending on the plain language, it can be concluded that a mortgagee becomes liable in virtue of the registration as the owner, but this indeed will create major problems for financial institutions when they are deemed to adopt responsibilities for merely registering themselves as registered owner, where the only assured protection is the document they possess as a security for their lent money. Even though the settlement of claims would not be the hardest issue at stake due to the insurance policies taken out by such financial institutions or mortgages in respective cases, it would be a problem concerning the attainment of actual liability for situations outside their physical participation. In turn, such implications would on one hand constitute grave fear on such entities and upheld invalidity of indemnity clauses on the other.[451] The same

448 See Int'l Convention on Maritime Liens and Mortgages, 1993

449 See S. Harwood, *Shipping finance*, 3rd ed., Euromoney Institutional Investor Plc, 2006 rep. 2008, pp 167–168; M. A. Sheppard, *supra* 374, pp 359–363

450 See M. A. Sheppard, *supra* n. 374, p. 286; Belize, Registration of Merchant Ships Act, Ch. 236, Rev. ed., 2006, http://www.belizelaw.org/lawadmin/PDF%20files/cap236.pdf; D. Lasok, *The Polish Merchant Shipping Code*, Int'l and Comparative Law Quarterly (1964), 13:1087–1093; R. Wacks, *The New Legal Order in Hong Kong*, Hong Kong University Press, 1999, pp 424–429; The Belgian Ships Register, http://www.brv.be/eng/download/BELGIAN%20REGISTER%20-%20Recommendations.pdf ; Merchant Shipping (Registration of Ships) Regulations 2009, Mauritius, http://supremecourt.intnet.mu/Entry/dyn/GuestGetDoc.Asp?Doc_Idx=2334775&Mode=Html&Search=No; D. J. Cremean, *Admiralty Jurisdiction: Law and Practice Australia, New Zealand, Singapore and Hong Kong* , The Federation Press, 5th ed., 2008, pp 49; G. de Costa, *Ship Registration in the Cayman Islands*, http://www.conyersdill.com/uploadedfiles/Article%20104%20Ship%20Reg%20in%20Cayman%20Islands.pdf; J. M. Joseph, *Ship Registration and Ship Mortgages in Malaysia*, http://www.josephandpartners.com/data/SHIP%20REGISTRATION%20AND%20SHIP%20MORTGAGES%20IN%20MALAYSIA.pdf

451 See A. M. Sheppard, *Modern Maritime Law and Risk Management*, 2nd ed., Informa Maritime and Transport, 2009, p 769

situation prevails with regard to CLC 69/92 and HNSC, but domestic legislations tend to hold differences in many jurisdictions. Especially in the US, the term 'operator' as a responsible person means to include mortgagee within the term 'lessor' under the OPA 1990.[452] In the UK, the matter would be settled on the test of 'actual control'. If a mortgagee exercise sufficient degree of control over such ship together with the operational aspect, he may become liable.[453] But the mere fact that he owns the ship as its registered owner would not impose the onus on him under the Merchant Shipping Act 1995.[454] Furthermore, holding possession alone would not suffice to attribute liability, unless he actually operates such vessel.[455] Therefore, the United Kingdom law on mortgagee's liability provides certain satisfactory groundings on the financial market, something which lacks at international level.

Situations of cessation and abandonment of ownership

The definition of 'registered owner' is autonomous under the Convention and do not relate to CLC 69/92 or HNSC in direct sense, as those Conventions used the term 'owner' in identifying the 'responsible party', although they bore the same meaning as is used under the Bunkers Convention. However, the Convention has not clarified the situation in the absence of registration, citing only within its meaning 'person or persons owning the ship'. This is somewhat misleading in the sense; how could it be possible to identify the actual person or persons owning the ship, without any clear documentary or practical evidence, especially where this Convention is intended to apply to all 'sea going vessels and seaborne crafts whatsoever'? This becomes a more serious issue in situations associated with abandonment of vessels, where the owner is ceased to be the owner upon abandonment.[456] Especially in circumstances where consequential loss or damage occur

452 Federal Register, Vol. 59, No. 126, 1st July 1994, Rules and Regulations, 34210, p. 34229, section 138.20

453 See G. L. Canfield, G. W. Dalzell, *The law of the sea: a manual of the principles of admiralty law for students*, pp 145–147; A. M. Sheppard, *supra* 451, pp 95–96

454 See paragraph 10 of Schedule 1. This is the same provision that was adopted way back in 1894 by the Merchant Shipping Act in its section 34

455 See *The Troubadour*, 16 L.T. 156

456 See A. E. Chircop, *Places of refuge for ships: emerging environmental concerns of a maritime custom*, Martinus Nijhoff, 2006, p. 255; F. G. Clarke, *The American shipmaster's guide and commercial assistant: being an enlargement of the seaman's manual*, Allen & Co., 1838, pp 349–350; S. Hodges, R. Carlile, *Case and materials on marine insurance law*, Cavendish, 1999, pp 629–633; T. Alderton, Seafarers International Research Centre, *The global seafarer: living and working conditions*

after being abandoned by the owner upon a given notice to that end to its insurer, creates a huge vacuum in instituting actions for constructive total losses suffered aftermath an incident. In such situations the vessel becomes *res nullius* lacking a proper owner or registered owner as to its ownership. Certainly, the law must ensure that innocent victims are not left out without compensation for damages caused upon abandonment of ships. Bunkers Convention in particular, need to fill this gap as there could exist of complex issues in attribution of liability and obtaining compensation in those circumstances. Although there could exist of various insurance schemes that serve this purpose,[457] and State intervention by way of public law remedies, both international and domestic private laws have not been able to provide a concrete solution, leaving it to be decided on principles of transfer and cessation of ownerships under respective national law under which such case appears.[458] The only justification would be to attach liability based on the concept of 'responsible party' alleging a flowing of ownership for presumption of tort in case of abandonment. In arriving at this conclusion one may consider the causative link between the pollution damage and the incident created by such abandoned vessel, with the relationship of the owner to the ship prior to such incident and after the actual abandonment. The OPA 1990 provides satisfactory ground to a certain extent by making provisions under Section 1001(26)(C) which defines 'owner or operator' meaning 'in the case of any abandoned offshore facility, the person who owned or operated such facility immediately prior to such abandonment'. Though, it provides some satisfactory tools for international legislators, an attempt has not been made to this end. Nevertheless, in the United States it seems to operate satisfactorily in domestic level within its national maritime jurisdictional boundaries.

(b) Bareboat charterer[459]

Unlike the CLC 69 & 92 regimes, the Bunkers Convention specifically list number of parties who would become responsible under the Convention, and against whom actions may be instituted. The common wording strategy used by the said CLC

in a globalized industry, ILO, 2004, pp 178–188; R. Gottschalk, *Problems in Israeli Maritime Law*, Int'l and Comparative Law Quarterly (1965), 14:302–308

457 E.g. Rule 32(2) of the Newcastle P&I Association Rulebook, 1997, which attaches responsibilities upon proof that the claim could not have been avoided after such abandonment. See also West of England Ship Owners' Mutual Insurance Association (Luxemburg), 1997 Rulebook, Class I, Rule 29

458 See C. J. Colombus, *Ships and Their Owners in the Prize Courts of France, Great Britain, Italy, and China*, Journal and Comparative Legislation and International Law, Cambridge, 3rd Series, vol. 2, no. 3 (1920), pp 289–297

459 See *supra* pp 83, 85

regimes is no longer applied under the Bunkers Convention, mainly because of certain shortcoming that were seen after their respective entering into force. One of the main difficulties associated in those regimes was whether the term 'owner' used in such, covers bareboat charterers as well. This was mainly due to the fact that bareboat registration[460] is somewhat unique in maritime activities and especially in maritime law.[461] Although, a registration as an owner or registration as a bareboat charterer, are two distinct concepts, the conduct of such charterer makes him an 'owner *pro hac vice*'. The only right the registered owner would have is to take back the ship upon the expiration of the contract and settlement of charter hire by the charterer.[462] In such context, the charterer controls the ship under his possession, but through the workforce provide by the registered owner.[463] According to the standard charterparty contract, the vessel should remain in the possession of the charterer under his complete control in every respect, for the disposal of all his purposes, during such period of time for which the charter was agreed upon.[464] Furthermore, he bares the responsibilities for all expenses throughout the operation of the vessel and costs of insurance.[465] Normally, the insurance coverage that is obtained by bareboat charterers does include potential liability cover in cases of pollution incidents, likewise in the case of policies granted for shipowners.

Although, attempts were made to include bareboat charterers under the 1984 protocol to the CLC regime, they ended up without any success.[466] This resulted in CLC 92 adopting the same definition for the owner as its predecessor, followed

---

460 Bareboat charter registration temporary permits a vessel to fly the flag of another country while ownership remains in the underlying registry. In such context, the vessel will prevail in the registry of the owner (so called underlying registry) as well as in the user's registry. Bareboat registration is recognized by virtue of the International Convention on Maritime Liens and Mortgages, 1993 and UNCLOS.

461 See A. Odeke, *Bareboat charter (ship) registration*, Kluwer, 1998, p. 28; ; S. Harwood, *supra* 449, pp 473–475

462 In *The Scaptrade* [1988] 2 Lloyd's Rep. 253, the House of Lords held that relief from forfeiture was not available where an owner exercised a right of withdrawal under a time charter. It further held that because the time charter was a contract for services and transferred no right to possession and because it was not a contract susceptible to an order for specific performance, equitable relief against forfeiture was not available.

463 See C. E. McDowell, H. M. Gibbs, *Ocean Transportation*, Beardbooks, (1954 Rep. 1999) p 186

464 E.g. Form BARECON A, Clause 8

465 See X. Chen, *Limitation of liability for maritime claims: a study of U. S. law, Chinese law and International Conventions*, Kluwer, 2001, pp 4–5

466 See O.R., International Conference on Liability and Compensation for Damage in Connection with the Carriage of Certain Substances by Sea, 1984 and the International Conference on the revision of the 1969 Civil Liability Convention and the 1971 Fund Convention, 1992, Vol.1, pages 296/297

by the HNSC. At domestic level, the Merchant Shipping Act 1995 follows the same trend as the CLC regime excluding bareboat registration from the ambit of that law by virtue of Section 17,[467] which does not regard such registration as falling within the meaning of 'owner' in the said Merchant Shipping Act 1995, and further, restricts actions against bareboat charterers.[468] All those described above have a common implication with regard to bareboat charterers, and the only way of suing such charterer rests with the right of recourse available for the shipowner as defined in those respective instruments.[469]

In this respect, the OPA 1990 has a different strategy, including within its meaning for 'responsible party', as 'any person owning, operating or demise chartering the vessel'.[470] Furthermore, the OPA 1990 notwithstanding any requirement of registration states in its Section 1001(26) that the term 'owner or operator' means *inter alia* 'in the case of a vessel, any person owning, operating or chartering by demise, the vessel …'[471]

However, the inclusion of bareboat charterers within the meaning of shipowner under the Bunkers Convention is a welcoming move in keeping with the developing aspects of maritime activities and the related law in general for purposes of effective implementation of the liability and compensation system.

(c) Manager

In the modern maritime market place, a new phenomenon is overriding classical doctrines of ownerships and related interests. Many ship-management entities[472]

467 It states that "the private law provisions for registered ships shall not apply to a ship registered by virtue of this section and any matters or questions corresponding to those for which the private law provisions for registered ships make provision shall be determined by reference to the law of the country of original registration"

468 Sec. 156(2)

469 For example see Contracts (Rights of Third Parties) Act 1999, United Kingdom. See also F. B. Cross, R. LeRoy Miller, *The Legal Environment of Business: Text and Cases*, 7th ed., South Western Cengage, 2009, pp 225–226; R. LeRoy Miller, G. A. Jentz, *Fundamentals of Business Law: Summarized Cases*, 8th ed., South Western Cangage, 2007, Ch. 11, p 222; D. Baker, *Law Made Simple*, 12th ed., Elsevier, 2007, pp 173–176; E. W. Huffcut, *Elements of the Law of Agency*, BiblioLife LLC, pp 178–204; Dr. A. D. Lazaraki, *Third party rights of suit in contracts for the carriage of goods by sea and the Contracts (Rights of Third Parties) Act 1999*, JIML [2008] 3, 208

470 Sec. 1001(32) (A)

471 Section 1001(26) (A)

472 See J. Lauritzen A/S, *Ship Management*, Annual Report 2008 available at http://www.j-lauritzen.com/upload/ship_management__annual_report_2008.pdf. See also A. E. Branch, *Elements of shipping*, 8th ed., Routledge, 2007, pp 460–461

are entering the common market in competing with the ever growing maritime interests. This has evolved in such ship-management companies, which are commonly known as managers of ships, take a lead role in commercial activities of shipping as a provider of transportation.[473] Spruyt[474] provides a definition to 'ship management' as;

> "The contracted and professional supply of all on-board services, together with their shore supervision, which would normally enhance a vessel from a bareboat into a time charter description, by a management company usually separate from the vessel's ownership."

The above definition poses many crucially important points that need to be clarified further for the best understanding of it. Spruyt takes a closer look at these special features and explain those points illustratively as follows;

The use of the term "Contracted" is aimed at having a clear contract between the owner and the manager in the form of what he states "crisply written". The importance of having proper standards and codes of conduct on the part of such managers is highly respected in terming them as "professionals" of the field. Activities that are carried out by managers can best be listed in the way "all on board services … enhance a vessel from a bareboat to a time charter description" together with the "shore supervision which maintains effective control and supervision of the crew, thereby safeguarding liability doctrines. He further describes the sentence 'management company usually separate from the vessel's ownership' as meaning shipowners with a fleet of ships to have single ship companies for each ship under their ownership that can be contracted with separate contract for each of such ships in facilitating both situations where, owners who manage and managers who sometimes seem to be owners in virtue of management contracts, of ships.[475] Willingale[476] provides a definition based on the earlier interpretation provided by Spruyt with certain modifications and in a much simple manner as;

> "Ship management is: 'The professional supply of a single or range of services by a management company separate from the vessel's ownership in support of the primary objectives of the shipowner' where: 'in support of the primary objectives of the shipowner' makes clear that the shipowner's main objectives are different."

---

473 See K. Mitroussi, *The role of organizational characteristics of ship owning firms in the use of third party ship management*, Marine Policy, vol. 28, issue 4, July 2004, pp 325–333; K. Mitroussi, *Third party ship management: the case of separation of ownership and management in the shipping context*, Marine Policy and Management, vol. 30, no. 1, Jan. 2003, pp 77–90(14)

474 John Spruyt, "Ship Management", Ch.1, 2ed, LLP 1998

475 *Id.*, pp. 18–19

476 Malcolm Willingale, "Ship Management", Ch. 2, 3ed, LLP 1998, p. 13

In view of the abovementioned definitions and practically visible situations with regard to ship-managements, it is clear that the managers perform a wide range of functions ranging from commercial to technical activities.[477] In performing such, they need to possess sufficient degree of professionalism, and in turn accept responsibilities for which the owners would have otherwise been bound by. Under the contract of ship management, the manager not only inherits most of the rights and privileges of the registered owner, but also liabilities that are connected with the operational aspects of such ships. Therefore, in the absence of the owner or registered owner in actual operation of the ship at a time of an oil spill incident, such manager under whom the actual conduct of the ship remains, shall become liable.[478] This seems to be the idea behind the inclusion of the term manager in the definition of 'shipowner' in the Bunkers Convention, a somewhat unique step, which cannot be seen under the interpretations provided in the CLC 69 & 92 and HNSC for 'owner'. Nevertheless, the jurisprudence of the CLC regimes unveil certain instances where this term was interpreted allowing the acceptance of the inclusion of 'managers' under the liability concept. Especially in the case of *Amoco Cadiz*, it was held by the court that the CLC 69 is not intended to exclude ship-management companies, unless they are considered to be servants or agents of the owners who fall outside the direct ambit of its sphere of application.[479] Furthermore, it developed a criterion of investigating the actual party upon whom the operation of the ship had been vested.[480]

Likewise in the case of bareboat charterer, the counter part provision of the Merchant Shipping Act 1995 restricts the attribution of liability to a manager of a ship,[481] unless in actions in recourse.[482] However, the OPA 1990 does not exonerate a ship manager being sued for actions in liability and compensation.[483] As

---

477 See also A. M. Sheppard, *supra* n. 451, pp 306–314

478 See N. de Sadeleer, *Liability for Oil Pollution Damage versus Liability for Waste Management: The Polluter Pays Principle at the Rescue of the Victims: Case C-188/07, Commune de Mesquer v. Total France SA [2008] 3 CMLR 16, [2009] Env LR 9*, Journal of Environmental Law 2009, 21(2):299–307

479 See [1984] 2 Lloyds Report 304 onwards

480 See K. Hofstetter, *The Ecological Liability of Corporate Groups: Comparing US and European Trends*, http://www.hottopos.com/harvard1/ecological_liability_of_corporat.htm; J. J. White, *Corporate judgment proofing: a response to Lynn LoPucki's 'The Death of Liability'*, Yale Law Journal, vol. 107, 1998; V. Tomljenovic, *New Conflicts Approach: Is It Necessary*, Maritime Torts, Yearbook of Private International Law, vol. 1 (1999), pp 249–298

481 See Sec. 156

482 See *supra* p 97

483 See H. J. Peters, *Strengthening Marine Pollution Control the International move towards Statutory Requirements*,http://siteresources.worldbank.org/INTTRANSPORT/Resources/336291-1119275973157/td-ps9.pdf

mentioned earlier, the wider application of the term 'responsible party' and 'operator' compels a ship manager to hold liability in case of an oil spill. Accordingly, an 'operator' of a ship would mean to include 'any person who is responsible or who agrees by contract to become responsible, for a vessel'.[484] Although, this does not provide a clear view whether a ship manager would fall within the definition of 'operator', it can be derived from Clause 12 of the Shipman Standard Ship Management Agreement[485] which explicitly declares *inter alia* that the operation of a vessel will be provided for by the ship managers and under optional clause 2.3 (x), they shall carry out the operation of the vessel as agents for and on behalf of the owner.[486] Therefore, it suggests that managers of ships are under the duty of holding responsibility concerning the ship and furthermore become ultimately liable in operating ships which are under their possession and control. Hence, the inclusion of the term 'agents' within such clause, enlarges the application of the OPA 1990 to agents, who are acting as operators, while being managers under the contracts of ship management.

(d) Operator

In reality, the best appropriate person upon whom liability should vest is such person who is in actual terms operating such ship. Due to various legal implications and lack of proper legal tools for such execution, it has been unable to impose liability on the actual operator of a ship.[487] Especially, in circumstances where the actual owner or the person who practice the controlling aspects of a ship in legal terms cannot be located in the real operation of the ship, it would be unfair to impose primary liabilities on its servants who are engaged in the operation of the ship in actual practice, under the commands and control of such respective owner. International maritime law has well respected the theories of primary and vicarious liabilities found in law of tort in all of its instruments dealing with the issues concerning liability and compensation. The Bunkers Convention in particular has deviated from the common trend of interpretation with regard to the liable party with a given list of persons who shall become liable. The term 'ope-

484 See Federal Register, Vol. 59, No. 126, 1st July 1994, Rules and Regulations, 34210, p. 34229, section 138.20

485 See http://www.simsl.com/Shipman98.html

486 See also Gauci, *supra* n. 130, p 94

487 See Xia Chen, *supra* n. 465, p 3; C. C. A. Voskuil, J. A. Wade, *Carriage of goods by sea: maritime collisions, maritime oil pollution*, T. M. C. Asser Inst., The Hague, 1980, pp 105–106; S. Mankabady, *The International Maritime Organization*, vol. 2, IMO, 1987, pp 84–90, A. M. Sheppard, *supra* n. 451, p 885

rator' within the definition of 'shipowner' seems to serve this said purpose and further strengthen the possibility of identifying the actual person so liable for oil spill pollution damages against whom, victims could institute actions. The term 'operator' itself cohabits with the other respective persons upon whom liability has been vested under that definition. That is to say that the owner, registered owner, bareboat charterer and manager could fall within the term 'operator' in relation to the specific activities such person executes.

In the absence of identifying the actual owner of a ship, it would be prudent to impose liability on the 'operator' who actually carries out the technical and administrative operations of such ship in question. In certain circumstances, it would not be possible in practical terms to locate such identity as to ownership due to various reasons. Especially in situations where the ship is under a contract of chartership, management or lease, the actual owner or registered owner would not possibly bare responsibilities for any wrong doing. As discussed earlier, when a ship is under a mortgage or other financial facility, such financial facilitator would not assume any potential future liabilities for tortures acts committed by the operator during the operation of the ship. Therefore, it is necessary to impose liability on the actual operator rather than on the owner on record.

Sometimes, there may exist of complex situations where both the owner and the operator jointly operate a ship.[488] Moreover, this could happen in virtue of an agreement entered between such parties as to this effect. The Bunkers Convention in this respect has followed the same pattern as the CLC 69/92 and HNSC regimes. In the foregoing discussion it would be prudent to analyse those situations in detail.

Joint and Several Liability[489]

The multiplicity of liable parties could arise as a potential barrier to legal actions in cases of oil pollution damage in many instances. Past experiences have shown

488 See D. Ulpian, *Damage caused by multiple tortfeasors* in *Digest of European Tort Law*, vol. 1, Springer, 2007, pp 265–352

489 See R. V. Randel, *Oil Pollution Deskbook*, Environmental Law Institute, 1991, p 290; D. R. Cooley, *Strict Joint and Several Liability and Justice*, Journal of Business Ethics, Springer, vol. 47, issue 3, Oct. 2003, pp 199–208; L. Bergkamp, *Liability and Environment*, Kluwer, 2001, p 306; K. Segerson, *Government regulation and compensations for environmental quality and natural resource use*, Contemporary Economic Policy, col. 15, 1997; M. T. Katzman, *Pollution Liability Insurance and Catastrophic Environmenatl Risk*, The Journal of Risk and Insurance, vo. 55, No. 1 (Mar., 1998), pp 75–100; R. G. Hansen, R. S. Thomas, *The efficiency of sharing liability for hazardous waste: effects of uncertainty over damages*, Int'l Review of Law and Economics, vol. 19, Issue 1, Mar. 1999, pp 135–157

critical impacts on institution of actions based on complex issues as to the identification of one exact person or several person having certain ratios or shares towards responsibility to an incident involving one ship. A concrete solution was therefore adopted at the very beginning of the related field of application once the CLC 69 came into existence.

The issue is two folded. There can be situations where more than one ship is involved in an incident leading to escape or discharge of oil thereby resulting in pollution damage been caused due to contamination.

Multiplicity of ships

Article 5 of the Bunkers Convention deals with a situation where more than one ship is involved, and states accordingly that;

> "When an incident involving two or more ships occurs and pollution damage results therefrom, the shipowners of all the ships concerned, unless exonerated under Article 3, shall be jointly and severally liable for all such damage which is not reasonably separable."[490]

Although there have been circumstances where shipowners were joined in several litigations concerning tanker disasters,[491] courts are extremely careful in applying this strategy in a given case due to the strict wording of this particular Article. A proper observation of what has been said accordingly will be highly respected in doing justice. There are several aspects that need careful consideration in this respect, including the interpretation of the term 'incident' with respect to the alleged oil spill. Apart from mere collision incidents where bunker oil could escape from one or more the ships involved,[492] one has to distinguish ships which fall into the CLC regime and Bunkers Convention regime separately, in order to proceed successfully. Furthermore, in a more practical view, it would not be easy to identify situations where two or more ships jointly contribute to escape of discharge of bunker oil from each ship, apart from a collision. But however, there

490 Identical provisions can be found in Article IV of CLC 69, Article IV of CLC 92 and Article 8.1 of HNSC.

491 E.g. *Amoco Cadiz* litigation, *supra* n. 34, and the case of *Tanio*, *supra* n. 357

492 See H. Schussler, *Process for monitoring ship traffic at sea while recognizing and oil spills and potential ship collisions*, 1997; P. T. Pederson, S. Zhang, *Effects of ship structure and size on grounding and collision damage distributions*, Ocean Engineer, vol. 27, Issue 11, Nov. 2000, pp 1161–1179; H. Ohtsubo, Y. Kawamoto, T. Kuroiwa, *Experimental and numerical research on ship collision grounding oil tankers*, Nuclear Engineering and Design, 1994, vol. 150, no. 2–3, pp 385–396; see also *Oil Pollution Liability of Colliding Vessels in China*, Lloyd's Maritime Law Quarterly, [2001] L. M. L. C. Q. 19

can be rare situations where such coincident exists, and hence, where mystery spills[493] occur without any substantial evidence as to the actual perpetrator. Therefore, based on the interpretation of 'incident' as provided by the Convention, one may assert the right to join the shipowners,[494] of all ships involved in pursuing, a successful case in such a situation leading to a mystery spill. However, that alone would not suffice in relying on this provision, as it is important to show the characteristic feature of all such damage that they are connected with, and in substance they are constituted in one incident 'having the same origin' plus such damage 'is not reasonably separable'. This seems to be the hardest task, as it deals with the link of the incident to all pollution damage caused by that particular incident. The terms 'reasonably separable' is nowhere clarified in the Convention. Therefore, the only solution would be to analyse its pure meaning based on the circumstance. Suppose that the pollution damage occurred due to two different oil spill incidents with a considerable degree of separating them for the convenience of instituting actions against those respective ships, the case is not that complicated. But if there was one incident in which oil escaped or discharged from all or some of those ships, then the pollution damage occurs on a contributory basis.[495] The pollution damage itself will not be able to separate physically into each others components, and traditionally they remain indivisible or in absence of the ability of apportioning liability among responsible parties. This difficulty constitutes the pollution damage to remain inseparable. Although actions in recourse may prevail for any or all of those shipowners in other forums, they shall hold joint and several liability under the Bunkers Convention. Understandably, any such incident involving ships of this Conventional regime and non Bunkers

---

493 See California Department of Fish and Game National Oceanic and Atmospheric Administration United Nations Fish and Wildlife Service National Park Service, *S. S. J. Luckenbach and Associated Mystery Oil Spills,* Nov., 2006, http://www.dfg.ca.gov/ospr/spill/nrda/luckenbach_final_darp.pdf; Z. Wang, M. Finags, L. Sigouin, *Characterization and identification of a "mystery" oil spill from Quebec (1999)*, A Journal of Chromotography, Elsevier, vol. 909, no. 2, Feb. 2001, pp 155–169(15); Z. Wang, S. A. Stout, *Oil spill environmental forensics: fingerprinting and source identification*, Elsevier Inc., 2007, p 226; Dr. H. D. Nguyen, Eng. M. D. Dao, H. Q. Nguyen, Eng. T. T. Nguyen, *Oil Spills in Vietnam – Facts and Challenges*, http://www.iosc.org/papers/2008%20010.pdf

494 See L. I. Kiern, *Liability, Compensation and Financial Responsibility Under the Oil Pollution Act of 1990: A Review of the First Decade*, Spring 2000, 24 Tul. Mar. L. J. 481; L. Bender, *Frontier of Legal Thought III: Feminist (Re) Torts: Thoughts on the Liability Crisis, Mars Torts, Power and Responsibilities*, 1990 Duke L. J. 848

495 See C. J. Giaschi, D. Williams, *The Maritime Liability Act: Parts 2, 3 and 4, Apportionment of Liability, Limitation of Liability and Carriage of Passengers*, Presented at Maritime Conference 2002 (Jan.), Toronto, Ontario, available at http://www.admiraltylaw.com/papers/MLA.htm

Conventional regime, the provisions of this Convention will only apply to ships of the Bunkers Conventional regime, as stated before. The other ships will fall under such specific regime to which those ships belong, such as for example the CLC 92,[496] under which liability would be assessed accordingly.

Multiplicity of 'shipowners'

The other situation with regard to the concept of joint and several liability lies on the issue of one incident where several persons could become liable due to coexistence within the term of 'shipowner'.[497] The CLC 69/92 and HNSC regimes have the same legal implication as regards to this situation and such penetration of this concept has evolved in accepting the same idea under the Bunkers Convention through its Article 3.2 which states;

"Where more than one person is liable in accordance with paragraph 1, their liability shall be joint and several."

This provision enables an oil pollution victim to institute action for compensation against number of persons who shall deemed to be liable, and who fall within the meaning of shipowner as defined under Article 1.3. As described earlier there could prevail situations where the term 'shipowner' consists of number of natural or juridical persons.[498] Especially, with regard to dual or multiple ownerships, partnerships or co-existence of joint operators or managers and etc., under a contract or other such arrangement, it is possible to join all such parties in one plaint seeking compensation for their tortuous acts in a given incident.

Convention being the supreme law

The other important characteristic feature is the restriction of instituting actions against the shipowner outside the Convention. Accordingly the victim lacks any right to sue the shipowner for compensation unless such action is made within and under the provided parameter. Article 3.5 states that;

"No claim for compensation for pollution damage shall be made against the shipowner otherwise than in accordance with this Convention".

496 According to Article 4.1 of the Bunkers Convention, CLC conventional ships are excluded from the application of the Bunkers Convention
497 See *supra* pp 79–94
498 See Art. 1.1

The pre-existed possibility of instituting actions under a respective national law prior to the adoption of this Convention against State Parties to it would cease to exist once the Convention becomes binding on such States upon ratification.[499] Furthermore, the victim will have no option than observing what is required under the Convention.[500] Therefore, claims for compensation against the shipowner in exact or wider meaning as described in the aforementioned discussions, will have to be based on the Conventional pre-requisites. This will not only restrict the victim in filing single claim against one or more of the defendants, but also lay additional pressure in composing one strong case. However, it does not mean that a victim is prevented from presenting more than one plaint, as obviously there can be situations of continuing occurrences spreading over a period of time, but to confine itself to one legal regime, namely in this respect, the Bunkers Convention.

However, a victim deprived of any damage that occurs due to an oil spill incident other than pollution damage would still have the possibility of instituting claims under such respective laws. He may in this respect rely on the *lex fori* in which, remedies would be available towards compensating such other damage. But it does not remedy any damage falling with the meaning of 'pollution damage' as described in this Convention, for which the shipowner is liable unless exempted from liability pursuant to the application of certain provisions which exempt such shipowner from liability under this Convention, and hence, after being unsuccessful in maintaining his (victim's) original claim.[501] Matters relating to presentation of claims and jurisdictions would not be dealt here, but will be discussed in detail in the next Chapter.

---

499 See Gaskell, Nicholas and Forrest, *Marine pollution damage in Australia: implementing the Bunker Oil Convention 2001, and the Supplementary Fund Protocol 2003*, Uni. of Queensland Law Journal, 27(2), 103–165, available at http://eprints.soton.ac.uk/66737/01/Marine_Pollution_Damage_in_Australia.pdf; Report of the Transport and Industrial Relations Committee, House of Representatives, New Zealand, *International Treaty Examination of the international Convention on Civil Liability for Bunker Oil Pollution Damage*, http://www.parliament.nz/NR/rdonlyres/C901FD34–146E-43A8-A8CD-3EE643C293CB/94261/DBSCH_SCR_4183_6372.pdf; M. Jacobson, *Bunkers Convention in force*, JIML 15 [2009] 1:21; R. D. Thomas, *Bunker oil pollution – a new liability convention*, JIML 14 [2008] 5:375. See also Merchant Shipping (Oil Pollution) (Bunkers Convention) Regulation 2006 (SI 2006/1244) amending Merchant Shipping Act 1995, Part 6, Chapter 3

500 See M. Mason, *Civil liability for oil pollution damage: examining the evolving scope for environmental compensation in the international regime*, Marine Policy, vol. 27, issue 1, Jan. 2003, pp 1–12

501 For further reading see Abecasis, *supra* n. 92 pp. 177–179

Implication of liability on related parties and 'responder immunity'

Both CLC 69/92 and HNSC regimes contain the similar provisions on the application of those respective regimes in relation to institution of actions. But however, a notable deviation from those regimes could be highlighted in the Bunkers Convention dealing with the same objective. Both former regimes have implications on the restriction of actions against certain parties who have personal or official relationship towards the shipowner or to the complete scenario of the oil spill incident. In concise terms, both such regimes, except CLC 69 other than paragraph (a) below, prohibit any actions made against;

(a) the servants or agents of the owner or the members of the crew;
(b) the pilot or any other person who, without being a member of the crew, performs services for the ship;
(c) any charterer (howsoever described, including a bareboat charterer), manager or operator of the ship;
(d) any person performing salvage operations with the consent of the owner or on the instructions of a competent public authority;
(e) any person taking preventive measures;
(f) all servants or agents of persons mentioned in subparagraph (c), (d) and (e);

> *unless, the damage resulted from their personal act or omission, committed with the intent to cause such damage, or recklessly and with knowledge that such damage would probably result* (emphasis added).[502]

The Merchant Shipping Act 1995 has same implications as the CLC 92 and HNSC regimes,[503] but without any limitation as to the application of those provisions only to CLC conventional ships[504]. The OPA 1990 on the other hand excludes only a certain category of parties from the above list.[505] However, inclusion of bareboat charterer, manager and operator of a ship in the definition of shipowner under the Bunkers Convention has short listed the parties upon whom due consideration is ought to be paid in the foregoing discussion. There are number of justifications towards the rule of non inclusion of this provision in the present Convention. Firstly, it was a legislative move motivated by several delegations during the law making process. Although, the provision was indeed visible in the foremost draft, as Article 3 (5), number of delegations were of the view that it

502 See Article III.4 of CLC 92 and Article 7.5 of HNSC
503 Sec. 156 (2) (a) – (f)
504 See Sections 156, 153 and 154
505 See Sec. 1002 (d) (1) (A) – (B)

should be kept within brackets for further consideration. During the 75th Session of the Legal Committee, they based their argumentation on inappropriateness of such provision in this instrument. The main reason was that the latter paragraph, which appears above in italics,[506] would allow the shipowner certain exemptions of liability, which he should not be left to invoke in a regime such as compensation for bunker oil pollution. By the 78th Session, many delegations held the same view and expressed their suggestions in deleting the whole section without leaving any possibility on behalf of the shipowner to enjoy benefits of a safety net.[507] They further contrasted the CLC/HNSC regimes from the present one, on the ground that those regimes possess funds to cover compensation for damage in the absence of actual shipowner liability. Furthermore, such exemptions would result in unsuccessful litigations against the insurer in direct actions.[508] The whole issue was not resolved by only addressing the exemption clause, as many delegations pointed out the importance of the concept of 'responder immunity'[509]. Especially, when 'pollution damage' cause as a result of measures carried out by salvors and other persons taking preventive measures, claims for compensation against such acts could still fall outside the Convention, in the absence of any provisions to that effect. Such situations would deter responders from taking any necessary and timely preventive measures, resulting in more damage.[510] There was in fact a consideration for the adoption of a modified text reading as;

"Subject to paragraph 6, no claim for compensation for pollution damage under this Convention or otherwise may be made against any person taking preventive measures unless the damage resulted from their personal act or omission, committed with the intent to cause such damage, or recklessly and with knowledge that such damage would probably result".[511]

---

506 See *supra* p 106

507 *Supra*, n. 378, p. 8

508 *Supra*, n. 206, p. 18

509 The term is used to express the limited immunity from civil liability provided towards the responders to an environmental accident whose acts or omissions result in worsening the environmental consequences so far as their conduct was within certain principles and that environmental worsening was not resulted due to their gross negligence or willful misconduct. See W. L. Peck, *Responder Immunity and Salvage*, http://www.iosc.org/papers/01852.pdf; Colin de la Rue, *Bunker Spill Risk*, http://www.incelaw.com/documents/pdf/News-PDF/Bunker-spill-Risk; Int'l Salvage union, *An exchange of view on responder immunity*, Salvage World, June 2007, available at http://www.marine-salvage.com/salvage_world/isu%20salvage%20world%20june%202007.pdf

510 *Supra*, n. 378, p. 2

511 *Id.*, p. 3

Many delegations including the co-sponsors and others apart from those who raised this issue[512], indicated lack of support for this argument, and therefore, a vital and critical issue has been left out, leaving it to be settled at national level.

Secondly, it is quite evident in practice that the proof of fault against a third party in an action for tort outside the scope of the Convention would impose additional burden on the victims, leaving them sometimes, without any considerable remedy. As a matter of fact, it would not be easy to find proof to that end in order to pursue with a positive cause of action in tort against employees of the shipowner, independent contractors or public authorities engaged in clean up and preventive measures[513]. As the present Convention deals with the issue of bunker oil spills, it would be more appropriate to impose liability on the 'shipowner' as the best person entrusted with the functions of a ship while holding actual or contractual ownership. Thirdly, there prevails an actual fairness to this end in terms of maritime law. In actual circumstance, the 'shipowner' is vicariously liable under the law of tort for any tortuous acts committed by its servants or agents.[514] But, that concept would not extend to such acts committed by independent contractors, such as salvors.[515] However, the salvors operate on contractual relationship with the shipowner, and enter into such contract *with consent*, before or after an incident occur. Both CLC and HNSC regimes exclude actions by victims against such persons operating salvage activities, in principle, and furthermore, the Mer-

---

512 Number of delegations and both IUCN and P & I Clubs favoured the inclusion responder immunity provisions and urged the Committee to adopt a similar provision to that of Article 4.III of the CLC regime.

513 See D. J. R. White, I. C. White, *Is the oil industry's tiered preparedness and response approach applicable to oiled wild life response planning?*, http://www.oilspillresponse.com/pdf/Published%20Papers/David%20White/paper_wildlife_tiered.pdf

514 See G. Williams, *Vicarious Liability and the Master's Indemnity (Continued)*, The Modern Law Review, vol. 20, No. 5 (Sept., 1957), pp 437–446; F. Berlingieri, *The Hamburg Rules: a choice for the EEC?*, European Institute of Maritime and Transport Law, Maklu Uitgevers, 1994, pp 169–171; R. Zimmermann, *The law of obligations: Roman foundations of the civilian tradition*, Oxford (1990 ed., 1996), pp 1136–1138

515 See S. Buaghen, *supra* n. 354, pp 308–309; S. Mankabady, *supra* n. 487, pp 249–256; M. W. D. White, *Australian maritime law*, 2nd ed., The Federation Press, 2000, pp 1–3; National Research Council (U.S.), Committee on Marine Salvage Issues, *Purposeful jettison of petroleum cargo*, National Academy of Sciences, 1994, pp 78–96. A. E. Chircop, O. Linden, *Places of refuge for ships: emerging environmental concerns of a maritime custom*, Maritnus Nijhoff, 2006, pp 289–295; G. W. Paulson, *A reassessment of the marine salvage posture of the United States*, National Academy of Sciences, 1994, p 72

chant Shipping Act 1995[516] re-affirms its predecessor's provisions,[517] while bearing similar effect. Therefore, a long existence of the idea seems to be apparent in upholding their non responsibility in maritime oil pollution incidents. The same effect extends to pilots and pilotage authorities[518], and furthermore, to any persons engaged in preventive measures. Suppose, any action against the above said persons was allowed in maritime law, the whole procedure would have become much complex, and as a result, many victims would have been left unattended with respect to settlement of claims. Although, the CLC and HNSC have allowed exemptions in cases of acts committed by such parties outside their contractual relationship with the 'shipowner', Bunkers Convention will not be able to tolerate such situations due to the above said reasons, and other justifications as to simplicity, speedy recovery of compensation and more importantly evading situations of non settlements due to the absence of a fund or indemnities of insurers. These end results would thereby enlarge the pure threat of being deprived by pollution damages at large.

Finally, the entirety of the instrument would provide satisfactory grounding with the help of national legislation to the effect of enacting clauses dealing with responder immunity. According to Article 3.1, the liability lies on the 'shipowner' as it is clear from the wording, "... the shipowner at the time of an incident shall be liable for pollution damage caused by any bunker oil on board or originating from the ship ..." Hence, Article 3.5 states that "No claim for compensation for pollution damage shall be made against the shipowner otherwise than in accordance with this Convention".[519] Therefore, it is obvious that the actions by victims in relevant circumstances have been restricted to shipowner and be made only under this Convention. Suppose the tortuous act is a result of the interference or commission of some other party, the shipowner remain liable towards the victims as a matter of holding responsibilities for his ship, but consequently, reserving a right to sue such other parties in another forum seeking any remedies for himself. This is clear from Article 3.6 which categorically allows right to recourse by stating "Nothing in this Convention shall prejudice any right of recourse of the shipowner which exists independently of this Convention".[520]

516 Schedule 4, Section 156/156(A)

517 Sec. 3 of Merchant Shipping (Oil Pollution) Act 1971

518 See IMO, *Resolutions and other decisions of the 23rd Assembly*, IMO, 2004, pp 210–212; *Pilotage under siege*, http://www.tankeroperator.com/pastissues/2002%20Jul/TO2002Jul%20p1-3%20Pilotage%20under%20siege.pdf. See also Pilotage Act 1983, U. K., available at, http://www.england-legislation.hmso.gov.uk/acts/acts1983/pdf/ukpga_19830021_en.pdf

519 Same as in Art. III. 4 of CLC 69, Art. III.4 of CLC 92 and Art. 7.4 of HNSC

520 Art. III.5 of CLC 69, Art. III.5 of CLC 92 and Art. 7.6 of HNSC

In practice, the above mentioned solution adopted by the Convention creates an extra cautious situation for the insurer or other person providing financial security, as he will not be able to benefit the same privileges enjoyed under the former regimes, especially in circumstances where the whole issue is dependant on the respective national law concerning responder immunity. Though it does not create any additional financial burdens in its outset, as recovery and reimbursement is possible under the recourse procedure as well; "how will the insurance market react?"- is a question purely based on future determinations by respective insurers or financial security providers, on this aspect.

The Victim

The Bunkers Convention has nowhere specifically mentioned a straightforward definition for 'victim' nor introduce such term, rather than referring to a broad spectrum by the use of the phrase, 'person who suffered the damage'.[521] There are several derivatives of this meaning. Firstly, it is a 'person' who has the right to institute action upon being deprived of a tortuous act. Accordingly, as discussed above, a person means 'any individual or partnership or any public or private body, whether corporate or not, including a State or any of its constituent subdivisions'.[522] Therefore, it includes all natural and juridical entities having a *locus standi* with a proper cause of action. The said cause of action should arise with a suffering experienced due to damage. The damage is further described under the Convention as meaning only "pollution damage" for which such actions could be brought against the liable party under this Convention. As discussed in the preceded Chapter[523], 'pollution damage' can be subdivided into different types and categories of damage that occur as a result of an oil spill incident. It further extends to 'preventive measures',[524] so that, persons taking such measures also fall into the category of 'persons who suffered the damage' subject to a proper *locus standi*. The expression 'suffering' may have a wider application, ranging from physical, mental, emotional[525] to somewhat theoretical[526] and pure econo-

521 Art. 3.4.
522 Art. 1.2.
523 *Supra* pp 53–59
524 *Supra* pp 65–71
525 Such as damage to reputation
526 See M. L. Honkasalo, *Grips and Ties: Agency, Uncertainty, and the Problem of Suffering in North Karelia*, Medical Anthropology Quarterly, Blackwell, vol. 23, no. 1, Mar., 2009, pp 51–69(19); D. L. Kahn, R. H. Steeves, *The experience of suffering: conceptual clarification and theoretical definition*, J. Adv. Nurs., 1986 Nov., 11(6):623–31

mic[527] means. Most importantly, a 'person who suffered the damage' is under an obligation to prove a causative link between the incident and the loss or damage suffered in circumstances where such person had somewhat direct proprietary rights or interests on which such damage or loss occurred.

Having those prerequisites satisfied would not suffice to make a successful case for compensation. There must exist of a proper jurisdiction under which such a cause of action could prevail. In overcoming this barrier, the victim has to lay his case on the applicable law of the State in which he pursue with such matter, which grant the possibility to proceed with a causable action. In the context of international law, once a State ratifies a certain Convention, such State implements the relevant provisions in its own national law.[528] The process might differ from one State to another so as its recognition, whether it fully implements or with reservations, and the applicable law will be based on its national legal system, i.e. it falls under the law of torts or by way of some other legislation implementing such provisions. Therefore, it is important on the part of the victim to peruse the actual national legislation or law which may grant him a cause of action.[529] Once that is established, the victim may proceed with his action in the prescribed or proper jurisdiction, namely in a court of law.

As a matter of understanding, there may exist of various and different types of victims or plaintiffs in terms of litigation, who could potentially proceed with an action based on the Bunkers Convention. Such a category could vary exhaustively, but to name a few; persons engaged in hotels and recreation facilities, fishermen who suffer from loss of earnings and other damages associated with their fishing gear and etc., persons suffering from physical and mental disorders or such other related issues, salvors employed by victimised States, local governments and governmental authorities for loss of earnings by way of tax and levies,

527 See E. K. Banakas, *Civil liability for pure economic loss*, Kluwer, 1996, pp 86–88; W. H. van Boom, H. Koziol, C. A. Witting, L. Bloch, *Pure economic loss* in *Tort and Insurance Law, Volume 9*, Springer, 2004, pp 115–140; E. H. P. Brans, *Liability for damage public natural resources: standing, damage and damage assessment*, Kluwer, 2001, pp 254–158

528 See W. Tetley, *Uniformity of International Private Maritime Law – the pros, cons and alternatives to international Conventions – How to Adopt an International Convention*, (2000) 24 Tul. Mr. L. J. 775–856; M. N. Shaw, *supra* n. 230, p 820; B. A. Boczek, *The Law of Treaties* in *International law: a dictionary*, Ch. 6, Scarecrow, 2005; J. O'Brian, *International law*, Cavendish, 2001 rep. 2002, p 336; L. Kurukulasuriya, N. A. Robinson, *Training manual on international environmental law*, UNEP, 2006, pp 5–7; E. Lauterpacht, *International Law Reports*, vo. 42, Cambridge, 1971 rep. 1994, p 249

529 For example, under the English legal system, a victim may have such a right under the legislations implementing provisions with regard to cause of action brought in terms of CLC regime

and clean-up, preventive and restoration measures undertook by such entities and interested groups.

Issues relating to actionable damages, allowed claims and recoverable will be dealt in detail in a later Chapter, and the issues dealing with jurisdiction and *locus standi* under the Bunkers Convention will be taken up in the next Chapter.

Chapter 4: Jurisdiction

One of the most important characteristic aspects of any legal instrument, whether national or international, is the provision dealing with jurisdiction laid down by the respective instrument. Many authors tend to address this aspect at the latter end of a discussion, but the practical implementation of any law proves it to be an important part of the whole document. Although, the actual adherence to an international legal regime depends on the administrative aspect, the smooth existence is deeply rooted on adjudication. The judicial view not only attracts the attention and conduct of such a regime, but it also considered to be one of the most critical decider of the well being of a designed legal order. The role of the judiciary being dispute resolution and law making to a certain extent by way of case law or 'judge made law', its pioneering piece of work has well been respected in both civil and common law jurisdictions. It would be true to say that; liability is regulated by means of legislative instruments, irrespective of being national or international in character, but the judiciary imposes liability upon adjudication. Therefore, the ascertainment of such liability is wholly dependent on the decisions of the courts. Having said that, it would be more appropriate and prudent to address the jurisdiction aspect of the Bunkers Convention,[530] before attending to the liability issue in a much detailed way. On the other hand, the exact aim of this entire discussion is to evaluate the implementation of liability for bunker oil pollution damage in a more practical manner, and therefore, adjudication falls second only to, the administrative aspect. Once pollution damage has been caused to the victim upon a bunker oil spill, and the respondent has been traced, the next step would be to file actions according to the prescribed formula. All other issues relating to the incident are matters that have to be addressed in such prescribed forum.

Jurisdiction as proclaimed

The Convention deals extensively with regard to this subject, and differs slightly with its other counter parts namely, the CLC[531] and HNSC[532]. Although, it ad-

530 See K. Takahashi, *II. External Competence Implications of the EC Regulation on Jurisdiction and Judgments*, Int'l and Comparative Law Quarterly (2003), 52:529–534

531 See Art. IX

532 See Arts. 38 and 39

dresses this issue in somewhat direct form via one main provision, there can be found certain interconnections with some other provisions dealing with related issues.[533] Therefore, the intention behind addressing the jurisdiction issue in this Chapter is confined to those direct issues concerning litigation, leaving miscellaneous issues relating to administrational aspects of the Convention, untouched.

Article 9 of the Convention expressly declares jurisdiction that needs to be implemented in accommodating legal actions. Its sub-paragraph 1 states that;

> "Where an incident has caused *pollution damage in the territory, including the territorial sea or in an area referred to in Article 2(a) (ii) of one or more States Parties*, or *preventive measures have been taken to prevent or minimize pollution damage in such territory, including the territorial sea, or in such area*, actions for compensation against the shipowner, insurer or other person providing security for the shipowner's liability may be brought *only in courts of any such States Parties*." (emphasis added).

Viewing at it from its outset, the relevant provision seems to be understandable, but complex to a certain extent. Therefore, it is advisable to break it into separate portions in clarifying for better understanding.

The causative link

Any action of litigation under the Convention is based on the joint cohabitation of 'incident'[534] and the 'pollution damage'[535] or 'preventive measures'[536]. The place in which the incident has occurred is of no importance. The vital element that needs to be established is the physical appearance of pollution damage, or preventive measures that have been taken upon an incident that caused such damage or led to the taking of such measures. Therefore it is clear that an oil spill incident alone would not suffice to file any actions under the Convention.

The geographical link

The causative link is further supplemented by the geographical aspect. The pollution damage that was caused or preventive measures that were taken upon an incident should have occurred within a one or more of the States' Parties' jurisdic-

533 Such as Time Limits, Article 8, Recognition and Enforcement clause, Article 10, States with more than one system of law and Article 13
534 See *supra* pp 63–65
535 See *supra* pp 53–59
536 See *supra* pp 65–71

tional ambit.[537] There may exist of simple situations where only one State Party is affected against much complex ones where more than one States' Parties are affected. In any such situation(s), it is of vital importance, in the process of obtaining compensation, to link the pollution damage caused or preventive measures taken upon an incident to the geographical extent to which any of such States Parties have effectual jurisdiction. Accordingly, the area which falls within the scope of application includes the territory, territorial sea and the exclusive economic zone or the area extending up to two hundred nautical miles from the baseline of such State, as discussed in Chapter 3 dealing with geographical scope of application of the Bunkers Convention.[538] Therefore, it is understood that any damage caused outside such limits, such as in the high seas, do not fall within the meaning of this Convention and no action for compensation therefore be laid under it.

The initial draft that was presented at the very beginning by the Sponsoring Delegations had certain extra implications on the issue of jurisdiction. Although it is not intended to open up a debate based on that, it would be worthwhile to underline few characteristic features that were in existence in that draft. As mentioned in Chapter 3 under 'pollution damage', there were some persuasions during the initial stage to include damages caused other than by contamination.[539] This resulted in proposing an effective dispute resolution process for such damages that caused outside the territory or territorial jurisdiction of any State[540]. Therefore, the drafters eyed for three options in accommodating such claims, as the situation lacks actual jurisdictional competency of any State in the absence of territorial jurisdiction. According to the relevant provision, it would have been possible to bring such claims in the courts of;

(a) the State of the ship's registry; or
(b) the State Party where the shipowner has habitual residence or principal place of business; or
(c) the State Party where a limitation fund has been established.[541]

537 See Vienna Convention on the Law of Treaties between States and International Organizations or between International Organizations 1986, available at http://untreaty.un.org/ilc/texts/instruments/english/conventions/1_2_1986.pdf

538 *Supra* pp 45–52

539 *Supra* pp 59–62

540 See K. M. Meessen, *Extraterritorial jurisdiction in theory and practice*, Kluwer, 1996, Ch. XI, pp 225–238; C. L. Brooker, *The Clash of Obligations: Exercising Extraterritorial Jurisdiction in Conformance with Transitional Justice*, available at http://works.bepress.com/christen_broecker/2/; R. Lefeber, *Frontiers of International Law: Counteracting the Exercise of Extraterritorial Jurisdiction*, Leiden Journal of Int'l Law (1997), 10:1:1–7

541 Draft Article 8.2, Annex, *supra* n. 51, p. 5

Certainly, this persuasion was based on Article IX of the CLC 92 under which the existence of a fund complemented recovery of compensation by the victim[542], and Article 38 of the HNSC under which damages caused outside the territory or territorial jurisdictions of any State were allowed in principle as a result of the existence of a similar fund provisions[543] that prevails likewise in the case of the CLC. Many delegations were of the view that this range of alternative jurisdictions would promote forum shopping[544], and furthermore, create chaotic situations if the limitation fund has been established[545] elsewhere than in the State in which the claim has been filed.[546] During the law making process the issue concerning damages outside the territory or territorial limits of State Parties did not receive any welcome so as the matter relating to jurisdiction in such circumstance. Furthermore, leaving the limitation of liability under the competence of the applicable law of a State which possesses the jurisdiction to hear such matters, would necessarily require such limitation proceedings to be instituted in that jurisdiction,[547] without paving ways for the establishment of multiple funds.[548] Therefore, using such strategy to address extra-territorial damages would create chaotic situations.

The present strategy used by the Bunkers Convention seems to deviate from the common law approach of physical presence of the defendant[549]. It furthermore, eliminates possible forums such as defendant's habitual residence as proposed during the legislating process and the place where the vessel of the defend-

---

542 "[3.] After the fund has been constituted in accordance with Article V the courts of the State in which the fund is constituted shall be exclusively competent to determine all matters relating to the apportionment and distribution of the fund"

543 "[5] After a fund under article 9 has been constituted by the owner or by the insurer or other person providing financial security in accordance with article 12, the courts of the State in which such fund is constituted shall have exclusive jurisdiction to determine all matters relating to the apportionment and distribution of the fund"

544 See A. O. Sykes, *Transnational Forum Shopping as a Trade and Investment Issue*, http://www.law.virginia.edu/pdf/olin/0708/sykes.pdf; A. S. Bell, *Forum shopping and venue in transnational litigation*, Oxford, 2003; M. L. Busch, *Overlapping Institutions, Forum Shopping, and Dispute Settlement in International Trade*, Int'l Organization (2007), 61:4::735–761

545 See Wu Chao, *supra* n. 107

546 *Supra*, n. 59, pp. 3–4.

547 See J. C. McHorse, *Admiralty Jurisdiction and Limitation of Liability in Single Claim Cases*, California Law Review, vol. 22, no. 5 (Jul., 1934), pp 526–534

548 *Supra*, n. 206, p. 16

549 See O. Schachter, *International law in theory and practice*, Martinus Nijhoff, Kluwer, 1991, pp 256–257; J. O'Brian, *International* Law, Cavendish, (2001 rep. 2002) pp 232–234; L. I. De. Winter, *Exclusive Jurisdiction in Private International Law*, Int'l and Comparative Law Quarterly (1968), 17, pp 706–720; E. W. O'Brian, *The Hague Convention on Jurisdiction and Judgments: The Way Forward*, The Modern Law Review, vol. 66, no. 4 (Jul., 2003), pp 491–509

ant is arrested. This approach is not peculiar to the Bunkers Convention, but was adopted way back in 1969 when the first CLC regime came into existence.[550] Wu Chao express her opinion about the elimination of the common law approach, as "may have resulted in a variety of decisions being made regarding liability and extent of damage, which would furthermore, delay the grant of compensation".[551]

The collective link

Once the causative link and the geographical link are realized, the parties may proceed with the right of action based on the cause, either pollution damage or preventive measures, or both. Under such circumstances, the existence of 'pollution damage' or carrying out of 'preventive measures' coupled with the territorial jurisdictional competency play a vital role, and neither, the place of the incident of the oil discharge occurred nor the nationality or domicile of the plaintiff or defendant are decisive factors. The multiplicity nature of this collective link in more than one States Party would make courts of all such States Parties gain competency to entertain claims for compensation. Thus, Article 9 imposes a certain restriction on instituting such actions, permitting such actions for compensation to be brought only in the courts of any such States Parties.[552] This furthermore, strengthens the concept of non-dependency on the above said categories of possible alternatives for the institution of claims in different jurisdictions. However, the Convention is ambiguous in citing the exact forum in which an action may be brought. As far as litigations are concerned the Convention uses the phrase "... only in the courts of any such States' Parties" in which actions may be brought, creating conflictions between its literal meaning and the real intention of the drafters. Does this mean that actions pertaining to damages caused in a certain territorial jurisdiction can only be filed in that particular States Parties' courts or in courts of any State Party which suffered damage? Illustrating the CLC 1969 regime which had the same wordings, Abecassis and Jarashow pose a similar question relying on the literal meaning while mentioning that an action be brought only in the courts of States Parties' in which each and every damage

550 *Supra*, n. 238, pp. 495–496

551 Wu Chao, *supra*, n. 95, pp. 72–73

552 "[1] Where an incident has caused pollution damage in the territory, including the territorial sea, or in an area referred to in article 2(a)(ii) of one or more States Parties, or preventive measures have been taken to prevent or minimise pollution damage in such territory, including the territorial sea, or in such area, actions for compensation against the shipowner, insurer or other person providing security for the shipowner's liability may be brought only in the courts of any such States Parties"

occurred. They in fact lay doubts on this aspect whether it corresponds with the intention of the drafters.[553] The said phrase however is associated with the rest of the wordings, alleging that such actions be brought in States Parties "Where an incident has caused pollution damage in the territory ... or preventive measures have been taken ..." and this does not include situations where actions are brought in States Parties which have not suffered damage. Therefore, the meaning of Article 9.1 implies to possible actions that may be brought outside the States' Party in which such damage has occurred, but only in another States Party which suffered damage out of the same incident. In such circumstance, the Convention would facilitate forum shopping to a certain extent, which is somewhat contrary to the intention of the drafters.

Another practical problem arises in such situations where claims are brought in several jurisdictions pertaining to damage caused or measures taken in those respective States Parties.[554] With this respect, the Convention lacks provisions in proposing one single forum or registration of *lis pendens*[555] in cases of multiplicity of actions in different States Parties, and further to prescribe anything in response to prevent complexity of total claims, time constraints and leaving the defendant in an inconvenient situation. Although, it would have been much practicable to prescribe a solution based on diplomatic arrangements, matters concerning limitation of liability seems to have overridden any such solution that could have evaded the weakness. Therefore, the Convention has adopted the same effects as in CLC 69/92 and HNSC which allow plurality as to institution of claims with regard to the forum.[556]

In addition the Convention has not prescribed the need for a constitution of a limitation fund as in the case of CLC[557] and HNSC[558], mainly due to optional conceptual right vested on the shipowner in determining the best regime. This in fact places the shipowner under more complex situations in circumstance where multiple litigations are to be dealt in more than one jurisdiction. The important question that arises here is; whether he is required under respective national laws to constitute multiple limitation funds in all such jurisdictions? Absence of any such provisions dealing with this aspect under the Convention presupposes the

553 See Abecassis, *supra*, n. 92, p 220

554 See D. Jielong, *Statement on Diplomatic Protection, International Liability for Injurious Consequences Arising out of Acts not Prohibited by International Law*, Chinese Journal of Int'l Law 2007, 6(1): 189–194

555 The procedure allows to stay the actions instituted in the rest of the courts other than the court in which an action was instituted first, till the latter court determines the jurisdiction in which the action be tried

556 Art. IX.1, Art. IX.1and Art. 38.1

557 Art. V.3

558 Art. 9.3

non requirement of a limitation fund, but to depend on respective limitation regimes in each and every case. Furthermore, the international private law lacks rules pertaining to a system of executing *lis pendens* in situations where actions are brought in more than one jurisdiction with respect to matters relating to one incident. The European Community law stands as an exception to this where Member States are bound by the Brussels[559] and Lugano Conventions[560] on Jurisdiction and the Recognition and Enforcement of Judgments in Civil and Commercial Matters, under which 'maritime claims' are thus included.[561]

Suing the appropriate defendant

The Convention is expressed in outlining the parties against whom action for compensation may be brought, unlike the CLC and HNSC regimes. The central scope of imposing liability on the shipowner has well been laid down in the jurisdictional provisions. Accordingly, the shipowner, insurer or other person providing security for shipowner's liability fall within the meaning of defendant in litigation. Obviously, the prime person who is bound to become the defendant is the shipowner, but in certain circumstances the plaintiff would have the possibility of instituting direct actions against the insurer or financial security provider as a matter of curtailing associated difficulties.[562]

---

559 Brussels Convention on Jurisdiction and the Enforcement of Judgments in Civil and Commercial Matters 1968, also known as EEX Convention. For full text of the convention, see http://www.jus.uio.no/lm/brussels.jurisdiction.and.enforcement.of.judgments.in.civil.and.commercial.matters.convention.1968/doc.html. See also J. A. Pontier, E. Burg, *EU Principles on jurisdiction and recognition and enforcement of judgments according to the case law of the European Court of Justice*, T. M. C. Asser, 2004, pp 243–244; I. Carr, P. Stone, *International trade law*, 3rd ed., Cavendish, 2005, p 599; S. Sime, *A Practical Approach to Civil Procedure*, 10th ed., Oxford, 2007, p 131

560 Lugano Convention on Jurisdiction and the Enforcement of Judgments in Civil and Commercial Matters 1968, also known as EVEX Convention. For full text of the convention, see http://www.ofj.admin.ch/etc/medialib/data/wirtschaft/ipr.Par.0022.File.tmp/260307_entw_lugano_convention-e.pdf. See also J. A. Pontier, E. Burg, *supra* n. 559, p 243; I Carr, P. Stone, *supra* 559, p 599–600

561 See Article 54 A.1 of Lugano Convention

562 Such as locating, identifying and other related difficulties

Direct actions

The Convention provides the victims a more flexible mode of instituting claims in events of unidentifiable owners of ships. Recalling the main pillars on which the entire regime was established, direct actions constitute one of these main pillars. Although it is not an actual transfer of liability on the insurers or financial security providers, under a true meaning of the law, direct actions make the procedure simpler and faster for the recovery of compensation.[563] The estimated risk calculated in terms of economy is the basic factor which builds the idea of such a transfer of responding. Therefore, it is the insurance cover or the financial security undertaking that supports the concept of direct action while holding the shipowner's liability intact. Neither the CLC 69 nor 92[564] contained such direct implications on those responsible parties, but however, the experiences of handling matters under those regimes in the past proved the necessity of upgrading the system via HNSC[565] and the Bunkers Convention. The response on the part of the insurers and financial security providers were negative to such moves, at the delegation level meetings of the Legal Committee, especially both the P&I Clubs and Comite Maritime International (CMI) being scared of consequential financial burdens that would put the industry in a less-convenient situation[566]. Formally, the main fear was centered on the issue of enormous administrative burdens on the Clubs, their respective members and also on the States, in observing higher measures for implementation of respective policies and issuance of port entry certificates. Hence, the insurance industry would need to assume high degree of risks in cases of policy breaches by their respective shipowners.[567]

Upon an adoption to this end, the P&I Clubs in particular will have to bear the ultimate outcomes. They laid many questions at the meetings, focusing towards a justification in the light of a demonstrated added benefit on the part of the claim-

563 See Wu Chao, *supra* n. 95; V. Petaccio, *Water Pollution and the Future Law of the Sea*, Int'l and Comparative Law Quarterly (1972), 21, pp 15–42; J. Wren, *Overview of the Compensation and Liability Regimes Under the International Oil Pollution Compensation Fund (IOPC)*, Spill Science and Technology Bulletin, vol. 6, issue 1, Feb., 2000, pp 45–58

564 Art. IX.1: "... actions for compensation may only be brought in the courts ..."

565 Art. 38.1: "... actions for compensation may be brought against the owner or other persons providing financial security for the owner's liability only in the courts ..."

566 See M. Jacobson, *supra* n. 449; Wu Chao, *supra* n. 95; See also LEG/95, Mar., 30-Apr., 4, 2009, available at http://www.imo.org/newsroom/mainframe.asp?topic_id=1772&doc_id=11233; LEG 94/11/2, http://www.igpandi.org/downloadables/submissions/imo/IMO%20LEG%2094%20Paper%20-%20October%202008.pdf

567 *Supra*, n. 68 at p. 9, para 60

ant in a direct action.[568] Due to the fact that all States nowadays require a valid insurance for entry into ports, it would be very rare that an insurer abstain from granting compensation consequent to a valid certificate of insurance. However, it would not be a point of justification in every circumstance concerning the effects of the Bunkers Convention. In a restrictive sense, the Convention need not leave any loop holes through which tortuous actions could penetrate without any positive solution.[569] In P&I point of view, the issue creates unnecessary complexities especially in circumstances where the pollution damage is relatively remote and the amount at stake is too small to justify proceedings.[570] The backing up of the International Chamber of Shipping (ICS) on this issue made significant impact in highlighting the obligation to carry certificate of entry in all ships of State Parties[571] together with the main concept of imposition of strict liability on the shipowner, which exclude the use of direct action.[572] Nevertheless, the delegations had no intention of reverting to several past situations where victims were left without remedies in the absence of identification of the responsible party, in particular the shipowner, rather proceeding with the inclusion of direct action concept, as a matter of considering it as one of the main pillars of this regime,[573] and more importantly, facilitating the victims with the benefit of speedy recovery and cost effectiveness. In view of the provisions, direct actions are feasible only

568 Bunker spills constitute less than 5% of the risks covered by Clubs

569 For the ultimate solution adopted by the Convention, see Art. 7.2: "[A] certificate attesting that insurance or other financial security is in force in accordance with the provisions of this Convention shall be issued to each ship after the appropriate authority of a State Party has determined that the requirements of paragraph 1 have been complied with. With respect to a ship registered in a State Party such certificate shall be issued or certified by the appropriate authority of the State of the ship's registry; with respect to a ship not registered in a State Party it may be issued or certified by the appropriate authority of any State Party. This certificate shall be in the form of the model set out in the annex to this Convention and shall contain the following particulars: (a) name of ship, distinctive number or letters and port of registry;(b) name and principal place of business of the registered owner; (c) IMO ship identification number;(d) type and duration of security;(e) name and principal place of business of insurer or other person giving security and, where appropriate, place of business where the insurance or security is established;(f) period of validity of the certificate which shall not be longer than the period of validity of the insurance or other security"

570 LEG 80/4/2, 9 September 1999, p. 2

571 Art. 7.5: "[T]he certificate shall be carried on board the ship and a copy shall be deposited with the authorities who keep the record of the ship's registry or, if the ship is not registered in a State Party, with the authorities issuing or certifying the certificate"

572 *Supra*, n. 206, p. 4

573 See *supra*, n. 277

with respect to ships having gross tonnage greater than 1000,[574] unless other ships which fall outside this scope maintain insurance policies. As the convention applies to all types of ships irrespective of its tonnage, small ships are not under the compulsory requirement of maintaining insurance or other financial security.[575] Nevertheless, it would be up to the respective national laws to regularize any requirements pertaining to such guarantees in foreseeing risk mitigation. However, the Convention does not lay any burden on small ships with a necessity to that effect.

As the compulsory insurance requirement is placed on the registered owner of the ship under the Convention through its Article 7.1, the victim is at a better position in pursuing with his case against the shipowner even in circumstance where the ship in question was under some other contractual arrangement. Article 7.10 specifically deals with the subject of direct actions against the insurer or other person providing financial security, stating in its first sentence that;

> "Any claim for compensation for pollution damage may be brought directly against the insurer or other person providing financial security for the registered owner's liability for pollution damage."

In accommodating the court with the fullest co-operation and avoiding unnecessary complications with regard to substantive and procedural questions in litigations, the insurer or financial security provider has the right to require the registered owner to be joined in the proceedings.[576] Therefore, as a matter of convenience and obtaining justice, the victim is enabled to proceed with the case in the presence of all such responsible parties, as the case may require. This provision clearly assembles all connected issues and persons upon which an action for compensation may lie.[577]

---

574 Art. 7.1: "[T]he registered owner of a ship having a gross tonnage greater than 1000 registered in a State Party shall be required to maintain insurance or other financial security, such as the guarantee of a bank or similar financial institution, to cover the liability of the registered owner for pollution damage in an amount equal to the limits of liability under the applicable national or international limitation regime, but in all cases, not exceeding an amount calculated in accordance with the Convention on Limitation of Liability for Maritime Claims, 1976, as amended"

575 See Ling Zhu, *supra* n. 52, pp 36–39

576 See Art. 7.10: "… [T]he defendant shall in any event have the right to require the shipowner to be joined in the proceedings"

577 See Jus. R. C. Corona, *Class Action, Public Interest litigation and the Enforcement of Shared Legal Rights and Common Interests in the Environment and Ancestral Lands in the Philippines*, available at http://www.aseanlawassociation.org/9GAdocs/Philipines.pdf; J. A. Page, P. Rothstein, *Brief of Amici Curiae of American Law Professors in Support of Respondents, Norfolk & Western Ry. Co. v. Ayers, No. 01–963(U.S. Aug. 19, 2002)*, U. S. Supreme Court Briefs, Georgetown

Prescription

Once the cause of action is realized under the Bunkers Convention, the Claimant should be aware of the time bar within which the claim has to be filed in the relevant court. Though he is vested with the right to compensation, it is vital that such Claimant exercise this right according to the clause dealing with prescription of actions. Normally, actions in tort at both national and international level lay high emphasis on the time limit.[578] The Bunkers Convention is no exclusion for such an approach, and outlines the provisions in Article 8. According to its first sentence;

> "Rights to compensation under this Convention shall be extinguished unless an action is brought thereunder within three years from the date when the damage occurred."

It is entirely clear that the action has to be brought within three years from the date when the pollution damage occurred irrespective of the date on which the oil spill actually took place. Here again, the oil spill incident does not have any bearing on the requirements, but the occurrence of pollution damage. Impliedly, it extends to the date of any preventive measures taken in an absence of any physical damage, as such measures fall within the meaning of 'pollution damage' and are permitted under the definition of 'incident' as meaning pure threat removal measures. In certain circumstances the original damage might not occur immediately when an incident takes place, such as in the event of a drowned vessel.[579] In such a situation, the damage might occur as a result of the salvaging effort or under natural circumstances while remain in the sea bed. Notwithstanding any

---

University Law Centre, Year 2002, available at http://scholarship.law.georgetown.edu/cgi/viewcontent.cgi?article=1059&context=scb; R. W. Wright, *The Vitality of Joint and Several Liability: Amici Curiae Brief, Norfolk & Western Railway Co. v. Ayers*, 123 S. Ct. 1210 (2003) (August 19, 2002), available at SSRN: http://papers.ssrn.com/sol3/papers.cfm?abstract_id=392786

578 See J. L. Mathews, *The Lawsuit Survival Guide: A Client's Companion to Litigation*, 2003, pp 5/38–5/39; E. S. Knutsen, *The Ultimate Limitation Period: Updating the Limitation Act*, Canadian Business Law Journal, vol. 39, no. 149, 2003.; J. C. Care, *Contract law in the South Pacific*, Cavendish, 2001, p 272; D. L. C. Miller, *Three of a Kind? Positive Prescription in Sri Lanka, South Africa and Scotland*, Electronic Journal of Comparative Law, vol. 12.1(May 2008), available at http://www.ejcl.org/121/art121–3.pdf; For respective national laws on limitations, see Limitation Act 1980, UK; Prescription and Limitation (Scotland) Act 1984; Limitation Act 1969, Australia; Limitation Act 1963, India; Limitation Act (Ordinance No. 57 of 1959) Chap. 163, Singapore; Limitation Act 2002, Chap. 24, Schedule B, Canada; Limitation Act 1950, New Zealand; Limitation Act 1975, Samoa; Limitation Act 1953, Malaysia; Prescription Ordinance No. 22 of 1871, Sri Lanka to name a few

579 See Colin de la Rue, *supra*, n. 316, p 115

of these events, the claimant is under an obligation to specify the date on which the actual damage caused to his interests in order to overcome possible objections by the defendant as to the time bar. Therefore, it is necessary that claimant acts in due course in preventing its claim from been dismissed. In actual occurrence, it is very rare that damage is limited in scope without extending further for an indefinite period of time causing subsequent effects. Once pollution damage takes place it could spread indefinitely with regard to the extent and time. Unless effective and timely measures are applied to prevent further damage, and to clear up the effected area restricting any type of spreading its effects thereafter,[580] it is uneasy to overcome connected problems.

However, the Convention does not allow any prolongation of institution of claims in the event of such on going process, unless the action fulfills the requirement laid down in the second sentence of this Article, which states that;

"... in no case shall an action be brought more than six years from the date of the incident which caused the damage."

Therefore, the Convention grants certain benefits in such events, but courts will examine the actual scenario with a high degree of attention to claimants' conduct. This implies that the claimant needs to provide a valid and indeed strong 'show cause' as to the prevalence of an ongoing characteristic behavior of damage. It is important to note that all such damages shall result from one occurrence or series of occurrences having the same origin in order to rely on the foregoing provision, as such pollution damage is directly linked to an incident which caused such damage, and include 'incident' as defined in the Convention.[581] The provision would not only bar undue actions based on the absence of due diligence, but also encourages States Parties and other authorities in taking speedy recovery acti-

580 See C. R. Chase, L. G. Roberts, *Advanced Detection Technology for Early Warning – The Key to Oil Spill Prevention*, available at http://www.slicksleuth.com/pdf/IOSC2008paper.pdf; C. R. Chase, S. van Bibber, *Utilization of Automated Oil Spill Detection Technology for Clean Water Compliance and Spill Discharge Prevention*, available at http://www.epa.gov/OEM/docs/oil/fss/fss06/chase_2.pdf; F. Lopez-Pena, R. J. Duro, *A Hyperspectral Based Multisensor System for Marine Oil Spill Detection, Analysis and Tracking* in *Knowledge-Based Intelligent Information and Engineering Systems*, Springer, vol. 3213, 2004, pp 669–676; J. F. Baxter Jr., *Oil Spill Detection System*, 1995, US Patent No. US5532679, available at http://www.freepatentsonline.com/5532679.pdf; J. P. Pederson, L. G. Seljelv, T. Bauna, O. A. Follum, J. H. Anderson, T. Whal, A. Skoelv, *Towards an Operational Oil Spill Detection Service in the Mediterranean? The Norwegian Experience: A Pre-operational Early Warning Detection Service using ERS SAR Data*, Spill Science and Technology Bulletin, vol. 3, no. 1, 1996, pp 41–46(6)

581 See in conjunction, Arts. 3 and 1.8

ons in completing the whole scenario of pollution damage caused and preventive measures taken with respect to an oil spill from bunkers.

Furthermore, an explanation with regard to the said six-year period has been provided in the third and final sentence of Article 8. According to it;

> "Where the incident consists of a series of occurrences, the six-years' period shall run from the date of the first such occurrence[582]."

Likewise many other provisions in the Convention, this relevant section is a repetition of the CLC regime having the same wordings. The HNSC follows somewhat different approach separating action brought against the owner in a court of law from that of action brought against the HNS Fund.[583] Although, its characteristic feature of having two separate forums under which actions could be brought, something which the Bunkers Convention lacks, subject to the three year period in event of direct results, and further the six-year period with regard to on-going damage strategy in the Bunkers Convention, is complimented with a ten-year period in the HNSC.[584] One may see this as a wide relaxation to the onus on the claimant and another may see this as an undue grant. It is worthwhile to note that referring to the subject matter it deals with, the HNSC ought to provide such concession,[585] unlike the Bunkers Convention.

Institution of Actions

As illustrated in the above discussions, any action instituted in virtue of the provisions of the Convention, need to satisfy conventional requirements. Rights and obligations that arise under international multilateral conventions such as the Bunkers Convention are contracts between States Parties as far as private international law is concerned. Therefore, matters arising out of a non-contracting State involvement fall outside the ambit of such Conventions. As evident from the practice of the CLC and HNSC regimes, the relevant provisions of Conventions entered into by States Parties are strictly applied for and against such States Parties' interests. Before instituting an action based on the Bunkers Convention, the claimant would need to ascertain whether the defendant is a person, natural or juridical, who is domiciled, flying the flag in the ship in question or carrying

582 Occurrence meaning occurrence of pollution damage as defined

583 See Arts. 37.1 and 37.2

584 See Article 37.3

585 See R. Force, M. Davies, *Jurisdiction and forum selection in international maritime law*, Kluwer, 2005, pp 172–174; M. D. Guner-Ozbek, *The Carriage of Dangerous Goods by Sea*, Hamburg Studies on Maritime Affairs, vol. 12, Springer, 2008

out the relevant business, in a State Party to the Convention or not. This is the most decisive factor in determining whether actions could be brought under the Convention according to its provisions.

Matters relating to non-States Parties

It is obvious that a successful action could only be instituted by and against persons of States Parties under a relevant Convention, as a basic rule of international law. But the situation is different in cases of matters involving non-States Parties' interests. The Vienna Convention on Law of the Treaties in its Article 34 clearly states that "A treaty does not create either obligations or rights for a third State without its consent".[586] Therefore, in such circumstance, the claimant is unable to depend on a Convention which lacks effects on such parties rather than sorting out other legal remedies. In such context, the identification of the applicable law would depend on choice of law aspect that is available for the claimant under the legal system of the State in which he intends to institute the action.[587] This may sometimes lead to situations where a claimant relies on foreign jurisdiction. A clear example is the litigation of *Amoco Cadiz*,[588] where French authorities relied on the jurisdiction of US courts, seeking full remedies. Although it could promote forum shopping for States seeking best remedies, one could not prevent such

586 See I. M. Sinclair, *The Vienna Convention on the Law of Treaties*, Manchester Univ. Press, 1973, p 10; E. W. Vierdag, J. Klabbers, R. Lefeber, *Essays on the law of treaties: a collection of essays in honour of Bert Vierdag*, Kluwer, 1998, p 98; E. W. Vierdag, *The Law Governing Treaty Relations between Parties to the Vienna Convention on the Law of Treaties and States not party to the Convention*, American Journal of International Law, vol. 76, 1982; James Crawford, UN ILC, *The International Law Commission's articles on state responsibility*, Cambridge, 2002, p 243

587 See. M. Weitzenbok, *Determining Applicable Law and Jurisdiction in contractual disputes regarding virtual enterprises*, Proceedings of the 8th International Conf. on Concurrent Enterprising: Ubiquitous Engineering in the Collaborative Economy, Rome, Italy, 17–19 June 2000, pp 27–34, available at http://folk.uio.no/emilyw/documents/S01_p04_Weitzenboeck.pdf; R. C. Force, A. N. Yiannopoulos, M. Davies, *Admiralty and Maritime Law, Volume 2*, Beard, 2005, pp 85–86; Admiralty, *Federal Court Applies Maritime rather than State Law in Diversity Suit for Injury sustained within State's Territorial Waters*, Harvard Law Review, vol. 63, no. 3 (Jan., 1950) pp 525–526; M. L. Hendrikse, N. H. Margetson, *Aspects of Maritime Law: Claims under Bills of Lading*, Kluwer, 2008, pp 330–342; R. Force, M. Davies, *Jurisdiction and forum selection in international maritime law: essays in honour of Robert Force*, Kluwer, 2005, pp 311–316; M. George, *Choice of Law in Maritime Torts*, Journal of Private International Law, vol. 3, no. 1, 2007, available at http://papers.ssrn.com/sol3/papers.cfm?abstract_id=1117942

588 1992 AMC 913

situation where the other party involved is a non-State Party. Many authors criticize these Conventions on the ground that shipowning interests are overvalued than the benefits for victims,[589] and therefore result in overlooking the regimes in cases involving non-States Party interests.[590] On the other hand, the shipowners will be left in a much disadvantageous situation in non-States Parties' forums, with additional burdens and liabilities in addition to the absence of choice with regard to limitation under the Bunkers Convention in particular, when actions are brought in such courts. Furthermore, the victims in non-State Parties will be in a disadvantageous situation as well, having required to institute the action depending on its laws, and relying on the jurisdiction which would either be based on the place on which the damage occurred meaning *lex loci delicti* or under the law of the jurisdiction in which the shipowner is domiciled or having its business, pursuant to the Rome Convention.[591] Even then there might exist of situations where such judgments are unenforceable and unrecognized in opposite States,[592] unless a bilateral or multilateral arrangement is in force[593]. In all these situations, it would be much beneficial on the part of the victimized States to strictly abide by its' arrest of ship procedure, but however, is unlikely to produce best results.

---

589 See M. G. faure, H. Wang, *Financial Caps for Oil Pollution Damage: A Historical Mistake?*, Marine Policy, 2008, available at http://papers.ssrn.com/sol3/papers.cfm?abstract_id=1116173; O. F. Kundsen, *Transport interests and environmental regimes: The Baltic Sea transit of Russian oil experts*, Energy Policy, Elsevier, 2009; N. Gaskel, C. Forrest, *supra* n. 499; G. Gauci, *supra* n. 130

590 For example see Wu Chao, *supra*, n. 95, p 262

591 See P. Sarcevic, Yearbook of Private International Law, Vol. IX, Staempfli, 2007, available at http://www.sellier.de/pages/downloads/9783866530713_leseprobe.pdf?code=28fa0049d3f20d01a9cbe47eabd022b2; P. Stone, *EU private international law : harmonization of laws*, Edward Elgar, 2006, pp 7–8

592 See Enforcement and Foreign Judgments: *ED & F Man Sugar Ltd., v. Lendoudis*, [2007] EWHC 2268 (Comm), [2008] 1 All ER 952, JIML 14 [2008] 1:17; J. Y. Gotanda, *Supplemental damages in private international law*, Kluwer, 1998, pp 218–219; S. M. Hope, *Recent Private International Law Developments Before the Supreme Court of Canada*, The Globetrotter, Ontario Bar Association – International Law Section Newsletter, vol. 13, no.2, 2009, available at http://papers.ssrn.com/sol3/papers.cfm?abstract_id=1342413; A. Bonomi, P. Volken, *Yearbook of Private International Law: Vol. X, 2008*, Sellier, 2009, pp 456–458; S. P. Baumgartner, *The proposed Hague Convention jurisdiction and foreign judgments: trans*, Paul Siebeck, 2003, pp 1–6

593 Since only a very few states have acceded to the Hague Convention on the Recognition and Enforcement of Foreign Judgments 1971, the concerned subject still based on relevant bilateral and multilateral arrangements such as the EC Convention on Jurisdiction and Enforcement of Judgments in Civil and Commercial Matters, EC Regulation No. 44/2001

Fairness Rule

In any civilized legal system the fairness of the proceedings stands as the most vital aspect. *Audi alteram partem* is strengthened by way of assuring that notice of an instituted action has well been served. The Convention stresses the importance of guaranteeing this requirement by enacting sub-paragraph 2 to the main Article dealing with jurisdiction. Accordingly;

"Reasonable notice of any action taken under paragraph 1 shall be given to each defendant".[594]

Although it again follows the traditional trend introduced by CLC[595] and HNSC[596] regimes, neither of these Conventions specifically describe what is meant by 'reasonable notice'. It is very much believed that the matter is left for national law of the States Parties in which actions for compensation are filed. In general it varies from one jurisdiction to the other with regard to the type, procedure and time frame. Unlike in a case where the defendant or defendants are domiciled or having the place of business within the geographical jurisdictional zone of a court or a State, a litigation of this nature involving foreign parties will employ additional burden on the claimants as well as the respective courts. In those circumstances, the procedure would most often be highly complex involving numerous diplomatic means of realization of the desired end costing enormous time and ofcourse, financial means. Due to the absence of a clarification as to 'reasonable' methods, a clear justification is unreachable. Therefore the courts would normally use the tool suiting its accepted norms of interpretation to that end in cutting short complex situations that might arise, enabling the victims to recover compensation upon a valid judgment. But however, sometimes it might not be the most suitable approach in considering the after effects of a trial *in absentia*.[597]

Once a judgment is delivered by a respective court of a State Party competent to hear such matter, it is left for enforcement against the defendants in their State of domicile. Suppose that the defendants were not present due to the fact that they have not received proper notice from the original court, the intended enforcement becomes impossible under Article 10.1.(b)[598]. Accordingly, any such judgment

594 Art. 9.2

595 See Art. IX. 1

596 See Art. 38.3

597 See Briefing note: *EU strengthens trials in absentia – Framework Decision could lead to miscarriage of justice*, available at http://www.openeurope.org.uk/research/tia.pdf; H. Schwartz, *Trials in Absentia*, available at http://www.wcl.american.edu/hrbrief/v4i1/schwar41.htm

598 Art. 10.1: "[A]ny judgment given by a Court with jurisdiction in accordance with article 9 which is enforceable in the State of origin where it is no longer subject to

that has been given under the circumstance "where the defendant was not given reasonable notice and a fair opportunity to present his or her case" would not be able to secure its aim. It is sometimes true and very important to ensure the existence of such a clause, but in certain circumstances would not make a difference rather than consuming time and related costs. If the court, which is moved to endorse or enforce the foreign court judgment is convinced that the provisions of Article 10.1(b) has not been met by such respective court, it will be reverted back to that original court for re-trial. The only way of preventing such complicated formalities is by way of ensuring that reasonable notices are served on the defendants in a credible manner.

The Forum

As far as the jurisdiction is concerned, the Convention lacks self executing character. This is evidenced by Article 9.3 which specifically requires States Parties to ensure the competency of its respective courts in entertaining actions for compensation under the Convention.[599] The actual meaning would not only be based on the forum, but also the procedures involved in issuing notices and other legal documents relating to the summoning of the defendant to the court ceased with such action. Different States apply different tools in establishing this requirement, especially depending on its respective legal system. That is to say, that common law countries use procedures that are common to their system of administration of justice[600] and in the meantime, civil law countries execute their obligations under respective civil or commercial codes or any other statutes giving effect to dispute resolution[601] under the Convention. Even before the coming into force of the Bunkers Convention, number of States practiced bunker oil pollution damage liability and compensation cases under their respective legal system and had enacted legislations to that effect granting jurisdiction to respective courts in

---

ordinary forms of review, shall be recognised in any State Party, except (a) where the judgment was obtained by fraud; or (b) where the defendant was not given reasonable notice and a fair opportunity to present his or her case"

599 "[E]ach State Party shall ensure that its courts have jurisdiction to entertain actions for compensation under this Convention"

600 See M. W. D. White, *Australian maritime law*, 2nd ed., The Federation Press, pp 2–3; D. L. A. Barker, *Law Made Simple*, 12th ed., Elsevier, 2007

601 See J. G. Apple, R. P. Deyling, *A Premier on the Civil Law System*, available at http://www.fjc.gov/public/pdf.nsf/lookup/CivilLaw.pdf/$file/CivilLaw.pdf; A. Nussbaum, *Proving the Law of Foreign Countries*, The American Journal of Comparative Law, vol. 3, no. 1 (Winter, 1954), pp 60–67

resolving such disputes.[602] Rules pertaining to judicial procedures vary in much diversity with respect to Common and Civil Law States, and therefore, it would be wise to analyze those different systems with a view of gaining a better understanding and further to discuss how the situation would operate under the Bunkers convention once it is ratified or acceded to by varying States.

The Common Law Systems' Strategy

In common law States, actions can be instituted in way of matters *in rem*[603] or *in personam*[604]. Even though, the Bunkers Convention clearly embodies a method based on actions *in personam*, States which practice common law system would still have the possibility for resorting to either of these two methods in practice. There are substantial and procedural differences in these two procedures although they are aimed at providing solutions to the claim.[605] In matters *in rem*, the action is instituted against the '*res*', in this case the ship,[606] whereas in matters *in personam*, the case lies against the defendant,[607] the shipowner or the person as defined under the Convention, to hold liability. Jackson points out that

---

602 For example, the Merchant Shipping Act 1995 in UK and OPA 1990 in US

603 Commonly known as action or Writ *in rem* where it refers to a thing (or *res*). In the United Kingdom it falls within Civil Procedure Rules 1998 (S. I. 1998/3132) and is instituted by a 'claim form in *rem*' under Practice Direction 61 (Admiralty Claims), para 3.1 and Form ADM 1 that has been amended upon promulgation pursuant to Part 61(Admiralty Claims), Rule 61.3 of Civil Procedure Rules of 1998 with effect from Mar., 25, 2002

604 Commonly known as action *in personam* and is instituted against a person including juridical. In UK the action is instituted by way of issuance of 'claim form' by court under the Civil Procedure Rules 1998(S.I.1998/3132) now governed by Practice Direction 61 (Admiralty Claims) paras 12.1 – 12.6 and in Form ADM1A

605 See G. H. Smith, *The Law of Private Right*, Biblio Bazar LLC, 2008, p 6; A. Berger, *Encyclopedic dictionary of Roman Law*, The American Philosophical Society, 2004, p. 346; A. Dahdal, P. S. Gillies, *Characterizing the Action in Rem in Australia and the Implications on International Commercial Arbitration*, Journal of Maritime Law and Commerce, vol. 40, no. 2, 2009

606 See Dr. S.C. Derrington, *The nature of the modern action in rem*, Paper presented to the Judges of the Admiralty Panel of the Federal Court of Australia, August 2007, available at http://www.docstoc.com/docs/16056932/The-Nature-of-the-Modern-Action-In-Rem---Federal-Court-of (last visited; Nov., 15, 2009); P. R. Barnett, *Res judicata, estoppel, and foreign judgments: the preclusive effects of*, Oxford, 2001, pp 77–83; A. M. Sheppard, *supra* n. 422, p 75–101; A. Dahdal, *supra* n. 605; D. J. Cremean, *Admiralty Jurisdiction: Law and Practice in Australia, New Zealand, Singapore and Hong Kong*, 3rd ed., The Federation Press, 2008, pp 21–24

607 See D. J. Cremean, *supra* n. 606

an action *in rem* could form a basis for an action *in personam* once the defendant appears in court upon an acknowledgment of service for an action *in rem* that is constituted against his interests,[608] the ship. This would result in the matter continuing as an action *in rem* and also *in personam.*[609] As a common law State, the United Kingdom in particular does not bar such a situation where concurrent procedure could exist of, as stated earlier. There are several advantages under *in rem* procedure. Even though the remedies are limited in proportion to the value of the '*res*' in absentia of the defendant, the claimant is able to obtain a writ to arrest the ship[610], which leads him to be considered, a secured creditor.[611] Upon a successful decree, compensation could be obtained by sale of the ship, and the buyer himself receives a clear title from the court. In failing to institute action *in rem* in the absence of the custody of the ship by the court, as happen much often in cases of pollution damage that occur due to incidents that take place outside the territorial jurisdiction of the damage suffered State, or in circumstances of a pending unenforceable or partially enforceable *in rem* judgment which is in force, the victim may resort to *in personam* procedure. In such an action, the plaintiff is able to obtain a Mareva injunction,[612] in addition to a writ served against the defendant, in order to have a hold on the interests of the defendant that rest in the jurisdiction of the court that issues.

608 D. C. Jackson, *Enforcement of Maritime Claims*, 2 ed., 1996, LLP, p. 177

609 See D. J. Cremean, *supra* n. 606, pp 191–193; J. J. O. Harwood, *Lessons in Lien – The Separate Nature of in personam and in rem Claims*, The Arbitrator, Society of Maritime Arbitrators, vol. 40, no. 3, April 2009

610 See "full text", International Convention Relating to the Arrest of seagoing Ships 1952, available at http://www.admiraltylawguide.com/conven/arrest1952.html; A. M. Sheppard, *supra* n. 422, p 128; R. Gillich, *Practical issues arising from the Arrest of Ships*, available at https://maritimejournal.murdoch.edu.au/index.php/maritimejournal/article/viewFile/96/140;

611 See S. S. Karic, *NOTE: Admiralty – Procedure for Maritime Action in Rem Found Unconditional*, 55 Tul. L. Rev. 936 (1981); G. Samuel, *Law of obligations and legal remedies*, 2nd ed., Cavendish, 2001; D. G. S. Chong, *Security in Actions in Rem*, Singapore Academy of Law Journal, vol. 1, 1989, p 81; R. C. Force, A. N. Yiannopoulos, M. Davies, *supra* n. 587, p 179–183; A. M. Sheppard, *supra* n. 422, p 69

612 Pursuant to the order of the case *Mareva Compania Naviera S.A v. International Bulk Carriers S.A.*, [1975] 2 Lloyds' Report 509. In UK it is now provided for by sec. 37(3) of the Supreme Court Act 1981, (UK 1981 c. 54). It is also known as 'freezing injunction'. However, Mareva Injunction does not exist in USA, but exist in commonwealth States. See K. Lipstein, *Conflict of Laws – Jurisdiction – Mareva Injunction as Basis for Service Aboard – R. S. C., Ord. 11, R.1(1)(i)*, The Cambridge Law Journal, 37, pp 241–244; J. B. Berryman, *Remedies: Cases and Materials*, 5th ed., Emond Montgomery, 2006, pp 938–951; K. M. Meesen, *Extraterritorial jurisdiction in theory and practice*, Kluwer, 1996, pp 42–59

The common law remedy is effectively used by States that have a common law system and the competencies of the respective courts in dealing with issues pertaining to liability and compensation for oil pollution damage need to be granted by way of national legislations enacted in virtue of the provisions of the Bunkers Convention once it is ratified or acceded to by such respective States, unless it has not made any reservations with respect to that issue, before such ratification or accession, or have certain conflictions with the new regime. For example, the English legal system is competent to apply bunker oil pollution damage that occur within its territorial jurisdiction according to substantive and procedural law under its provisions laid down in the Merchant Shipping Act 1995,[613] and Supreme Court Act 1981[614] as amended and further supplemented by Rules of the Supreme Court,[615] respectively, although further transitional provisions have come into place with the enactment of Merchant Shipping (Amendment) Act 2008 and Merchant Shipping (Amendment) Act 2009, such as matters concerning the enforcement and recognition of foreign judgments as proclaimed by the Bunkers Convention likewise the enactments of the CLC regime.[616]

The Civil Law Systems' Strategy

The civil law system on the other hand operates on a much simpler manner than that of the common law system. Basically, delictual actions are brought according to the statute or code which grant the right for cause of action and dealt in way of a civil litigation. Remedies are normally based on the respective civil procedures and relevant statutes that grant such rights, and remedies pertaining to such actions.[617] In most instances, the statute which implements cause of action and the procedure that should be followed, expressly confer jurisdiction to a relevant civil court to hear the matter. Therefore, the plaintiff is required to follow the requisites laid down in such statutes and plead the case accordingly in the relevant court of law entrusted with the power to determine liability and compensational issues arising out of the subject matter. In this respect some countries legislate such issues, as to pollution damage, in their environmental legislations and some include these in other regimes such as transport law. [618]

---

613 See Sec. 154 and 166

614 Subsection 5 of Sec. 20

615 Order 75, Rule 2

616 See Merchant Shipping Act 1995, Section 166 (4) and Supreme Court Act 1981 as amended, subsection 5(a) of Section 20

617 See M. G. Faure, J. Hu, *Prevention and compensation of marine pollution damage: recent developments in Europe, China and the US*, Kluwer, 2006, pp 328–330

618 See M. Faure, J. Boyd, *Deterrence, insurability, and compensation in environmental liability: future developments in the European Union*, Vol. 5, Springer, 2003, pp

United States, for example, had enacted its laws pertaining to these issues in a different regime prior to 1990, but upon the enactment of the OPA 1990, all relevant issues are compiled in that document. The said document not only deals with the relevant law for damage resulting from oil spills from ships,[619] but also deals considerably with regard to the jurisdiction[620] in terms of actions that fall under that regime. Therefore, it would be interesting to analyze certain important characteristics concerning jurisdiction as proclaimed in such regime.

Jurisdiction under the OPA 1990

Section 1017 (b) dealing with the jurisdiction issue expressly states that;

"Except as provided in the subsections (a)[621] and (c)[622], the United States district courts shall have exclusive original jurisdiction over all controversies arising under this Act, without regard to the citizenship of the parties or the amount in controversy. Venue shall lie in any district in which the discharge or injury or damages occurred, or in which the defendant resides, may be found, has its principal office, or has appointed an agent for service of process. For the purpose of this section, the Fund ..." [623]

Furthermore, subparagraph (c) deals with the jurisdictional competency of State Courts with regard to the issue and states that;

"A State trial court of competent jurisdiction over claims for removal costs or damages, as defined under this Act, may consider claims under this Act or State law and any final judgment of such court (when no longer subject to ordinary forms of review) shall be recognized, valid and enforceable for all purposes of this Act."

As the OPA 1990 presently covers all types oil spills[624] and related damages including removal costs[625], as used under the Act,[626] and therefore governs the regime

100–111; F. Cafaggi, H. M. Watt, *The Regulatory Function of European Private Law*, Edward Elgar, 2009, pp 138–143

619 Title I – Oil Pollution Liability and Compensation
620 Sec. 1017 of Title I
621 Review and Regulations
622 State Court Jurisdictions
623 33 USC 2717
624 Sec. 1002 (a)
625 Sec. 1002 (b)
626 See *Rice v. Harken Exploration Co.*, [2001] WL 422051 (5th Cir. Tex.); M. Mason, *Transnational Compensation for Oil Pollution Damage: Examining Changing Specialties of Environmental Liability*, available at http://www2.lse.ac.uk/geographyAndEnvironment/research/Researchpapers/rp69.pdf; I. Kim, *A comparison between the international and US regimes regulating oil pollution liability and com-*

of bunker oil spill damages, the jurisdictional provisions are helpful in determining the way that system operates in the United States. Presently it provides the claimant certain optional forums, in which a case may be instituted depending on one of the grounds specified therein.[627] A question could be raised, whether it would change these provisions if US ratifies or accedes to the Bunkers Convention, but its absence in the CLC regime would lead to the belief that the present OPA provisions would prevail as it is for the future.

locus standi *of the parties*

Once an action for compensation for pollution damage has been instituted in a competent court as described under the Bunkers Convention, the court which is ceased with the action will certainly look into the issue of *locus standi* of the parties involved in the litigation.[628] In most cases, the defendant himself lodges primary objections on this issue seeking determination of the court. Situations with regard to proper *locus standi* arise mainly, where claims are filed by parties who lack direct involvement to the matter, based on the absence of clear title or legal right to damaged property or preventive acts that has been taken by such parties. Normally, the principal claimant and associated other direct claimants do not constitute a problem of this nature. Especially in a case where environmental damages constitute a considerable degree of compensation based on impairment, clean up and restoration, the *locus standi* of nature conservation organizations and pressure groups[629] that have coordinated during such process would create huge turmoil in the process of litigation.

---

pensation, Marine Policy, vol. 27, issue 3, May 2003, pp 265–279; A. C. Homan, T. Steiner, *OPA 90s impact at reducing oil spills*, Marine Policy, vol. 32, issue 4, July 2008, pp 711–718; *D.* Stalcup, G. Yoshioka, B. Kaiman, A. Hall, *Effects of OPA 90 Response Planning on Reported Spills*, available at http://www.iosc.org/papers/00640.pdf

627 See M. Mason, *supra* n 626; C. A. Jones, *Compensation for Natural Resource Damages from Oil Spills: A Comparison of USA Law and International Conventions and Pollution*, available at http://papers.ssrn.com/sol3/papers.cfm?abstract_id=1656; R. C. Helm, R. G. Ford, H. R. Carter, *The Oil Pollution Act of 1990 and Natural Resource Damage Assessment*, available at http://www.marineornithology.org/PDF/34_2/34_2_99–108.pdf

628 See also *supra* pp 69–70

629 See A. Geddes, *Locus standi and EEC environmental measures*, Journal of Environmental Law (1992), 4(1), pp 29–39; D. Wilkinson, *Moving the boundaries of compensable environmental damage caused by marine oil spills: the effect of two new international protocols*, Journal of Environmental Law (1993), 5(1), pp 71–90; G. S. Nijar, *Public Interest Litigation: a matter justice an Asian perspective*, available at http://www.aseanlawassociation.org/9GAdocs/Malaysia.pdf; G. Betlem, *Li-*

Victim being the central point at issue in resolving any dispute concerning the *locus standi* of the claimant, this discussion need to focus on the rights of such person in instituting actions. Failure of the Convention to define the victim in clear terms rather than referring to 'the person who suffered the damage' creates greater difficulties in this aspect.[630] Furthermore, providing the right of action to persons who take preventive measures being one of the important pillars of this compensation regime should have been complimented with much better clarifications.[631] All such failures have contributed heavily towards leaving those issues for respective national courts to deal under such national law. Although there exists of a definition for 'person', the rights of such 'person' with regard to the damaged property are not clearly set out in the Convention. Even though title or documentary rights that exist on the claims are clear evidence to establish the *locus standi* of a party to an action, hidden and unverified rights cause many difficulties to this end. There are two main instances under pollution damage and preventive measures respectively, where a thorough determination is needed, namely with respect to claims for natural resource damage and actions by environmental organizations and pressure groups.

Claims for natural resource damages

Classical theories of Roman Law describes two types of properties that fall outside the concept of real ownership, namely *res nullius* such as wild life and *res naturale* such as the sea, water and air.[632] Justinian's classification on these components of properties fall within the purview of State ownership accepted under modern legal principles. The doctrine has furthermore penetrated into international law since its birth with due recognition under customary rules. Ranging from conventions to case law based on international customary law rules concerning territory to airspace[633] extending to seashores[634], the international law provides

---

ability for Damage to the Environment in *Towards a European Civil Code* [A. S, Heartkamp et al. (eds.)], Kluwer, 1998, pp 473–491; W. Anderson, *Environmental Law Enforcement: the role of the judiciary*, available at http://www.inece.org/conf/proceedings2/58%20-%20Enviro%20Lw%20Enforcement.pdf; E. M. Basse, *Environmental Liability – Modern Developments*, available at http://www.cenneth.com/sisl/pdf/41-2.pdf; P. Hassan, A. Azfar, *Comment: Securing environmental rights through public interest litigation in South Asia*, 22 Virginia Envtl. L. J. 215

630 See also *supra* pp 111–113

631 See also *supra* pp 69–70

632 See R. W. Lee, , "The Elements of Roman Law", 4 Ed, at p. 110

633 See Article 1 of the Chicago Convention on International Civil Aviation

634 See UNCLOS 1982

due regard to so-called open access resources[635] that lie within respective territorial jurisdictions. The public law perspective of recognizing the rights of States' to its own natural resources,[636] confirms the idea of ownership on one hand and trusteeship[637] on the other. Irrespective of the kind of such titles, States in the international legal order posses clear rights with regard to properties of public nature. Furthermore, there have been instances where States are considered as *parens patriae* of a public resource in situations of common injury to its citizens.[638] Therefore, the rights to these public properties are exclusively vested upon States, and any action pertaining to pollution damage or claims on preventive measures, arising out of the Bunkers Convention need to be instituted by the affected States Party or its constituent sub-divisions having proper legal status. Rights of States to obtain compensation for environmental damage occur as a result of pollution damage and for the preventive measures taken under the CLC 69, were upheld by the Court of Genoa in the case of *The Haven*[639] laying great recognition to the concept.[640]

The OPA 1990 on the other hand allows designated trustees ranging from the President or authorized representatives of any State, the so-called State trustees, Indian tribe to foreign government, to present claims on behalf of such public, in order to recover damages to the natural resources.[641] In addition, it provides a basic structure on types of missions that all such claimants need to implement upon obtaining compensation, such as to develop and implement a plan for the restoration, rehabilitation, replacement or acquisition of the equivalent, of the natural resources under their trusteeship.[642] The wide encompassing details in the OPA 1990 provides an extensive understanding of the rights of parties in actions for natural resource damage and applied preventive measures, within the scope of

---

635 See T. Tietenberg (Lead Author); S. Niggol Seo (Topic Editor), 2006, *Open access resources* in *Encyclopedia of Earth*. Eds. Cuttler J. Cleverland (Washington D. C.: Environmental Information Coalition, National Council for Science and the Environment) published in the Encyclopedia of Earth, Dec., 11, 2006 [Retrieved Nov., 18, 2009], http://www.eoearth.org/article/Open_access_resources

636 See N. Schrijver, *Sovereignty over natural resources: balancing rights and duties*, Cambridge, 1997, pp 276–278

637 As is used by the law of the United States

638 See Gauci, *Oil Pollutoin at Sea – Civil Liability and Compensation for Damage*, John Wiley & Sons, 1997, p. 254

639 The Cypriot VLCC (very large crude carrier) named "the Haven" anchored off the coast of Genoa loaded with 144,000 tonnes of crude oil caught fire and exploded with one part sinking on the spot and other two sinking while towing. The Italian authorities took steps to burn the majority of the oil to prevent further damage

640 FUND 1971/EXC. 50/17, paras 3.2.10 – 3.2.13

641 Sec. 1006 (33 USC 2706)

642 Sec. 1006 (C)

liability and compensation for oil pollution damages, and especially the requirement of *locus standi.*

Status of environmental organizations and pressure groups

Locus standi of these organizations and interested persons fall into much of the controversies in connected litigations.[643] The main reason being the environment is subjected to non appropriation by way of private ownership, which is rather vested with the concept of State trusteeship. Eventhough it possesses a legal personality in the context of modern international law, courts are very reluctant to grant any right to private organizations or persons who represent the rights of such an entity[644]. Courts of some States have indeed facilitated such motivations in numerous instances, but a true legal proclamation via legislative enactments is absent in most States. In particular, the Council of Europe grants such rights to these organizations and interested parties to a certain extent enabling them to state the case as a matter of directing the liable party in carrying out its obligation of reparation and prevention in cases of oil pollution damage, but prevent itself from declaring any legal rules as to obtaining compensation.[645] Many States tend to prevent themselves from enacting such laws, due to various constraints ranging from disagreement as to delegation of common public ownership to private interests, to possible conflicts of dual rights that could exists between State authorities and these organizations. In such circumstance, a court would find lack of authority in granting *locus standi* on such entities due to the absence of proper proprietary interest on which a claim of maritime lien could rests. Such absence would in no way justify a cause of action for attainment of compensation upon the attribution of liability.

---

643 See *supra* pp 138–139

644 See B. A. Chokor, *Environmental pressure groups and habitat protection in the developing world: The case of Nigeria*, The Environmentalist, Springer, vol. 12, no. 3, Sep., 1992, pp 169–180; A. P. Y. Ho, *The Organizational and Political Resource of Environmental Pressure Groups in Hong Kong*, Asian Journal of Public Administration, available at http://sunzi1.lib.hku.hk/hkjo/view/50/5000318.pdf; P. Rawcliffe, *The challenge of Rio: environmental pressure groups and the UNCED process* in *Environmental pressure groups in transition*, Manchester Univ. Press, 1998, pp 185–218T. Bernauer, L. Caduff, *In whose interest? Pressure group Politics, Economic Competition and Environmental Regulation*, Journal of Public Policy, 24, pp 99–126; L. Tamiotti, M. Finger, *Environmental Organizations: Changing Roles and Functions in Global Politics*, Global Environmental Politics, vol. 1, no. 1, Feb, 2001, pp 56–76

645 See Article 18 of the Lugano Convention on Civil Liability for Damage Resulting from Activities Dangerous to the Environment, 21 June 1993

Therefore, at present the issue is highly dependent on the decisions of the court in securing a *locus standi* for these organizations and persons, who act solely on behalf of the environment in defending its rights[646] in the wake of pollution damage. Obviously, the work of such organizations is greatly appreciated by many in the absence of proper State intervention, but their ability to make potential claims within the framework of the law is highly sensitive and subject to severe constraints. Nonetheless, it is important that respective national laws grant legal rights to these organizations in the future in pursuing an effective policy on environmental protection in the aftermath of oil spills, especially focusing on issues relating to the proclamation of proper *locus standi*.

646 See J. L. Sax, "Defending the Environment", 1971, pp. 56–57

Chapter 5: Requirements of Liability and Admissibility of Claims

The success of an action for liability and compensation will truly depend on the fulfillment of requirements laid down under the particular law that establish the relevant legal regime. The foregoing chapter focuses on this aspect that generates highly controversial issues pertaining to different systems of law and respective requirements for liability in such different jurisdictions. The success of an international regime such as the Bunkers Convention would truly depend on common adherence of States Parties in applying unified rules[647] on the subject. Satisfaction and fulfillment of these aspects would certainly promote, enhance and guarantee the effectiveness of the regime, which is the primary goal of the Convention. The final outcome of the Convention has appeared to have covered most of the conceptual requirements of strict liability that are visible in most if not all similar regimes. Article 3 extensively deals with the liability aspect of the shipowner as discussed previously in Chapter 4 together with the defences that are normally accorded under the rule of strict liability[648]. Limitation of liability is dealt in Article 6 in coping with conventional goals of reducing the burden based on certain accepted criterion, and which will be discussed later in Chapter 7.[649] Article 7 which deals with compulsory insurance and financial security plays an important role in supplementing the financial risk factor of the shipowner which is based on strict liability, but will not be discussed in essence in this piece of work due to its non-direct bearing on the main topic.[650]

---

647 See A. Kiss, D. L. Shelton, *Strict Liability in Environmental Law*, GWU Legal Studies Research Paper No. 345, available at http://papers.ssrn.com/sol3/papers.cfm?abstract_id=1010478; I. Kim, *A comparison between the International and US regimes regulating oil pollution liability and compensation*, Marine Policy, vol. 27, iss. 3, May 2003, pp 265–279; C. Bernasconi, *Civil Laibility and Transfrontier Environmental Damage* in *Hague Yearbook of International Law* (J. G. Lammers ed., 1999) Kluwer, pp 45–136; M. Boile, S. Theofanis, F. Mikiki, *Oil Spills in Maritime Transport*, Transportation Research Record: Journal of the Transportation Research Board, vol. 1909, 2005, pp 100–107

648 Art. 3.3 and 3.4. See *infra* pp 191–230

649 *Infra* pp 233–280

650 For a full discussion, see, Ling Zhu, my fellow colleague at Max-Planck Research School for Maritime Affairs, *supra* n. 131

The Requirements

The most central point in establishing liability under law of tort rests on the provision of proclamation of liability and the necessity of fulfilling the requirements. In statutory law, it is the provision that establishes liability and requirements stated therein. In effect, the Bunkers Convention provides a straight forward approach towards this aspect by way of its Article 3.1. It is quite evident from an analysis of this provision that the requirements stand purely visible as in the case of similar tort law regimes in maritime law, both at national and international level. For the convenience of this discussion, it would be wise to separate these requirements into a more legal language. In such context, there exists of three main statutory requirements that need to establish liability under this regime, namely, the aspect of operation of the ship, causation and damages. These three seem to rule this entire liability and compensation regime imposing immense tension on the whole issue thus implying the necessity of common supra national laws on those topics. The absence of commonly accepted rules on specific issues such as causation and damage create many complications in litigations under common international platform of this nature.

Operation of the ship

Firstly, it is the shipowner at the time of the incident shall be liable. As shown earlier in the preceded chapters, the shipowner is vested with the legal responsibility for tort under the duty to guarantee to remedy any violations of rights of third parties. In order to satisfy this requirement, it is important to show that the ship in question was under his control. This would draw the attention in building up a control test around the shipowner with a view of ascertaining the nature of his command in actual operation of the ship and especially in economic terms. In such context, the person who had control of the ship is in the best position to bear liability and compensational costs for damage caused as a result of the things he carried and acts of other persons who were under his control, which caused such damage. It is far clearer that this control test in strict liability concept overshadows vicarious liability[651] in making the shipowner answerable to all torts

651 See, *Do the provision and work equipment regulations 1998 mean what they say?* April Pl Focus vol. 19, issue 6, available at http://www.hughjames.com/pdf/PWER%20-%20Cenric%20Clement-Evans%20-%20July%202009.pdf; C. Connal, *Of Chairs, Trains and Escalators*, The Journal Online, Jun. 15, 2009, available at http://www.journalonline.co.uk/Magazine/54-6/1006655.aspx; V. Harpwood, *Principles of Tort Law*, 4th ed., Cavendish, 2000, pp 347–349

committed by his subordinates in an event of such damage. Unlike in private ownership where the controlling aspect is wholly carried out by the owner without the involvement of skilled labourers such as independent contractors, it would be sometimes truly impractical for a shipowner to apply his control over all personnel employed in his ship.[652] In that respect, many legal experts lay critics on the control test citing it as an unjustifiable approach towards helpless shipowners. In such failure, one could build up a relationship-based test as developed by Cane[653] under which shipowner becomes liable for tortuous acts of all personnel including independent contractors with whom he had some kind of employment relationship upon contract or other means. In addressing this issue under the Bunkers Convention it is possible to see the existence of both these tests correctly included in order to avoid further clashes. In order to lay liability on the actual tortfeasor as interpreted under the Convention,[654] a wide spectrum of potential persons and entities that fall within the definition of 'shipowner' enables to simplify the procedure in identifying the suitable defendant[655] based on ownership or conduct.[656]

Secondly, the behavioral aspect of the shipowner should be further strengthened by identifying the substance that falls as the subject matter for the alleged damage and its appearance in and within the defendants' conduct in the operation of the ship in question. This requirement has been clearly addressed in Article 3.1 indicating the nature of the substance as bunker oil that originates or remain on board the defendants' ship. Although, it seems to be much easier in fulfilling this aspect *prima facie*, much litigation worldwide have proved that the situations aren't always straight forward. Many practical reasons appear to be prime causes in such scenarios including matters affiliated with identification of the type of oil, whether 'bunker oil'[657] or any other kind, identification of the exact ship from which such oil has escaped and other related complications as to proof. Therefore, an active role of States in mutual coordination in and outside their respective ports, proper maintenance of log books on natures of oil, oil inputs and outputs from all ships entering and leaving such ports, possessing scientific and technical capabilities especially with DNA testing and frequent checks for the detection of any suspicious substances are highly required and prescribed in order to reap the benefits of this regime.[658]

652 See D. Ferriere, *Corporate Responsibility: How Ship Owners can prevent Oil Spills*, available at http://www.iosc.org/papers/IOSC%202003%20a161.pdf

653 P. Cane, "The Anatomy of Tort Law", Oxford, Hart Publishing 1997, pp.45–47

654 Art. 1.3

655 See *supra* Ch. 3

656 I.e.: persons other than the owner or registered owner meaning bareboat charterer, manager or operator of the ship

657 See *supra*, pp. 31–37

658 See Bunkering Procedure in Singapore, available at http://www.shipowners.co.uk/lossprevent/pdfs/10-08%20BUNKERING%20PROCEDURES%20IN%20SINGA-

Causation

Law on liability and compensation frequently deals with causation as one of the main requirements for the establishment of liability. Although, causation is not always an essential criterion of liability, especially in strict liability regimes, the causal relationship between liability-inducing act and damage makes it an important ingredient in litigation.[659] According to Von Bar, causation is no more than the connection between a defendant's misconduct or that of a person for whom the defendant is responsible, or the existence of a source of danger on the one hand and compensable damage resulting from such on the other.[660] Both international regimes and national tort laws in general upholds its importance as a means of compensating the victims. The causation in maritime liability law of this nature as similar to the CLC and Fund regimes is the link between the alleged incident and the damage. By their respective definitions, it is clear that the damage must have caused by contamination of bunker oil as a result of an incident created upon the operation of the ship.[661] Once the conventional regime is absorbed into different national legal systems, the requirement of causation would be taken up differently according to such respective law on causation prevailing under such national jurisdictions, in addressing admissibility of claims and the extent of allowed compensation. In addition, the absence of a supra national compensating body here, such as the IOPF, leaves the victim solely dependant on respective national legal system. Therefore, a basic survey would enable to understand how this requirement would be met in certain developed systems of law in finding solutions for tortfeasing.

Although most of the States recognize causation as a requirement of tortuous liability, many of them utilize different tools in their respective law of causation in determining the extent of applicability. In general, causation can be categorized as a matter of fact and as a matter of law. The factual matter purely deals with the existence of an act or omission which cause the damage whereas in law,

---

PORE.pdf; Bunkering: Code of Practice, Peel Ports, Oct., 20, 2008, available at http://www.merseydocks.co.uk/assets/pdf/Code-of-Bunkering.pdf; Nanaimo Ports Authority, *Practices and Procedures to be followed by Ships in the Port of Nanaimo*, updated May 7, 009, available at http://www.npa.ca/PandPmay72009.pdf

659 See A. Honore, *Causation in the Law*, Stanford Encyclopedia of Philosophy, Oct., 12, 2005, available at http://plato.stanford.edu/entries/causation-law/; R. W. Wright, *Causation in Tort Law*, in *The International Library of Essays in Law and Legal Theory, Tort Law* (E. Weinrib ed.), Dartmouth, 1991; P. Widmer, W. H. van Boom, *Unification of Tort Law: Fault*, Kluwer, 2005

660 Von Bar, "The Common European Law of Torts", Volume 2, Oxford University Press, p. 441

661 See in conjunction, Arts. 1.8 and 1.9

there may exists of other reasons for the actual acceptance or rejection for the imputation of liability based on causation, depending on various issues. This two step approach gains a remarkable recognition in many jurisdictions in applying the best solution for the attribution of liability and settlement of compensation for the tortfeasor's tortuous conduct.

Factual Causation

In factual causation, the existence of a *conditio sine qua non* requirement plays one of the central roles in determining liability.[662] This requirement which has been hailed for many periods of time in law of tort is considered to be one of the main decisive factors.[663] In Belgium for example, it gathers all attention in upholding it as the sole requirement for the establishment of causation.[664] In *conditio sine qua non*, it is the act or omission which is regarded as the cause for the damage or loss, and the absence of such act or omission would not contribute any effect towards an existence of a tortuous outcome. Simply, it is the act or the omission that fulfills one of the basic requirements for creating the damage. According to Article 1.9 of the Bunkers Convention, it is clear that in first instance the loss or damage must have occurred as a result of escape or discharge of bunker oil from the ship. Therefore, *conditio sine qua non* in this respect relates to the operation of the ship. Suppose that the escape or discharge did not take place as a matter of the operation of the ship, the loss or damage would not have occurred, and therefore the question of liability and compensation would not arise.

As Spier points out, there may be other reasons for the formation of factual causation in relation to damage.[665] The first formula deals with a judgment based on science and relevant experience of a specific occurrence. Accordingly, the act or omission is subjected to a judgment on the basis of the best science and experience of the moment pursuant to materials of evidence in determining whether the damage could occur in such absence. Although such situations are very rare in tortuous acts one may not forget the damage that could occur as a result of oily

662 See *supra*, n. 660, pp. 437–441

663 See European Group on Tort Law, Chap. 3, *Principles of European Tort Law*, pp 43, 53, 59–60; B. Winiger, H. Koziol, B. A. Cock, R. Zimmerman (eds.) *Essential cases on natural causation (Digest of European tort law, vol. 1)*, Springer, 2007; D. Hodgson, *The Law of Intervening Causation*, Ashgate, 2008, p 5; D. Ulpian, *Essential Cases on Natural Causation* in *Digest of European Tort Law, Vol. 1*, Springer, 2007, pp 9–101

664 See Cousy and Vanderspikken, *Causation under Belgian Law* in *Unification of Tort Law: Causation* (J. Spier eds.), Kluwer, 1996, p.23

665 J. Spier, *Unification of Tort Law: Causation*, Kluwer, 1996

smoke balls consisting bunker oil substance that vaporize from a ship.[666] The second formula is related to the equivalence theory that is in existence in tort law[667], allowing all causal links without any distinctions as to their adequacy, directness or normalcy to be weighted as equivalent in the formation of factual causation. Here, the application of *conditio sine qua non* together with all surrounding issues enables to establish liability against the tortfeasor.

The *conditio sine qua non* requirement is accepted as a weaker test to be applied alone in determining causation.[668] This is mainly due to the fact that it is unable to distinguish causative and non-causative events based solely on such test. Involvement of other necessary elements towards the damage that might exists in addition to a *conditio sine qua non* requirement, especially in strict liability cases, proves the importance of developing a better test.

Legal Causation

Although conditio sine qua non is accepted as the basic requirement in law of causation, many jurisdictions have shown reluctances in applying it as the sole test. Therefore, a construction of legal causation[669] based on number of other relevant factors seems to elaborate a wide spectrum in the field of tort law. Nevertheless, it does not disallow factual causation completely, but introduce a more liberal approach as a second step for the benefit of the victim. These approaches vary in different jurisdictions, but in certain circumstances there are similarities in those respective laws. Mainly, one could identify four different types of approaches applied in developed systems of law, as legally accepted theories of causation of such respective States. These together with their application under the Bunkers regime are discussed below in detail.

666 In *Aegean Sea* incident the Executive Committee of IOPC Fund determined that the damage caused by fire and damage caused by contamination from the oily smoke resulting from the fire are all damage caused by contamination. See FUND/EXE. 38.9, 11 February 1994, sec. 3.3.3

667 See K. N. Hylton, *Missing Markets Theory of Tort Law*, 90 Nw. U. L. Rev. 977 (1995–1996); J. Borgo, *Casual Paradigms in Tort Law*, The Journal of Legal Studies, vol. 8, no. 3 (June 1979), pp 419–555; J. S. Johnston, *Punitive Liability: A New Paradigm of Efficiency in Tort Law*, 87 Colum. L. Rev. 1385 (1987)

668 See D. Loefgren, *Factual causation in Delict: a look at law of Scotland and Sweden*, European Young Lawyers, Eurodevils, 2002, available at http://www.law.ed.ac.uk/eyl/02repdl.htm

669 See D. Ozonoff, *Legal Causation and Responsibility for Causing Harm*, American Journal of Public Health, vol. 95, no. S1, July 2005, S35-S38; J. C. Knobel, *The Feasibility of the co-existence of concrete negligence and legal causation*, Electronic Publishing (Online Journal), vol. 124, issue 3, 2007, pp 579–599

(a) ***conditio sine qua non*** *as the sole cause*

This equivalence theory as pointed out in the above discussion relating to factual causation penetrates from the idea that the tortfeasor's act or omission serves as the sole cause in determining liability. Although it provides a more user-friendly attitude, damage do not always fall in straight forward ways. Use of this approach as the main test would certainly leave accountable remote tortfeasors in a much favourable position by opting them out of liability due to the absence of a visible *conditio sine qua non* requirement in presence. Therefore, it is obvious that the application of this test alone would not provide satisfactory base on establishing liability. In a situation where escape or discharge led to a contamination, but immediately such oil caught fire and the damage caused as a result of fire,[670] the shipowner would not become liable under the Convention, as the *conditio sine qua non* is the escape or discharge that resulted in fire, which lack the causal link required under the Convention. Victims in States that apply the sole test approach will be unable to establish liability on such shipowner eventhough; contamination did play a role in the damage prior to the burning.

(b) proximate cause test

The English Common Law strategy is also applied in the United States', but the latter uses a much wider scope in determining causation.[671] The classical doctrine of taking into consideration causes which are only 'immediate' or 'near' to the act of tort have been replaced in the English system with the introduction of foreseeability of harm. This provides a much wider scope in making the tortfeasor liable, except in cases where the cause is very remote. Although it operates as a central issue, English case law proves[672] the acceptance of other factors such as fairness, reasonableness, proximity, policy and the protective purpose of the violated rule[673] as basic tools in judging the effectiveness of the proximity cause test. The United States' apply the test with much liberty, openly using tools such

670 See *supra* pp 59–62

671 See D. Harris, *Tort Law Reform in the United States*, Oxford Journal of Legal Studies (1991), vol. 11, no. 3, pp 407–415; D. M. Engel, M. McCann, Introduction: *Tort Law as Cultural Practice* (July 8, 2009), Buffalo Legal Studies Research Paper No. 2009–15. Available at SSRN: http://ssrn.com/abstract=1431501; G. E. White, *Tort Law in America: an intellectual history*, Oxford, 1980

672 See W. V. H. Rogers, *Causation under English Law* in *Unification of Tort Law: Causation* (J. Spier eds.), Kluwer, 1996, p.40

673 See W. L. Prosser, W. P. Keeton, D. B. Dobbs, R. E. Keeton, D. G. Owen, *Prosser and Keeton on Torts*, 5th ed., West Publishing, 1984, section 42, p.273

as foreseeability of harm, directness, proximity in space and time and policies such as closing the floodgates in compensating and disproportionate portions on liability as to the wrongdoing.[674]

In applying the proximate cause test by a respective State in a matter falling under this Convention, such court would pay much attention to the relevant criterion used in that jurisdiction in ascertaining proximity of the cause. For example, in a case where fish of the surrounding waters become unhealthier to eat due to contamination, the fishermen fishing in such waters could eventually impose liability on the shipowner for loss of income.[675] But however, this would not extend to issues such as recovery of unpaid lease installments for such period that were due to a lessor of a boat.

(c) adequate causation test

Certain States apply the adequate causation test as the second step towards the formation of legal causation. Germany, France and Greece in particular, respect the adequacy test although; French courts have shown dependence on the equivalence test as well.[676] In essence, it is the probability of occurrence of the cause, such as expectation and foreseeability of the consequence, even under remote circumstances. Although, German law is much restrictive in applying the said theory based on the degree of probability,[677] the French law applies tools such as direct, immediate and pre-visible. In general, this approach operates in a much strict and inflexible manner in many States, basically with a view of facilitating limitation of liability thereby hindering the establishment of a cause to a reasonable extent. In such context, only direct and much foreseeable causes will have the possibility of attracting the courts leaving out less remote causes on the grounds of low probability. Even though some States would still allow such situations to exist in the absence of a normal degree of probability, interpretation of 'adequacy' in different States would pressurize the issue immensely.

674 See G. T. Schwartz, *Causation under US Law* in *Unification of Tort Law: Causation* (J. Spier eds.), Kluwer, 1996, p.123. See also M. F. Grady (2002), *Proximate Cause Decoded,* UC Los Angeles: UCLA School of Law. Retrieved from: http://escholarship.org/uc/item/2313h061

675 See R. V. Randle, *supra* n. 283, p 88; V. P. Goldberg, *Recovery for Economic Loss following the Exxon "Valdez" Oil Spill, The* Journal of Legal Studies, vol. 23, no. 1 (Jan., 1994), pp. 1–39

676 See Galand-Carval, *Causation under French Law* in *Unification of Tort Law: Causation* (J. Spier eds.), Kluwer, 1996, p.54

677 See U. Magnus, *Causation under German Law* in *Unification of Tort Law: Causation* (J. Spier eds.), Kluwer, 1996, pp 64–66

Under this test, damage or loss to the beach due to contamination by oil may be judged as an adequate cause for loss of business for nearby hotels[678], but the jurisdictions which apply a stricter adequacy test may not uphold loss of business for shops in that city's centre on grounds of distance from such touristic area. As mentioned earlier this would however depend on the fact; how such respective courts or its law interprets the term 'adequate'.

(d) flexible approach test

There are some States which apply a flexible approach in establishing liability by way of adopting a much liberal strategy. The South African legal system follows such an approach in determining whether there is a sufficient relationship between the cause and the consequences.[679] The causal link is judged on grounds of reasonableness, fairness and justice although these policy considerations lack firm and uniform meanings. It implies that the whole process may function together with the above discussed legal causations of such other category in allowing flexibility. The Dutch system also favours this test based on reasonableness, and according to the Dutch Civil Code 1992, liability is related to the defendant's act or omission taking into account, the nature of the damage and liability.[680] The same approach is adopted in Italy, where the so-called 'suitable causality' criterion is applied to achieve desired results.[681]

Alternative approaches used by States

Although, the double test approach is used by many legal systems, several States show tendency of not to consider it as conclusive. Some States go even beyond a two test approach by welcoming moves of facilitating a third approach. Courts in Germany tend to rely on certain policy considerations enabling an execution of a flexible approach by way of upholding 'protective purpose of the rule'

678 See P. Aspden, *Shipping law faces Europe: European policy, competition and environment*, Maklu Uitgevers, 1995, p 139; I. C. White, *Factors that determine the cost of oil spills*, available at http://www.itopf.com/_assets/costs03.pdf; V. P. Goldberg, *supra* n. 675

679 See Neethling, *Element of Causation in South African Law of Delict* in *Unification of Tort Law: Causation* (J. Spier eds.), Kluwer, 1996, pp 94–96.

680 See R. J. B. Boonekamp, *Schadevergoeding* art. 98

681 See Busnelli and Comande, *Causation under Italian Law* in Unification of Tort Law: Causation (J. Spier eds.), Kluwer, 1996, pp. 81–82

principle.[682] France on the other hand favours in deviating a strict adequacy test approach by considering other related factors although it does not apply in the German doctrine.[683] This rule is even compatible with the objects of the Bunkers Convention, where the drafters were fully concerned of the harmful effects that would appear as a result of a spill which violates both the public and private law aspects of protection of different natures. Furthermore, this issue of causation would drastically turn into a matter of high importance, both in the context of causation and as a breach of statutory duty to guarantee; the base of strict liability concept. Therefore, the doctrine of legal causation used by different States under respective law of causation evidently stretch its wings with the development of the law in particular. This is widely seen in the wake of important roles played by international conventions on liability as a matter of forming uniform rules on causation, under which all States could play a common role.

Apart from the basic rules, mostly all States in particular tend to reject the concept of expanding liability, labeling it as an unjustifiable scenario that disrupt the smooth growth of the whole liability system in tort law. Nevertheless, it is sometimes wise to impose hard obstacles for the tortfeasors as a matter of mitigating, preventing and cracking down serious wrongdoings of tortuous acts which are greatly harmful to the masses. Especially in strict liability regimes there seems to exist of such circumstances where disproportional approaches are respected as means of mitigation. As far as oil pollutions are concerned, the extent of the dangerous nature of the oil that escaped or discharged would certainly increase concerns of using this technique. Though it lacks explicit or implied authority in some States like South Africa, France and England, the disproportionate loss test is virtually visible in practice in States such as The Netherlands, Italy and Sweden[684]. In the United States it operates somewhat indirectly inside the 'policy argument'. Furthermore, concerns such as nature of the loss and nature of the liability also gather certain relevance when it comes to alternative requisites for causation. As far as extra ordinary factors such as pure economic loss, mental pain and suffering, reputation and emotional setbacks are concerned, nature of loss can be considered as a valid imputation, likewise intentional or grossly negligent act or omission gather some weight with regard to nature of liability. But however, these grounds should always be subjected to justifiable limits so that imputation of liability on the tortfeasor does not extend unreasonably. For example, loss of business for hotelier aftermath an incident of contamination of the beach by oil, should not extend to issues such as loss of reputation of his hotel, due to

682 See *supra*, n. 677

683 See *supra*, n. 676

684 See J. Spier, *Unification of Tort Law: Causation*, Kluwer, 1996, p.134

the fact that the hotel would not loose it reputation, but business, till the beach is restored and tourism is promoted. But however, an expected drastic decline of the tourist population once it is so restored would become a valid ground for pure economic loss arguments.

Causation in events of concurrent tragedies and in recurrence

Being one of the most important issues in maritime pollution liability, incidents that are joint and concurrent on the one hand and subsequent recurrences on the other, need a proper illustration in the context of law of causation. Especially in situations where the application of Articles 5 become relevant with regard to joint and several liability[685] and Article 3 implication on recurrence[686], courts will utilize the prevailing tools under such jurisdiction in imputing liability.

Concurrent Causes

According to Article 5, shipowners of all ships concerned shall be liable jointly and severally for all such reasonably inseparable damage. This is a result of concurrency of incidents that constitute the relevant cause.[687] In an event of a common incident, the act or omission of all connected parties are contributory towards the damage so as to causation by means of *conditio sine qua non*. In essence, the absence of this requirement would not have led to the creation of any harm, and therefore, lacks any imputation of liability. Imposition of joint and several liability on all tortfeasors does not merely define their respective portions, rather making everyone individually liable on constructive basis. Italy, Germany, Greece, France, Australia and South Africa in particular adhere to a strict principle of imposing entire liability equally on each and every tortfeasor. The Dutch law will go further in ascertaining whether the damage is divisible or each tortfeasor's contribution was on the same outcome. In the event of an inseparable damage, no legal system seems to recognize fractional liability portions even in situations of one's minimal contribution. It may be argued that such an approach produce unreasonable outco-

685 "...liable for all such damage which is not reasonably separable"

686 "...if an incident consists of a series of occurrences having the same origin, the liability shall attach to the shipowner at the time of the first of such occurrences"

687 See E. J. Weinrib, *Tort Law: cases and materials*, 2nd ed., Emond Montgomery, 2003, p 231; A. A. Ehrenzweig, *Releases of Concurrent Tort-Feasors in the Conflict of Laws: Law and Reason versus the Restatement*, Virginia Law Review, vol. 46, no. 4 (May, 1960), pp 712–716; J. Murphy, H. Street, *supra* n. 117, pp 94–97; P. Cane, n. 653, 1997, p 200

mes for low contributors and especially for remaining defendants in a case where one or more defendants become insolvent. Although, respective courts might use a strict approach in this regard, States which apply flexible tests[688] would grant some sought of relief to defendants with less contribution, taking into account of the relationship between his conduct and its consequences.

A further consideration need to emphasize situations with regard to subsequent and simultaneous occurrences where cause for concurrent damage is to be dealt carefully. The situation is treated differently in many systems of law based on the nature of the damage and existence of *conditio sine qua non*. Here, South African law denies causation, whereas in United States and in England the position is uncertain, though the English law would still consider characteristic feature of the whole damage whether indivisible or progressive. The Dutch law on the other hand uses a proportional approach[689] while German, Greek, French, Belgian and Italian laws impute full liability on the initial tortfeasor. The latter approach would be user-friendly for the victim in cases of shipowner-advantageous mystery spills aftermath an initial oil spill, but it would expose initial shipowner with less gravity as to damage, into a very unreasonable position.

Recurring Cause

Different natures of effects create possibilities of expanding damage in scope and time. Especially in oil pollution, damage is not visible or does appear at once or straight, due to many reasons varying from characteristic to ecological features. Marine environment in particular is subject to expanding after-effects upon oil spills. Even though Article 8 has prescribed a time limit, these recurring damages are judged with a much wider perspective in law of causation coupled with a system of providing reasonable compensation having due regard on future losses. [690] A true justification would be that a recurring damage sails along with the initial causation. In such a case, the tortfeasor becomes liable not by means of establishing causation at each point, but as a result of the penetration of the foremost causation to all recurring damage. Most importantly, the recurrence should bear

688 See E. J. Weinrib, *supra* n. 687, pp 221–222; M. G. Ross, *Concurrent liability in tort and contract*, Structural Survey, 1987, vol.5, issue 3, pp 288–299; V. Harpwood, *supra* n. 651, p 33; G. Stephenson, *Sourcebook on Torts*, 2nd ed., Cavendish, 2000, pp 183–184; A. Gray, *Flexibility in conflict of laws multistate tort cases: the way forward in Australia*, University of Queensland Law Journal, 2004, vol. 23, issue 2, pp 435–463

689 See J.Spier, , *Sluipende schade and Mon*, Nieuw Burgerlijk Wetbeok – 36, nr.52

690 See *supra* pp 125–128

the same origin as required by Article 3 in order to impute liability on tortfeasor responsible for the initial cause. In addressing such an issue, basically all States hold similar convictions.

Uncertain Cause

Lastly, the discussion should focus on the uncertainty character of a cause[691], which is not a stranger in oil pollution incident. Three such situations are based on uncertainty as to the parties, implying; uncertainty as to the tortfeasor who has committed the damage, uncertainty as to the victim whose rights are violated and uncertainty as to both. The first is sufficiently related to mystery spills where it is impossible sometimes to identify the exact tortfeasor, and as a result, been unable to establish the proper cause and impute liability in the absence of traceable evidence. The second situation is closely related to circumstances where the marine environment, outside a States' legal jurisdiction, is affected. Lack of national sovereignty or enforcement jurisdiction in such areas creates obstacles in bringing legal actions, especially due to the absence of a proper *locus standi*. These situations together with the third are not left out due to such ailments, but courts do apply suitable solutions such as having alternative claimants and defendants added in order to alleviate torts. In addition there may exist of another uncertainty as to whether the claimant actually suffered a visible or invisible loss or damage. Although, an adoption of a certain alternative approach to identify a damage or loss could play some role in arriving at a favourable situation on this matter, it would solely depend on evidence. In concluding, it could be said that uncertain causations of such nature have to be dealt in conjunction with the issue of damage, particularly referring to the type of damage.

Existence of Loss or Damage

Being one of the most paramount issues in compensational regimes, one must purely identify damage that cause as a result of bunker oil spill within the meaning of the Convention.[692] Thereafter the law of damages in which the claim for com-

691 See J. Spier, *Uncertain Partial Causation* and *Uncertain Causes in the Victim's Sphere* in *Principles of European Tort Law: text and commentary* (European Group on Tort Law eds.), Springer, 2005, pp 54–58; H. A. Causy, *Risks and Uncertainties in the Law of Tort* in *European Tort Law 2006* (H. Koziol, B. C. Steininger eds.), Springer, vol. 2006, 2008, pp 2–26

692 See *supra* pp 53–58

pensation lies would have to ascertain its character in order to determine whether such damage is attributable under such law or not, in arriving at decisions on compensation. Here again, the total issue is dependent on type of law applied in the jurisdiction in which such actions have been filed and the hearings take place.[693] As this issue primarily focuses on granting compensation rather than as a matter of imputing liability on the tortfeasor, different jurisdictional opinions would strive in diversity. Such diversities would not only create conflicting theories of compensable damages but also enhance forum shopping.

Compensation under the Bunker Oil Convention deals with existence of 'pollution damage' as interpreted therein. As discussed previously, it is three folded with the first category being mainly concerned of contamination. For a proper understanding, it would be intelligent to divide the on-going discussion into two parts, namely loss or damage caused by contamination and further loss or damage caused by preventive measures on one side, and the costs of preventive measures on the other. Furthermore, admissibility of various claims under those categories would be taken up with respect to certain developed systems of law in order to ascertain the types of damages that are recoverable under such respective jurisdictions, and provide a base for a proper identification in other respective jurisdictions applying similar systems in their respective laws of torts.

Types of damage

With respect to many modern systems of law on damages, it would not be easy to develop a specific set of recoverable damage. But however, in most of those systems there exist of different types of damage that are primarily categorized for the convenience of understanding their attribution in matters of compensation. Referring to the loss or damage falling under the Bunkers Convention, following types of damage are considered to be of paramount importance.

Direct and indirect damage

The characteristic importance of damage is sometimes weighed with respect to its relationship. It varies from space, time, probability and foreseeability to remoteness,

693 See U. Magnus, *Unification of Tort Law: Damages*, Kluwer, 1997; S. D. Sugarman, *Tort Reform Through Damages Law Reform: An American Perspective*, available at http://works.bepress.com/cgi/viewcontent.cgi?article=1009&context=stephen_sugarman

in specifically determining the scope.[694] One may see a clear association of this aspect with causation[695] as well as with the issue of *locus standi*. Unlike in law of contracts, tort laws in general do not tend to strictly abide by direct or indirectness of the damage.[696] There have been many instances in varying jurisdictions where reasonable remoteness has been facilitated with remarkable degree of concern.[697] In cases of oil pollution at sea, loss of income and profit of hotel owners of adjacent polluted beaches have well been accepted even though these pure economic losses sometimes lack a proper link.[698] Furthermore, courts have sometimes allowed claims by unrelated parties, such as environmental pressure groups, considering their interest and role especially in situations of clean-up and threat monitoring processes.

Actual and future damage

A further classification based on visibility and expectation of damage gains much recognition in many jurisdictions. The difference between *damnum emergens* and *lucrum cessans* play an important role in ascertaining compensable damage, except in common law jurisdictions.[699] Especially in maritime oil pollution casu-

---

694 See J. Cartwright, *Remoteness of Damage in Contract and Tort: A Reconsideration*, The Cambridge Law Journal, vol. 55, 1996, pp 488–514

695 For example in Italy, the distinction between direct and indirect damage is rooted in Arts. 1223 c.c. and 2056 c.c. section 1, which has great relevance on causation. See F. D.Busnelli, G.Comande, *Unification of Tort Law: Damages*, Kluwer, 2001, p.123

696 See H. Koziol, B. C. Steininger, *European Tort Law 2002*, Springer, 2003, p 251; B. S. Markesinis, M. Coester, G. Alpa, A. Ullstein, *Compensation for personal injury in English, German and Italian law* , p 23; W. V. H. Rogers, W. H. van Boom, *Unification of Tort Law: Multiple Tortfeasors*, Kluwer, 2004; G.Calabresi, *Toward A Unified Theory of Torts*, Journal of Tort Law, vol. 1 : iss. 3, Article 1. 2007, available at: http://www.bepress.com/jtl/vol1/iss3/art1

697 See J. Thompson, *Remoteness of Damages – Extending the Doctrine?*, Edinburg Law Rev., vol. 8, iss. 3, 2004, pp 405–407; C. C. Turpin, *Roman-Dutch Law Remoteness of Damage. Wagon Mound Not Followed*, The Cambridge Law Journal, vol. 23, no. 1 (Apr., 1965), pp 34–38; R. Cooke, *Remoteness of Damages and Judicial Discretion*, The Cambridge Law Journal, vol. 37, 1978, pp 288–300

698 See X. Liu and K. W. Wirtz, *The economy of oil spills: Direct and indirect costs as a function of spill size*, Journal of Hazardous Materials, Volume 171, Issues 1–3, 15 November 2009, pp 471–477

699 See O. Moreteau, *Revisiting the Grey Zone between Contract and tort: The Role of Estoppel and Reliance in Mapping out the Law of Obligations* in *European Tort Law 2004* (H. Koziol, B. C. Steininger eds.), 2005, pp 60–80; Marcos, Francisco, S. Graells, Albert, *Towards a European Tort Law? Damages Actions for Breach of the EC Antitrust Rules: Harmonising Tort Law Through the Back Door?* , October 26, 2006, available at SSRN: http://ssrn.com/abstract=1028963

alties the subject is highly controversial and indeed sensitive with regard to the quality and nature of oil. Claims would vary from physical to immaterial damage that could last for a long period of time. Much of its prevalence in a legal context would certainly depend on evidentiary sources as in the case of French, German, Greek and Italian laws; where in those systems standard of proof is usually reduced. In Belgium, recoverable future damage must be the normal outcome, development, consequence, completion or recurrence of an actual situation,[700] whereas under Dutch law, both actual and future damage may be recoverable, but however, depending on the assessments.[701] The mixed law system of South Africa recognizes this distinction for mere analytic and practical importance in order to quantify prospective loss, and lays upon certain grounds; i.e. recoverable actual damage are damage suffered up to the date of trial and future damage are confined to loss of profit and prospective loss.[702]

In most cases, the courts apply a method of compensating actual damage based on the evidence of proper existence, whereas compensation for future damage are calculated on the basis of probability of consequential effects that can be proved by evidence for mere satisfaction of claims. This would not only comfort the claimants in obtaining suitable compensation for damage that might appear in the future, but also to avoid prescription problems involved in such procedures. As an example, it would not be possible to exactly measure future effects on living natural resources, concerning their breathing capabilities, but applying a method of compensating to a reasonable degree of expected aftermath damage would suffice the objected goals.

Pecuniary and Non-Pecuniary Damage

These types of damages are common in many tortuous liability laws including strict liability regimes, and are no exception to maritime casualties which are under consideration here. Many States employ different tools in addressing such claims, but most commonly, assessment theories seems to be somewhat similar.[703]

700 See H. A.Cousy, A.Vanderspikken, *5.6.2., Unification of Tort Law: Damages* in *Principles of European Tort Law* (U. Magnus ed.), Kluwer, 2001, p.38

701 See M. H.Wissink, W. H van.Boom, 5.6.21., *Unification of Tort Law: Damages* in *Principles of European Tort Law* (U. Magnus ed.), Kluwer, 2001, p.148

702 See J.Neethling, 5.2.1., *Unification of Tort Law: Damages* in *Principles of European Tort Law* (U. Magnus ed.), Kluwer, 2001, p.162

703 See H. Cousy, D. Droshout, *Non-Pecuniary Loss under Belgian Law* in *Damages for non-pecuniary loss in a comparative perspective* (W. V. H. Rogers ed.), Springer, 2001, pp 28–52; E. Karner, H. Kozoil, *Non-Pecuniary Loss under Austrian Law* in *Damages for non-pecuniary loss in a comparative perspective* (W. V. H. Rogers

Pecuniary damage is compensated on the basis of objective market price, whereas non-pecuniary damage is subjectively measured.[704] Though French law does not provide any rules for the assessment of non-pecuniary losses, its courts seem to act by its own judgment and furthermore have developed so-called 'judicial tariffs'.[705] Greek courts also enjoys such freedom of discretion,[706] while in Italy, courts apply a test based on 'reasonableness', taking into account the balancing of contrasting interests at stake.[707]

The above mentioned main types of damage can further be categorized into subsections considering their existence in oil pollution incidents. They would vary from direct to indirect, actual to future and allowed and excluded types of damage within the context of law of damage and compensable claims under different systems of law. However, the French law distinctions on different types of losses depending on their nature and other distinctive features are mere de-

ed.), Springer, 2001pp 4–27; W. V. H. Rogers, *Non-Pecuniary Loss under English Law* in *Damages for non-pecuniary loss in a comparative perspective* (W. V. H. Rogers ed.), Springer, 2001, pp 54–86; S. G. Carval, *Non-Pecuniary Loss under French Law* in *Damages for non-pecuniary loss in a comparative perspective* (W. V. H. Rogers ed.), Springer, 2001, pp 87–108; U. Magnus, J. Fedtke, *Non-Pecuniary Loss under German Law* in *Damages for non-pecuniary loss in a comparative perspective* (W. V. H. Rogers ed.), Springer, 2001, pp 109–128; K. D. Kerameus, *Non-Pecuniary Loss under Greek Law* in *Damages for non-pecuniary loss in a comparative perspective* (W. V. H. Rogers ed.), Springer, 2001, pp 129–134; F. D. Busneli, G. Comande, *Non-Pecuniary Loss under Italian Law* in *Damages for non-pecuniary loss in a comparative perspective* (W. V. H. Rogers ed.), Springer, 2001, pp 135–154; M. H. Wissink, W. H. van Boom, *Non-Pecuniary Loss under Dutch Law* in *Damages for non-pecuniary loss in a comparative perspective* (W. V. H. Rogers ed.), Springer, 2001, pp 155–172; M. Nesterowicz, E. Baginska, *Non-Pecuniary Loss under Polish Law* in *Damages for non-pecuniary loss in a comparative perspective* (W. V. H. Rogers ed.), Springer, 2001, pp 173–191; M. M. Casals, J. Ribot, J. Sole, *Non-Pecuniary Loss under Spanish Law* in *Damages for non-pecuniary loss in a comparative perspective* (W. V. H. Rogers ed.), Springer, 2001, pp 192–244; I. Gilead, *Non-Pecuniary Loss – The Israeli Report*, Appendix 1 in *Damages for non-pecuniary loss in a comparative perspective* (W. V. H. Rogers ed.), Springer, 2001, pp 297–300; P. Tercier, *Short comments Concerning Non-Pecuniary Loss Under Swiss Law* in *Damages for non-pecuniary loss in a comparative perspective* (W. V. H. Rogers ed.), Springer, 2001, pp 301–310; D. B. Garrow, K. L. Ayre, *The Recovery of Non-Pecuniary Damages in Canada: The Cap on Recovery, Jury Trials, and other Unique Considerations for General Damage Awards*, available at http://monmexique.com/5-Garrow.pdf

704 Germany, England, The Netherlands and South Africa employ such schemes

705 See Carval, *Unification of Tort Law: Damages* in *Principles of European Tort Law* (U. Magnus ed.), Kluwer, 2001, p.81

706 See Kerameus, *Unification of Tort Law: Damages* in *Principles of European Tort Law* (U. Magnus ed.), Kluwer, 2001, p.111

707 Corte Cost., 22 November 1991, n. 420, [1991, II] Cons. Stato 1827

scriptive rather than establishing different rules for various kinds of losses. Damage or loss is recoverable on conditions of being certain, personal, direct and legitimate.[708] Concerning the interpretation of 'pollution damage' in the Bunkers Convention, States would employ their respective laws in determining what compensable claims are and what are not. Those potential claims could be categorized in the following manner in order to have a broad idea.

Damage to Property

Damage that cause as a direct result of the oil pollution incident have been limited by way of restricting them to contamination. Therefore, claims with respect to direct or consequential loss or damage caused by contamination must establish proper causation and existence of damage as a result of such incident together with a right of action arising out of ownership or legally acceptable interests. There can be many claims of this nature varying from damage to fishing gear, fish farms, aqua-cultural fish stocks, ships, boats, other navigating objects, desalination plants, power stations to even real estate properties that can become victims of contamination due to oily winds and smoke.[709] In the case of further loss or damage caused by preventive measures, many claims could still arise concerning property issues.

In England, damage to property is based on diminution in value taking into consideration of the repair costs, but in circumstances where such property is destroyed, it is valued on grounds of replacement costs of a substitute of the same age and condition or equivalent.[710] Under German law, damage to property bears no particular weight in the field of damages and is compensated under principles of tort law, once liability has been established.[711] It compensates for property loss under strict liability statutes such as *inter alia*, water resources[712] and environment liability. The Dutch law follows the same trend, and damages to property are fully recoverable.[713]

---

708 See Carval, *supra* n. 705, pp 80–81

709 For a detail discussion see Colin de la Rue, *supra*, n. 316 at Ch. 11

710 See W. V. H. Rogers, *Unification of Tort Law: Damages* in *Principles of European Tort Law* (U. Magnus ed.), Kluwer, 2001, p.62

711 See U. Magnus, J. Fedtke, *Unification of Tort Law: Damages* in *Principles of European Tort Law,* Kluwer, 2001, p 95

712 See, Sec. 22 of the Water Resources Act and Sec. 1 of *UmweltHG.*

713 See M. H Wissink, W. H van.Boom, *supra*, n. 701, p.147

Damage to Person

With regard to these types of damage which are very rare in marine oil pollution incidents, especially concerning contamination at first sight, one must prove deteriorated health conditions that he suffered, by medical evidence.[714] Claims of this nature would arise more under further loss or damage caused by preventive measures, where personal injury are somewhat elevated due to slipping from rocks and other and consequential effects. These types of damage could further be categorized into physical and non-material damage.

Physical Damage

In almost all States, physical damage suffered as a result of direct and consequential events are recoverable subject to proof of causation. Under the Greek Civil Code, claims for personal injury result in bodily harm, may extend to all future expenses as well.[715]

Non-Material Personal Damage

Claims concerning pain, mental suffering and damage to reputation fall into this category of damages that is less weighted in oil pollution incidents.[716] Even though certain rare and exceptional situations could still exist with regard to pain and mental suffering aftermath an oil spill, especially when one's private ownerships or interests are directly affected, the issue is extremely remote and senseless to pay much attention. As far as the reputation is concerned, it would be a matter

---

714 A personal injury claim was instituted before the IOPC Fund under the Fund Convention 92 aftermath the *Braer* incident but was subsequently rejected on grounds of lack of proof of causation between the incident and the alleged injury. See FUND/EXC. 37/3, 18 October 1993, section 4.2.11

715 CC 929

716 In response to a letter forwarded by South Mainland Action Group concerning the suffering underwent by the population of South Mainland of Shetland as a result of the *Braer* incident and requesting a payment to that end as a 'one-off humanitarian payment', the Director of the IOPC Fund disagreed with such payment on the ground that the claim was not quantifiable and furthermore, as falling outside the conventional meaning of 'pollution damage'. See FUND/EXC. 35/5/Add. 1, p.2. It was further endorsed by the Executive Committee stating that exposure to health risks, anxiety and loss of environmental amenities would not fall within such meaning. See FUND/EXC. 35/10 sec. 3.4.25

that needs to be tried along with causation. The *Amoco Cadiz* litigation involved certain questions on reputational damage, and the Appellate Court in the United States took the view that one must establish a loss to oneself in order to qualify for such suits, but a mere injury to his reputation as a moral damage would not suffice, as in the case of the French law.[717]

Pure Economic Loss

There are many instances where pure economic loss[718] concerns have been considered in oil pollution incidents.[719] Although many jurisdictions tend to disallow such claims, there are situations where such claims however have been entertained by several States utilizing different tools of their respective laws despite of the general principles of tort laws. The reason behind rejection of such claims centers on the issue of opening the 'flood gates' to which, many States have shown tremendous degree of reluctance.[720] It is totally improper to exclude them completely from a compensational sphere due to the fact that many claims relate to direct consequences of patrimonial damage. Therefore, the issue tends to gat-

717 See *supra*, n. 588, pp 963 onwards

718 See M. Bussani, V. V. Palmer, *Pure Economic Loss in Europe*, Cambridge, 2003, available at http://assets.cambridge.org/97805218/24644/sample/9780521824644ws.pdf *R.* Martin, *Recovery of Pure Economic Loss in Tort*, available at http://sunzil.lib.hku.hk/hkjo/view/14/1400126.pdf; E. K. Banakas, *Civil Liability for Pure Economic Loss*, Kluwer, 1996; F. Blobel, *European Tort Jurisdiction and Pure Economic Loss*, The European Legal Forum (E) 3-2004, pp 187–191, available at http://www.simons-law.com/library/pdf/e/507.pdf; F. Gomez, H. B. Schafer, *The law and economics of pure economic loss: Introduction to the special issue of the International Review of Law and Economics*, Int'l Rev. of Law and Economics, vol. 27, iss. 1, March 2007, pp 1–7; H. Bernstein, *Civil Liability for Pure Economic Loss under American Tort Law*, The American Journal of Comparative Law (1998), vol. 46, p 111; W. H. van Boom, H. W. Koziol, C. A. Witting, *Pure Economic Loss*, Tort and Insurance Law, vol. 9, Springer, 2004

719 See *supra* n. 348; I. C. White, *supra* n. 513; Wu Chao, *supra* n. 107; A. W. McThenia, J. E. Ulrich, *A Return to Principles of Corrective Justice in Deciding Economic Loss Cases*, Virgenia Law Rev., vol. 69, no. 8 (Nov., 1983), pp 1517- 1535; O. Thebaud, D. Baily, J. Hay, J. Perez, *The cost of oil pollution at sea: an analysis of the process of damage valuation and compensation following oil spills*, available at http://www.aerna.org/Documentos_trabajo/Prestige7Hayetal..pdf

720 See F. D. Busnelli, G. Comande, *Some Remarks on the Borderline between Contract and Tort Liability* in *Unification of Tort Law: Wrongfulness* (H. Koziol, F. D. Busnelli eds.), Kluwer, 1998, p 84; P. Cane, *Taking Disagreement Seriously: Courts, Legislatures and the Reform of Tort Law*, Oxford Journal of Legal Studies, vol. 25, no. 3, pp 393–417, 2005, available at SSRN: http://ssrn.com/abstract=821686

her much of the attention with regard to losses possessing links to any physical loss to the same interest concerned. In this sense, the question of compensating pure economic loss is strongly dependent on the issue of causation, foreseeability and remoteness.

There are many potential pure economic loss claims that could well move into litigations in marine oil pollution incidents. They may vary from loss of profit or income of fishermen, hoteliers and other persons having special interests on sea and costal related activities to loss of revenue of local governmental authorities. In common law jurisdictions, the issue gains less significance, but the development of the law has seem to allow certain instances where these claims are less restricted to achieve its purpose. On the other hand, civil law jurisdictions tend to support such claims on grounds of sufficient directness and certainty of causation. Although many jurisdictions do not provide straight forward solutions for positive grants, they in turn allow such claims indirectly. This approach would allow courts to use more liberal view on the subject deviating from classical narrow limits to build up flexible demarcations.

In general tort law concerning strict liability, the situation is multi-faceted apart from the fact that some States expressly pave way for claims on pure economic loss and some States expressly denying. There indeed exist of a middle situation where pure economic loss is non-recoverable in cases of strict liability. In Spanish law for example, the concept is unknown, although it allows consequential losses ensuing from personal injuries such as loss of earnings or from damage to property under tort liability, irrespective of its nature.[721] Belgium, France and South Africa[722] favour such recovery under their respective strict liability law regimes. Under the Belgian law, these types of damages are recoverable subject to their certainty and non-existence of anything to the contrary in any specific legislation or case law.[723] The Civil Code of France in its Article 1384 subs. 1 provides possibility for the recovery of pure economic loss, and found to be a well accepted rule of its courts.[724] One may wonder why *Amoco Cadiz* litigation that applied French law in assessment of claims ended up with a much less compensation within this prevailing parameter, but the answer lies not on the issue of granting pure economic loss, but difficulties of proof that existed against such claims.

Being common law jurisdictions, Australia and Canada apply a much less restrictive move in allowing claims on pure economic loss. Australian courts use the proximity test in relation to the parties or between the causation and the dam-

721 See, Martin-Casals, Ribot, Sole, *Unification of Tort Law: Strict Liability*, Kluwer 2002, p. 311

722 See Neethling, *supra*, n. 679, p. 277

723 See Cousy, *supra*, n. 700, p. 69

724 See Civ. 2, 19 March 1980 in [1980] D., somm. comm. 414

age as justification,[725] and Canadian courts on the other hand hold similar convictions, but favour the development of this area of law taking into consideration of various other factors that might influence the grant, depending on its nature.[726]

A negative approach is visibly highlighted in German, Dutch, Israel and Polish legislations. In general, German strict liability statutes do not grant opportunity for pure economic losses, except in certain rare situations.[727] These situations are justified by means of acknowledging public property interests recognized by its constitutional law.[728] Exclusions in Israel extend only to circumstances of consequential economic loss and non-pecuniary loss as a result of death or bodily injury and further to pure economic loss involved in consumer protection and pure non-pecuniary loss factors in defamation.[729]

As a prime figure in common law systems, United Kingdom does not recognize compensatory policies for pure economic loss,[730] even in situations where the claimant is entitled to enjoy the use of the damaged property in virtue of a contractual right[731]. Nevertheless, there have been situations, where claims of this nature were accepted for policy reasons such as negligent misstatements cases in which plaintiffs rely *bona fide* on advices of defendants.[732] The Merchant Shipping Act has shown positive deviations from restrictive rules to somewhat liberal provisions. Its Sections 170 and 181 specifically includes 'loss' within the meaning of 'damage' without any restrictions. This compelling reason supports the idea that it recognizes the recovery of pure economic loss.[733] It does not however provide satisfaction as to whether any pure economic loss, irrespective of being directly connected to the patrimonial physical damage, is covered or not, but

725 See, *Caltex Oil (Australia) Pty. Ltd. v. The Dredge "Willemstad"* (1976) 136 CLR 529

726 See, *Norsk Pacific Steamship Co. Ltd. v. Canadian National Railway Co.* 91 DLR 4th 289

727 For example, sec. 22 of the *Wasserhaushaltsgesetz* (Water Resource Act 1986) however grants compensation for any kind of harm caused by pollution of natural water resources including pure economic loss

728 Magnus and Fedtke, *supra*, n. 711, p. 167

729 See, Gilead, *Unification of Tort Law: Strict Liability*, Kluwer 2002, p. 202

730 In *Spartan Steel & Alloys Ltd. v. Martin & Co. (Constructors) Ltd.*, (1973) 1 QB 27, CA, the court granted foreseeable financial damage, but prevented from granting compensation for any pure economic loss. Even then, only the owner could bring such claim

731 See, *Candlewood Navigation Corporation Ltd. v. Mitsui OSK Lines Ltd.* [1985] 2 Lloyds' Rep. 303, PC

732 See, *Hedley Byrne & C. Ltd. v. Heller & Partners Ltd.* [1964] A.C. 465, HL

733 According to the Hansard of 11.01.93, p. 610, the Secretary of State for Transport has stated that "Compensation under the CLC and FUND Conventions ... is also available for economic loss."

certainly, it will depend on issues of causation and of course, the decision of the court based on principles of tort law.

United States' on the other hand as a common law jurisdiction rejects pure economic loss claims under its respective tort law.[734] Applying a similar approach as United Kingdom, American courts in particular tend to follow the idea "in an indeterminate amount for an indeterminate time to an indeterminate class"[735] in support of such rejections. Much of the arguments stem from Justice Holmes judgment in the *Robins* case[736] which denied the recovery of pure economic loss claim resulted from physical damage to property lacking proprietary interests. This rule commonly known as the "bright line rule" in the US has been extensively used in American jurisprudence even in actions by claimants possessing such proprietary rights, but to the exception that the claim has no direct relationship with the physical damage.[737] Therefore, it is clear that the courts tend to deviate from this rule in certain respects, especially in circumstances where the claim directly attributable to a damage to a proprietary interest.[738] Hence, this deviation is very much apparent in maritime pollution cases upon the enactment of the OPA 1990 under which various claims including that of loss of earnings of commercial fishermen have been awarded by courts, on grounds of policy reasons. The "bright line rule" has thereafter been avoided in cases of maritime oil pollution casualties in considering pure economic loss claims without a requisite physical damage. Although the OPA 90 provides an extensive coverage through its definition for damage while including damage to revenue, profits and earning capacity within such meaning,[739] it does not however extends to recoveries for damage lacking real or personal property[740] rights of the Claimant.[741] The only exception, being recovery of economic losses arising from damage to natural resources,[742] which will be discussed later.

---

734 See H. Bernstein, *Civil Liability for Pure Economic Loss under the American Tort Law*, 1998 supplement, 46 The American Journal of Comparative Law, p 111

735 Pronounced by Justice Cardozo in *Ultramares Corp. v. Touche*, [1931] 255 N.Y. 170, 179

736 *Robins Dry Dock and Repair Co. v. Flint*, 275 US 303, 1928 A.M.C. 61 (1927)

737 See H. Carty, *An analysis of the economic torts*, Oxford, (2001 rep. 2003), p 255; E. J. Weinrib, *supra* n. 687, p 424; R. F. Blomquist, *Re-Enchanting Torts*, 56, South Carolina Law Review, 481 (2005)

738 See, *Golnoy Barge Co. v. M/T Shinoussa*, 1994 A.M.C. 1378–1379

739 See, Sec. 1002

740 Real and personal property is interpreted in the OPA as "Damages for injury to, or economic losses resulting from destruction of, real or persona; property, which shall be recoverable by any claimant who owns or leases that property"

741 See, USC 33 Sec. 2702 (b) (2) (B)

742 USC 33 Sec. 2702 (b) (2) (E)

Varying claims on Pure Economic Loss and their Admissibility

Claims by Commercial Fishermen

Drafters of the Bunkers Convention specifically drew the attention on rights of fishermen to institute actions for compensation. This issue has gained tremendous momentum in both domestic and international legislations due to the lack of proprietary rights such fishermen posses in public law context. States' regulations on granting of license to commercial fishermen[743] have eased complexities resulting in granting legal protection via doctrine of proprietary interests. A clear demarcation is however made between these types of occupational fishermen, directly depending on earnings out of such activities, who posses right of action, with persons fishing for mere interests or joy who are normally deprived of such right of action.[744] Moreover, most of the States' domestic laws have well accepted claims for loss of profit encountered by such fishermen upon oil spill incidents resulting in damage to natural resources and the marine environment. As described above, the respective States, which allow recovery for pure economic loss, will employ its tests on causation, remoteness and foreseeability, as suitable, in assessing types of damage that is recoverable and amounts of compensation for such loss of earnings and profits, to the permitted extent under such respective law. For example, both Merchant Shipping Act 1995 and US law[745] provide favourable outcomes in this respect.

743 See R. L. V. Eisma, P. Christie, M. Hershman, *Legal issues affecting sustainability of integrated coastal management in the Philippines*, Ocean and Coastal Management, vol. 48, iss. 3–6, 2005, pp 336–359; *Commercial Fishing License Update*, State of California, The Natural Resources Agency, Department of Fish and Game, Feb., 2009, available at http://www.dfg.ca.gov/licensing/pdffiles/fg1554a.pdf. See also, *Takahashi v. Fish and Games Comm'n*, 334 US 410 (1948)

744 This is a well accepted practice of the IOPC Fund which has allowed such actions in many instances. See, IOPC Fund Annual Report 1988, p.61 and IOPC Fund Claims Manual, 5th Edition, pp. 26–28. The IOPC Fund has however rejected claims by non-commercial fishermen who fish for mere livelihood, in settling of claims for damage aftermath the *Braer* incident. See, FUND/EXC. 38/9, 11 February 1994, sec. 3.4.6/7. US courts on the other hand have also rejected such claims under Exxon Valdez incident, 1995 AMC 1440, 1997 AMC 940

745 For example, in *Burgess v. M/V Tamano* case (1973) AMC 1939, the court upheld the claim of commercial fishermen on the ground that they had special interest to take fish from coastal waters. Furthermore, *In Re Glacier Bay* judgment (1991 AMC at p. 750) has elaborated the acceptance of economic loss claims despite proof of physical damage

Claims other than by Commercial Fishermen

Experiences in past litigations of this nature have proved extensive flow of claims pertaining to different types of damages been put forward by various claimants. They vary from pure private and quasi-private to entirely public in nature. Mostly, the courts employ specific tools to measure the admissibility of all such claims by using respective rules on liability. Overwhelmingly, these claims fall as borderline cases between pecuniary and non-pecuniary damage, where courts in respective jurisdictions lay careful considerations. These complex issues could be categorized in the following manner in order to gain a better understanding.

(a) Pure loss of use

Connected chiefly with the notion of private ownership by means of actual or contractual title, claimants legally possess the right to obtain compensation for pure loss of use suffered as a result of contamination. Under the German law, it extends only to essential things of lifestyle including *inter alia*, ships[746], and do not cover lost use for such other things that do not fall within the said criteria.[747] In England, they are recoverable in principle.[748] Laws such as French,[749] United States,[750] Greek[751] and South Africa[752] follow the same tradition without paying extra consideration whether a substitute had been used or hired or profits were lost or not. In most cases, calculation of lost use would depend on replacement or rental charges. The Italian law however dos not support such attitude leaving such claims in the category of non-compensable non-pecuniary damage.[753]

(b) Pure loss of earning capacity

This issue is mainly related to persons directly depended on coastal activities, such as hoteliers and businessmen operating shops and other recreational facilities for tourists. Most States respect this concept in principle, but use different

746 BGH[1980] VersR 418

747 See Magnus and Fedtke, , *supra*, n. 711, p. 97

748 See Rogers, *supra*, n. 710, p.64

749 See, Galand-Carval, *supra*, n. 705, p.82

750 See, Schwartz, *Unification of Tort Law: Damages* in *Principles of European Tort Law* (U. Magnus ed.), Kluwer, 2001, p.177 and Schwartz, *Unification of Tort Law: Strict Liability*, Kluwer, 2002, p.358

751 See Kerameus, *Unification of Tort Law: Damages* in *Principles of European Tort Law* (U. Magnus ed.), Kluwer, 2001, p.112

752 See Neethling, *supra*, n. 702, p.165

753 See Busnelli and Comande, *supra*, n. 681, p. 128

modes of assessment based on objective value. Some States such as France,[754] Germany[755] and South Africa[756] requires strict financial loss of actual or future income in order to allow such claims.

(c) Pure loss of holiday or free time

Although the matter has only a minor importance in the field of damages due to other prevailing remedies[757] such travel insurance, claims of this nature cannot be totally excluded from the context of shipowner liability. This is mainly due to consequential outcomes, such as pain and suffering that might encounter the claimants coupled with frustration. The Belgian law favourably allows compensation in such situations for any loss of holiday or free time under the heading of 'moral damages',[758] while in South Africa it forms part of loss of amenities in life,[759] and whereas in Greece it is observed in theory.[760] In the US, the question will be left for the discretion of the court juries.[761]

(d) Costs of un-repairable things

In general, costs of repair of the damaged property are unarguably recoverable under most jurisdictions. In events of damage to property that could not possibly end up in reparation, the law needs to assess reasonable grounds on which compensation can be granted. Sometimes there may exist of situations where reparation cost is much higher than replacement. In such circumstances, it would be undoubtedly prudent to grant compensation based on real market value, whereas in situations in which the real value cannot be easily determined due to diminution of value resulting in its use, many States evidently use methods based on objective-abstract formula taking into consideration, the current market value of a new similar thing. However, the mode of granting compensation for private or public property which has real market value does not pose much difficulty to courts, unlike claims of environmental damage. This notion which has gained much of the controversies in claims of this nature undoubtedly falls within the competency of public ownership of States, under both national and international legal contexts, and this aspect needs special and separate attention.

754 See Galand-Carval, *supra*, n. 705, p. 82
755 See Magnus, Fedtke, *supra*, n. 711, p. 97
756 See Neethling, *supra*, n. 702, p. 165
757 For example, the European Directive No. 90/314/EEC on travel contract law
758 See Cousy and Vanderspikken, *supra*, n. 700, p.33
759 See Neethling, *supra*, n. 702, p.166
760 See Kerameus, n. 751, p 113
761 See Schwartz, *Unification of Tort Law: Strict Liability*, Kluwer, 2002, p.178

Environmental Damage, Costs of Restoration and Reparation

These types of damages are associated with natural resources that rest on the land and that lie beneath States' sovereign and jurisdictional coastal waters. They include *inter alia*, land based vegetation and species, beaches, rocks, fish, seabirds, marine species and other natural living and non-living resources of varying types. Taking into consideration, the causation factor coupled with probability and foreseeability, the most vulnerable in the face of an oil spill are costal related natural resources. In particular, beaches and rocks remain as repairables, subject however to some degree of reasonable compensation for clean-up and lost value, whereas, damage to both living and non-living natural resources are mostly unrepairable to a large extent. Therefore, such damage that occurs in a physical deterioration can only be compensated in terms of economic means. Furthermore, their assessments would depend on economic calculations based on certain accepted reasonable criteria.

As discussed previously, definition of "pollution damage" under the Convention allows compensation only to costs of reasonable measures of reinstatement actually undertaken or to be undertaken as a result of impairment of the environment,[762] except loss of profit from such impairment. Accordingly, there exists of two types of compensational claims dealing with impairment of environment. One deals with loss of profit, and the other with environmental impairment which is limited to costs of reasonable measures taken or to be taken. The former can only be calculated in abstract-objective manner while, the latter would need evidentiary proof of measures with respect to restoration and reparation. The practical aspect of this process is chiefly depended on action plans concerning preventive and clean-up measures that will be followed later in this discussion on requirements of liability, but in present context, the matter would be left for the benefit of pure understanding of compensational claims.

Most States do not have specific rules on this concept in their respective general tort law provisions. But however, in many instances these can be found in specific legislations dealing with environmental liability derived from international treaties. For example, Article 37, Act of 20 January 1999 of Belgium provides somewhat restrictive provisions in this respect enacting that; every damage to or disturbing of the marine environment following an accident or a violation of the existing legislation must be restored by the one who caused the damage or disturbance, even if he did not commit a fault.[763] Furthermore, somewhat new

762 Many States tend to follow this strategy in order to balance the reasonableness of the amount of compensation with extremely high costs of restoration while protecting the interests of both the affected parties and the industry

763 See F. Douever, F. Maes, A. Vanhulle, J. Schrijvers, *The role of marine spatial plan-*

and innovative phrase of disturbance has been defined as; every adverse influence to the marine environment which does not constitute damage. The German *UmwelthaftungG* provides special rules on environmental liability and damage allowing claims which are even unreasonably high or disproportionate to the value of the damaged thing.[764] It has gone a long way in prescribing limitless levels of compensation for environmental damage including that of living species, as far as restitution is concerned. The Dutch Civil Code provides certain rules in general for environmental damage in addition to specific legislations, covering compensation for specific expenses such as cost of maintenance of installations. These costs are considered as falling within the meaning of patrimonial damage.[765] Meanwhile, US legislations have shown a great tendency of favouring environmental damage through its American Superfund Legislation that provides compensation for the purposes of both caused harm and clean-up.[766]

The practice in the US with regard to assessment and valuation of environmental damage is quite remarkable and unique. Both Natural Resource Damage Assessment Regulations (NRDA)[767] in respect of the former[768] and Contingent Valuation Methodology (CVM) for the latter[769] have proved successful outcomes. In particular, NRDA regulations[770] have evolved to assess and redress environmental damage claims far more effectively than adopting theoretical formulae or gathering expert evidence which are used by most of the States. On the other hand, the CVM is used to assess amounts of lost value through positive investigations and surveys carried out by public authorities and under their authorization.

---

ning in sea use management: The Belgian case, Marine Policy, vol. 31, iss. 2, Mar., 2007, pp 182–191

764 See Sec. 16 par. 1

765 See Article 6:184

766 See Schwartz, *supra* n. 761, p.355

767 See http://restoration.doi.gov/

768 See L. B. Burlington, *A Update on Implementation of Natural Resource Damage Assessment and Restoration under OPA*, Spill Science and Technology Bulletin, vol. 7, issues 1–2, June 2002, pp 23–29; C. A. Jones, *Compensation for natural resource damages from oil spills: a comparison of US law and international conventions*, International Journal of Environment and Pollution 1999 – vol. 11, no.1 pp. 86–107; P. D. Boehm et. al., *Advances Chemical Fingerprinting for Oil Spill Identification and Natural Resource Damage Assessments*, available at http://www.iosc.org/papers/00033.pdf; A. W. Ando, et. al., *Natural Resource Damage Assessment: Methods and Cases*, available at http://www.istc.illinois.edu/info/library_docs/RR/RR-108.pdf

769 See J. B. Loomis, *Contingent Valuation Methodology and the US Institutional Framework* in *Valuing Environmental Preferences: Theory and Practice of the Contingent Valuation Methods in the US, EU, and Developing Countries* (I. J. Bateman et. al. eds.), Oxford, 1999, Ch. 18

770 See http://restoration.doi.gov/laws.html

In addition to compensating environmental damage caused by an oil spill, the liability of the shipowner extends to costs of restoration and reinstatement carried out after the clean-up operation.[771] Likewise the former category, restoration and reinstatement involve tremendous degree of assessment and valuation due to unlimited scope of the work. Therefore, the marginal appreciation imposed by the Convention[772] is highly important to close the floodgates.[773] However, the real meaning or accepted claims under this margin is undefined, unclear and left for respective national courts for their own interpretations. Nevertheless, there may exist of several acceptable claims that would suffice the requirement of reasonableness under most developed systems of law. Naturally, these include claims for contingency planning carried out by means of intervention by human, technological or both sources to encourage and accelerate *natural* recovery, and costs of in-depth study on issues directly relating to the disaster, damages, environmental impact assessments, deterioration process of natural resources and consequential effects.[774]

Need of Prevention

Being a requirement of imputation of liability under the Convention, the need of prevention is also a well respected theory in law of damages. Although connected with the definition of "pollution damage" under the Convention, actions for pre-

---

771 See D. S. Etkin, *Estimating Cleanup Costs for Oil Spills*, #168, 1999 Int'l Oil Spill Conf., available at http://www.environmental-research.com/site_files_base/publications/content_pdf/paper6.pdf; C. A. Jones, *Compensation for natural resource damages from oil spills: a comparison of US law and international conventions*, Int'l Jour. of Environment and Pollution, vol. 11, no. 1, 1999

772 "... costs of reasonable measures..."

773 IOPC Fund has in several occasions rejected claims based on abstract or theoretical assessment schemes pointing out the necessary requirement of reasonableness, as prescribed by the CLC regimes. Massive claims lodged aftermath the Soviet *Antonio Gramsci* tanker disaster and their inadmissibility led to the adoption of Resolution No. 3 of the 1971 Fund (FUND/A/ES.1/13) through which it addressed the issue by stating that "the assessment of compensation to be paid b y the Fund is not to be made on the basis of an abstract quantification of damage calculated in accordance with theoretical models". The 1981 Assembly further endorsed a report submitted by the Fifth Intersessional Working Group (FUND/A.4/10) which emphasized the importance of entertaining only claims of quantifiable economic loss

774 In the *Braer case* (UK, 1993), the IOPC Fund rejected a claim for costs of environmental impact assessment studies carried out by a local authority, on the ground that such studies were general in nature without much detailed report on the particular incident. See IOPC Annual Report 1995, pp. 60–66

vention could exist even in situations where no actual damage has been caused, but where future damage can be expected. The rational for such conviction is based on the interpretation of 'incident' under the Convention, which specifically relates to situations that create grave and imminent threat. Therefore, preventive measures applied before or after the pollution damage has occurred or both, fall within the meaning of recoverable claims varying their application from pure threat removal to clean-up. There seems to be certain requirements that need fulfillment in order to support such claims within the meaning of 'legally acceptable'. According to the definition of "preventive measures" the idea is clear that it should fulfill the substantive requirement of 'reasonableness' and procedural requirements of 'preventing' and 'minimizing' pollution damage. Once all these grounds are established, one may successfully claims costs he incurred towards employing methods for prevention. The Convention specifically deals with the personal scope of application for the execution of such methods, recognizing the rights of action of any person and their consequential claims.

Shipowner's Response

The above discussed conventional definition meaning 'any person' allows the shipowner himself to take preventive measures and qualify for reimbursement of such expenditure through relevant claims. In actual practice, the shipowner gains the opportunity to forward his claims based on his response towards the prevention purpose against his insurers or financial security providers.[775] It is a well accepted practice that the shipowner is in the best position to take leading role in prevention process due to the commanding knowledge about his own ship. Many coastal States have even enacted national legislations to this end imposing mandatory involvement of shipowners with respect to such response upon directed by relevant public authorities. This would not only support the idea mentioned before as to the employment of his knowledge, but also protect such authorities from self-imputing of future liabilities on themselves that may be associated with further damage or loss.

Execution of preventive measures by the shipowner is mainly considered as a measure in kind, although in certain circumstances it may be compelling. In turn, he executes his service in kind as a settlement for compensation. Most States are in favour of this procedure except in rare situations where shipowners' response is found to be aimed at securing his interests such as recovering his cargo and the

775 Under the CLC regime Article V.8, the shipowners' expenses with regard to preventive measures rank equally with other claims against the Fund

ship rather than the prevention of pollution damage. Sometimes, the said recovery might well be compatible with the goals of prevention, but not without reasonable expectations. Due to these uncertainties coupled with shipowners' reluctances to undertake liability on themselves at first sight, most States tend to deviate from the idea of response in kind and claim for compensation in money for carrying out preventive measures through their respective public authorities. These may be well founded under their respective tort law principles that enable such authorities to act accordingly.

Costs incurred by Governmental Authorities

As the subject of marine environment falls within the sovereignty and trusteeship of respective States, plans on preventive measures are basically dealt by their respective governmental authorities.[776] In carrying out those functions, they employ different types of expertise including the hiring of work by independent contractors. In order to avoid confusions, it is the idea of this discussion to separate these two types of involvement by addressing the pure governmental handling under this heading and leave the latter to the next.

There are various claims based on costs incurred by governmental authorities in prevention process. They involve publicity, handling of the matter in paper by way of formulating combating plans through expert personal and extend up to the practical aspect of clean-up in minimizing any future damage. Apart from the normal process of monitoring marine environmental protection[777] and safety

776 See E. Amato, N. Cardellicchio, R. Giangreco, T. H. Moller, *European Local Authorities role in preparedness and response to oil pollution to safeguard seafood production: state and perspectives*, Report of the European Workshop, 28–31 October 2003, Tarentum, Italy, available at http://ec.europa.eu/environment/civil/marin/reports_publications/taranto_final_report.pdf; D. J. Blackmore, *The Australian Oil Spill Response Program*, AMOSC, available at http://www.pcs.gr.jp/doc/esymposium/12171/97_don-j-blackmore-e.pdf; EPA Office of Emergency and Remedial Response, *Responding to Oil Spills: The National Response System*, Ch. 7 in Understanding Oil Spills and Oil Response, available at http://www.epa.gov/oem/docs/oil/edu/oilspill_book/chap7.pdf

777 See J. S. Gray, D. Calamari, R. Duce, J. E Portmann, P. G Wells, *Scientifically based strategies for Marine Environmental Protection and Management*, Marine Pollution Bulletin MPNBAZ, vol. 22, no. 9, Sept., 1991, p 432–440; B.-E. Bengtsson, T. L. Coombs and M. Waldichuk, *Biological Variables, Especially Skeletal Deformities in Fish, for Monitoring Marine Pollution [and Discussion], Philosophical* Transactions of the Royal Society of London. Series B, Biological Sciences, vol. 286, no. 1015, The Assessment of Sublethal Effects of Pollutants in the Sea (Aug. 8, 1979), pp. 457–464

of navigation in their respective waters, these public bodies are expected to react effectively in emergency situations concerning risks in the wake of oil spill incidents. In such circumstance, these public bodies will have to utilize additional costs in ensuring proper and effective response.

Fixed and Additional Costs

Due to the extra workload that has to be handled, it is obviously important to obtain services of additional personnel, plant and other equipment, especially in clean-up operations. These steps result in extra financial burdens on the respective authorities. Therefore, additional costs incurred in such a manner are generally recoverable. However, the issue of fixed cost, which is related to capital costs of maintaining machinery and personnel in any event even in the absence of such insurgencies, is somewhat problematic and forms many legal argumentations. On the other hand it would not be easy to draw a line of demarcation for such an issue. In *Maritime Services Board of New South Wales v. Poseidon Navigation Inc.* (1982), it was held that the claimant is not entitled to make profits from clean-up operations or to make a 'commercial charge' for the work done, but to calculate actual cost bourn by it of using its own employees, including the overheads and administrative charges, actual cost of using own plant and its subsequent overheads, for which it is entitled.[778] The *Amoco Cadiz* litigation dealt with similar issues as to fixed costs where the plaintiffs were entitled for recovery without any reductions.[779] Due to these varying opinions on the subject and the absence of a clear interpretation to the term 'costs' in the Convention, most national jurisdictions would apply their own interpretations in overcoming this issue.

Further distinction is surrounded by the issue of actual costs and overheads with regard to personnel and machinery. Overheads relating to the former category extend drastically in terms of economy from issues such as expenses for payroll tax, worker's compensation and insurance, leave entitlements to other related administrative charges. In the abovementioned case of, *Posiden Navigation Inc.*, the court upheld such costs for personnel and machinery overheads as recoverable,[780] but disallowed any possibilities of gaining profits from such entitlements[781]. The *Amoco Cadiz* litigation is also supportive in this respect where the court has allowed claims based on entire committed cost of maintaining machinery on stand-

778 See the judgment of Yeldham J, [1982] NSWLR 72

779 See In *the Matter of: Oil Spill by the Amoco Cadiz off the Coast of France on March 16, 1978*, 954 F.2d 1279 (1992 US App., 7th Cir.)

780 See [1982] NSWLR 72, at 91

781 *Id*

by.[782] However, the courts are very restrictive in determining the questions on overheads and are reluctant to proceed with claims for remote overheads. Therefore, it can be proposed that only directly linked overheads should be considered as a matter of policy in order to avoid junk claims and unjust enrichment. In accommodating justice, a court may take into consideration of the accounting techniques of valuation of machinery costs, use, depreciation of efficiency and value coupled with their relationship to the desired commitment. With regard to personnel overheads it may consider the most direct, reasonable and necessary ones.[783]

Costs incurred by Independent Contractors

Although the works of the independent contractors are directly linked with the competency of governmental authorities, the matter needs a separate discussion due to certain reasons. Firstly, there are several instances where governmental authorities become insolvent leaving the costs of these contractors unattended. In such situations the aggrieved party would resort to action against the shipowner seeking for recovery on behalf of the original claimant or as his agent under the contract between the parties.[784] Secondly, the work carried out by such contractors involves various expertises in specialized areas of interests. Thirdly, their work is specific purpose-oriented such as salvaging, rehabilitation of endangered species and alike rather than general clean-up. For the sake of clarity it would be better to consider these different fields one-by-one.

Salvage of the Ship

Normally, salvage including towing is carried out by the shipowner through salvors employed by him due to many technical reasons. But however, public authorities take over this activity by employing independent salvors in circumstances where shipowner identification difficulties encounter, in order to protect the marine environment from further damage and in events where the shipowner tries to

782 *Supra*, n. 779

783 The IOPC Fund has adopted the practice of settling such claims by way of accepting only expenses which correspond closely to the clean-up operations, excluding remote overheads charges. In Swedish vessel *Thuntank 5* case, the IOPC Fund (IOPC Fund Annual Report, 1989, p.23) has disallowed such claims based on indirect overhead charges

784 For example, see the case of *Akari* tanker of Panama, IOPC Fund Annual Report 1991, p 31

protect his cargo and the ship without paying much interest on relevant damage, and for other reasons such States consider it necessary. Nevertheless, there is much danger in handling the salvage process by the respective State due to further damage which could occur from negligence of its' side and ultimately have to suffer without being able to obtain compensation for such loss or damage. In such event, the shipowner might attempt to escape liability for any consequential damage that would occur in salvage operation due to the negligence or carelessness of the salvors, other than to incidents connected with the normal recovery under expected competency. Nevertheless, the Bunkers Convention is specific in imposing all such liabilities on the shipowner as a result of the first incident,[785] which in turn impute all such consequential liabilities on such shipowner by virtue of the definition provided for 'pollution damage', which has a wide coverage even to include such above mentioned types of damage. In any case, the Convention has well guaranteed the right of recourse that exists for the shipowner outside the Convention.[786] But however, a clear line of demarcation could still be drawn between damage caused as a result of taking preventive measures and damage caused due to mere salvage of the ship and its cargo, which sometimes might have nothing to do with prevention of pollution damage. As the Convention does not deals specifically with clauses on 'responder immunity',[787] the issue is thus left for decisions of the national courts based on its accepted jurisprudence.[788] Undoubtedly, such circumstances would generate high tensions, especially in proving such matters with expert evidence, as to salvaging of ships.

Oil Recovery Process

This forms one of the most important steps towards controlling the spreading of loss or damage. Specialized in the field, boats and specially equipped crafts are used for the recovery process upon which recovered oil will then be directed for sewage.[789] Costs are incurred for the carrying out of these activities plus harbour fees, which is more a governmental type of claim.

785 See Art. 3.1

786 See Art. 3.6: "[N]othing in this Convention shall prejudice any right of recourse of the shipowner which exists independently of this Convention"

787 See *supra* Ch. 3 at p 112

788 This would depend on the interpretation of the instance by such court, whether to entertain the claim under the Convention or under its general law applicable to such event

789 See D. S. Etkin, *Clean up Costs for Oil Spills in Ports*, Oil Spill Intelligence Report, Arlington, MA, USA, available at http://www.philippines.canurb.com/projectsunrise/files/pdf/paper5.pdf; L. Clarke, *Oil Spill Fantasies*, available at http://

Spraying of Dispersants

The process is highly scrutinized in the area of marine environmental protection, but is widely used in order to protect the marine environment from oil and to accelerate the process of degrading. Costs relating to these measures are recoverable as steps taken in prevention,[790] and the shipowner is held liable for such and further costs in terms of strict liability.

Rehabilitation of Wildlife

Expert knowledge is used in most cases for such rehabilitation in reducing the vulnerability of the wildlife to adverse and side effects of the spill.[791] Costs of scientific research, plans for implementation and actual carrying out of work by persons specialized in scientific and environmental fields would certainly have to be compensated by the liable party. However, courts will employ rules such as directness, fairness and appropriateness in all such moves in determining the admissibility of all these claims.

---

leeclarke.com/docs/clarke_atlantic_b&w.pdf; W. M. Sullivan, M. J. Mulhare, *Oil Spill Prevention, Administration and Response (OSPAR) Fund, Annual Report FY 2006*, Rhode Island Department of Environmental Management, available at http://www.dem.ri.gov/pubs/ospar06.pdf

790 See ITOPF, *Dispersants*, available at http://www.itopf.com/spill-response/clean-up-and-response/dispersants/; ITOPF, *The use of Chemical Dispersants to treat oil spills*, Technical Information Paper, available at http://www.itopf.com/_assets/documents/tip4.pdf; J. Bywater, *Government response to marine pollution from ships*, Marine Policy, vol. 19, iss. 6, Nov., 1995, pp 487–496; J. Guyomarch, O. Kerfourn, F. X-Merlin, *Dispersants and Demulsifiers: Studies in the Laboratory, Harbor and Polludrome*, Cedre, France, Paper ID #283, 1999 Int'l Oil Spill Conference, available at http://www.iosc.org/papers/00541.pdf; A. Lewis, F. Merlin, P. Daling, M. Reed, *Applicability of Oil Spill Dispersants Part I: Overview*, European Maritime Safety Agency (EMSA), Aug., 2006, available at http://www.emsa.europa.eu/Docs/opr/emsa_manual_applicability_oil_spill_dispersants_2006.pdf

791 See B. Kathy, *Preparedness to oiled wildlife incidents in Belgium*, available at http://www.interspill.com/PDFs/UK/2006/wildlife_oiledwildlife_doc.pdf; IPIECA, *Objectives of Oiled Wild Life Response* in *A Guide to Oiled Wildlife Response Planning*, IPIECA Report Series, vol. 13, at http://www.ipieca.org/activities/oilspill/downloads/publications/reports/english/Vol13_Oiled_Wildlife.pdf

Participation of these associations and organizations are sometimes welcome by respective States due to varying degree of support they provide towards prevention and clean-up procedure. Sometimes, such grants are made for pure transparency role of the government in carrying out the work effectively and as a matter of political pressure. Especially, with regard to international agencies such as the Greenpeace, their involvement is much supportive towards gaining the knowledge of new and modern techniques that can be executed for speedy recovery for caused harm and also as ambassadors of goodwill who could take the matter into international arena.

The possibility of forwarding claims by any 'person', under the meaning of the Convention, creates certain extra burden on respective jurisdictions. Subject to requirements such as '*locus standi*' and other legal doctrines as to recognition of natural or legal personality respected by States' own legal system, any person involved in prevention process is entitled to seek compensation for its part. The said '*locus standi*' in this context would only be linked to the rightful claiming of such compensation based on involvement. The duty of the court to investigate whether the claimant actually has a right of action is simply surrounded by the issues of direct involvement and reasonableness of claims. In general, national laws of all States do not completely adhere to these kinds of claims in their respective tort law principles as a matter of lack of jurisdiction. For example, non-governmental agencies or groups linked to environmental protection purposes are under the obligation under the Belgian law to establish that the measures they took against possible damage are personal to such association.[792] Upon proving such, those organizations become eligible to obtain compensation as a matter of symbolic importance.[793] In England, any claimant in principle must posses proprietary or possessory interests in order to file a claim, but special statutory provisions in the environmental law sector has allowed certain exceptions in accommodating such actions.[794] The French law which has combined tort law principles with specific procedural law in cases of environmental harm is supportive towards this approach, where associations have the possibility of recovering compensation for so called '*dammage moral*' on the condition that they act for pure environmental purposes.[795] Under the Convention on Civil Liability for Damage

792 See A. Lebrun, *Aménagement, environnement, urbanisme et droit frontière (1990)*, Brussels, 2 Nov., 1989, p.42

793 For a comprehensive discussion on this see A. Carette, *Herstal van en vergoeding voor aantastingen aan niettoegeeigende milieubestanddelen (1997)*, p. 99

794 See Rogers, *supra*, n. 710, p 66

795 See Galand-Carval, *supra*, n. 705, pp 83–84

Resulting from Activities Dangerous to the Environment[796], the so-called Lugano Convention adopted by the Council of Europe, these organizations have no right to claim compensation rather than persuading liable party to carry out preventive measures.[797]

However in essence, the Bunkers Convention's allocation of right to claim compensation on these organizations would undoubtedly open up new frontiers in liability and compensation regimes. In practice, national courts should maintain a quite reasonable yardstick in determining which claims are to be considered justifiable without leaving any room for unnecessary jamming of irrelevant claims that might abuse the due and fair process, safeguarding the conventional definition in the meantime.

Miscellaneous Costs

There is no intention here to discuss all other types of costs that may be associated with the entire scenario of bunker oil spills, but to deal with most important issue of litigation costs. In any legal forum there would involve high litigation costs for all parties concerned. Especially, in these types of litigation, victims have to bear huge amounts as legal fees, judicial costs, travelling and so fourth,[798] upon been victimized originally by the pollution incident itself. Therefore, claims for compensation usually include a separate clause for costs incurred towards the judicial process, which are indeed recoverable under most national jurisdictions.

---

796 The full text of the Convention adopted in Lugano, 21.VI.1993, can be found at , European Treaty Series No. 150, Council of Europe, http://conventions.coe.int/Treaty/en/Treaties/Word/150.doc

797 Art. 18(1): "[A]ny association or foundation which according to its statutes aims at the protection of the environment and which complies with any further conditions of internal law of the Party where the request is submitted may, at any time, request: (a) the prohibition of a dangerous activity which is unlawful and poses a grave threat of damage to the environment; (b) that the operator be ordered to take measures to prevent an incident or damage; (c)that the operator be ordered to take measures, after an incident, to prevent damage; or (d) that the operator be ordered to take measures of reinstatement"

798 See *supra* D. S. Etkin, n. 789; D. Helton, *Putting Response and Natural Resource Damage Costs in Perspective*, Paper ID #114, 1999 International Oil Spill Conference, available at http://www.darrp.noaa.gov/library/pdf/costsofs.pdf

Litigation costs including Professional Fees

Litigation costs are normally regulated by provisions of the procedural law. National courts of most States follow its' own civil procedure or case law doctrines in awarding these types of costs. For example, in Germany, it is regulated by Section 91 of the Civil Procedure Code (ZPO) under which such costs are recoverable. However, the losing party has to bear its costs,[799] while solicitors' costs outside court proceedings are awarded on the condition that such costs are reasonable and associated with a justified damages claim or in defence of an unjustified claim.[800] Under the Dutch Civil Code, expenses of legal proceedings are recoverable within the meaning of judicial costs, and awarded as fixed amounts based on the nature and the length of the proceedings. These costs do not cover the real expenses and are awarded to the prevailing party. However, costs for court appointed experts are fully recoverable.[801] Although, national laws of respective States provide favourable approach towards the settlement of litigation costs, especially in cases which fall under the meaning of the Convention, courts would apply its discretion in determining the extent and amount based on reasonable tests, in the absence of any relevant legally accepted formulae.[802]

Plaintiffs' Standard of Proof

Unlike in general tort law remedies for damage such as negligence, trespass or nuisance, strict liability regimes impose the claimant with a lesser degree of proving his case against the liable party. It is important to note that there exist of a certain degree of proof on the part of the claimant even though the defendant is answerable in a stricter form. Although, proof based on circumstantial misconduct of the defendant gather no recognition in such cases, claimant is vested with the burden of proving the actual scenario with much evidence to his knowledge in order to impose liability on the defendant. These requirements are discussed below with a view of understanding the standard of proof expected of him.

799 Sec. 91 par.1 sent. 1 ZPO

800 BGHZ 111, 171; BGH [1986] NJW 2244

801 See Wissink and van Boom, *supra*, n. 701, p. 147

802 The IOPC Fund in several instances has allowed claims on professional fees emphasizing the importance of settlements on case by case basis. The Executive Committee has expressed it views that such claims will be examined on grounds of necessity as to use of expert advice, the usefulness of the work carried out by the expert, the quality of that work, time and duration of the work under normal circumstances. (FUND/EXC. 37.3, 8 October 1993, sec.4.2.21) It has also upheld the justifiable claim issue (FUND/EXC. 50/17, 23 October 1996, sec.3.12.19)

Identification of the Respondent

In bringing up a claim for compensation based on liability, it is highly important to identify the actual tortfeasor. Therefore, the claimant must know the person responsible for the incident, namely the 'shipowner' as defined in the Convention, with much reliable evidence. Although straight forward situations would not create much of the controversies, claimants would be left in less favourable positions in cases of unidentifiable or mysterious types of spills. As the majority of claims are normally lodged by governmental authorities or States in general, identification process is left to the most specialized sector having the best technical tools. Absolutely, the State intervention is very important in this process due to complexities that surround the whole issue, especially with regard to the work of its own harbours and coast guards. In respect of spills that occur in the high seas, effective coordination of all coastal States is highly essential, in monitoring activities and movements of ships that operate between different ports of those States. Furthermore, States' coordination in administrative matters, especially with regard to issuing of certificates and license to ships, enables to identify respective insurers against whom direct actions could be filed. Therefore, the identification of the tortfeasor is considered to be the main and the starting point of attributing liability under this conventional regime.[803]

Certainty as to Causation

Once the respondent is identified by the claimant, claim for compensation should satisfy the causation requirement in order to attribute liability on such respondent. The plaintiff in such circumstance must prove the link between the incident to which the respondent is connected and the pollution damage which occurs or steps of preventive measure are taken as a result.[804] Uncertainties that encounter with such causation would end up in failing to impute liability on the respondent.[805] Therefore, it is important on the part of the claimant to prove his case with a considerable degree of certainty as to causation in succeeding with his claims for compensation.

803 See Z. Wang, M. Fingas, D. S. Page, *Oil spill identification*, Journal of Chromatography A vol. 843, issues 1–2, 28 May 1999, pp 369-411; M. G. Faure, W. H. van Boom, Concluding Remarks of *Shifts in Compensation Between Private and Public Systems*, Springer, 2007, pp 219–236

804 See *supra* pp 146–153

805 See *supra* pp 157–158

Certainty as to Damage

In a successful claim for compensation based on damages, the claimant is required to maintain strict proof of the existence of damage and its extent, without any reasonable doubt.[806] As a principle in civil law, damage has to be proved on balance of probabilities, but the required degree of such probability varies from one jurisdiction to another.[807] Therefore, the standard of proof on certainty as to damage depends on the required degree of probability in the jurisdiction where such matter is tried.

In Germany and Belgium, courts require a probability next to certainty.[808] The German law requires the judge be convinced of the facts and be sure that no reasonable doubts remain,[809] whereas in Belgium, such certainty must hold the likelihood that a judge cannot seriously consider the contrary.[810] The English law apply a test of probability of the kind 'more likely than not',[811] and the Dutch law on the other hand requires some proof of data concerning the occurrence and extent of the damage which must be plausible.[812] South African law is much percentage oriented requiring fifty percent of the probability with regard to certainty.[813]

Even though actual damage does not pose much threat in this respect, claims on future damage tends to reflect the importance of ascertaining the required elements of respective national courts to a very great extent. The Belgium strategy goes further in holding the 'likelihood' criteria for such future damage in circumstances where expected damage may never actualize even in theory.[814] In Germany, the burden of proof is reduced in issues of lost profits, and the probability instead of certainty suffices, that the profits would have ensued.[815] The French law requires absolute certainty as to future damage, and in success it compensates

---

806 See *id*

807 See R. Brillinger, *Canadian Business Law*, Edmond Montgomery, 2007, p 72; J. Steele, *Tort Law: Text, Cases and Materials*, Oxford, 2007, pp 134–135; A. Porat, A. Stein, *Tort liability under certainty*, Oxford, 2001, p 16; J. Hodgson, J. Lewthwaite, *Tort Law Textbook*, 2nd ed., Oxford, 2007, p 256; E. L. Sherwin, K. M. Clermont, *A Comparative View of Standards of Proof*, American Journal of Comparative Law, vol. 50, 2002, p. 243; U of San Diego Public Law Research Paper No. 32. Available at SSRN: http://ssrn.com/abstract=285832 or doi:10.2139/ssrn.285832

808 See Magnus, *supra*, n. 711, p. 204

809 BGHZ 53, 245; BGHZ 61, 169

810 See Cousy and Vanderspikken, *supra*, n. 700, p. 38

811 See Rogers, *supra*, n. 710, p. 67

812 See Wissink and van Boom, *supra*, n. 701, p. 153

813 See Neethling, *supra*, n.702, p. 167

814 See *supra*, n. 810

815 See *supra*, n. 808

the claimant without any 'discount of contingencies'.[816] The South African law is somewhat relaxed in the sense it does not require proof if uncertainties prevail, again applying a percentage based criteria, this time with a requirement of a forty percent probability of the chance that damage may result.[817] In the United States, the issue has not received much attention, and there doesn't seem to be a uniform practice with regard to certainty as to future damage, where some courts apply reasonable certainty requirement, whereas some other use the 'more likely than not' test.[818]

Necessity and Reasonableness in taking Preventive Measures

Claims for expenditures incurred towards execution of preventive measures and clean-up do not have to be necessarily accompanied by proof of damages under the conventional interpretation. As preventive measures taken prior to the existence of 'pollution damage' are compensable under the Convention, claims of such nature are legally allowed to be entertained by courts. However, claimants must maintain a certain standard of proof in this respect, not merely as to apportion their expenses but as a matter of policy consideration in avoiding unnecessary claims been lodged. Therefore, a successful claimant must demonstrate compelling needs for the execution of preventive measures based on necessity and reasonableness.

The requirement of necessity arises with the fact that such preventive measures are important to avoid pollution damage been caused or grave and imminent threat of causing such damage is created. Once it is discovered that such a ground is apparent, government authorities in particular apply measures in prevention, which ultimately become compensable upon been attributed to the liable party. However, demonstrating the grounds for necessity is highly sensitive and falls into much controversies. In proving such a cause in the absence of actual pollution damage to the end result, the plaintiff is expected to present reliable evidence to establish the conventional requirements as legislated in the definitions of 'preventive measures', 'incident' and 'pollution damage'. The hardest obstacle would be to rationalize situations of 'grave and imminent threat', which lack a clear interpretation. In view of such, a court would usually employ a test at its will or pursuant to objections by the defendant at first sight, in determining whether the claimant has fulfilled this requirement.

816 See Galand-Carval, *supra*, n. 705, p.84
817 See *supra*, n. 702
818 See Schwartz, *supra*, n. 750, p. 179

The requirement of reasonableness depends on the nature of the oil spill and varies extensively from one type to another. Reasonableness is measured by way of taking into consideration; many aspects related to the spill, mainly, type of oil spilt, state of the environmental dynamics, sensitivity of the affected area and the behaviour of the climate. These conditions would influence the measures that the affected parties are expected to carry-out using scientific, technical and other types of operations. Measures which produce adverse end results would normally be considered improper, adding their failure as a negative approach to determine them as unreasonable in this context. In certain circumstances, preventive measures are highly optimized in order to achieve accelerated results that end up in negative forms. Sometimes, the media is used in the wrong manner to spread awareness unnecessarily, creating chaos among the public that would bring unpopularity to the affected area as well as disrepute the shipping industry. Therefore, reasonableness that prevails in compensable claims must be associated with positive, effective and correct modes of application of preventive measures. Furthermore, the monetary character needs to be fair and reasonable, especially being proportionate to the actual measures applied and expenses incurred towards the whole process.

Chapter 6: Available Defences and Exoneration from Liability

Since its origination in tort law pursuant to the rule in *Rylands v. Fletcher*, strict liability regime has undergone extensive recognitions of tortfeasor's defences.[819] Although, a breach of statutory duty is weighted heavily on grounds of inexcusable wrongdoings in both criminal and civil structures of law, certain permitted exceptions have always been welcomed by civilized systems in maintaining justice and reasonableness on the part of the wrongdoer. Unlike in other areas of tort law such as negligence, trespass and nuisance, defences under strict liability have always been limited to extreme circumstances that create extra-ordinary burdens on the liable party having been unable to remedy by itself. In other words, most strict liability regimes complement tortfeasors in defending themselves from liability in such circumstances where the tortuous outcomes are far apart from their contributions towards such end result. Although, these defences differ with respect to the jurisdiction and the nature of the tort, one could always identify similarities in isolation, as being part of the so-called *act of god*[820], *force maeju-*

819 See J. R. Mario, *Law amid Flux: The Economics of Negligence and Strict Liability in Tort*, The Journal of Legal Studies, vol. 9, no. 2, Change in the Common Law: Legal and Economic Perspectives (Mar., 1980), pp. 291–318; A. E. Richard, Defenses and Subsequent Pleas in a System of Strict Liability, The Journal of Legal Studies, vol. 3, no. 1 (Jan., 1974), pp. 165–215; J. Steel, *supra* n. 807, p 707; I. Englard, *Can Strict Liability be Generalized?*, Oxford Journal of Legal Studies, vol. 2, no. 2, 1982, pp 245–256; *A.* Calnan, *Strict Liability and the Liberal Justice Theory of Torts*, March 2, 2007, available at http://works.bepress.com/cgi/viewcontent.cgi?article=1000&context=alan_calnan; R. Grantham, C. Rickett), *A Normative Account of Defences to Restitutionary Liability*, The Cambridge Law Journal, vol. 67, 2008, pp 92–125

820 See K. Kristl, *Diminishing the Divine: Climate Change and the Act of God Defence*, Widener Law Review, Forthcoming; Widener Law School Legal Studies Research Paper No. 09–21, April 8, 2009, Available at SSRN: http://ssrn.com/abstract=1375115; C. S. Switzer, L. A. Bulan, *CERCLA: Comprehensive Environmental Response, Compensation, and Liability Act (Superfund)*, American Bar Association, 2002, pp 41–42; J. F. Berry, M. S. Dennison, *The environmental law and compliance handbook*, The McGrow Hill, 2000, p 385; D. Delaney, *Law and nature*, Cambridge, 2003, p 151; R. V. Randle, *supra* n. 283, p 5; D. Riesel, *Environmental Enforcement: Civil and Criminal*, 12-54.3 – 12-54.4

re[821] and *volenti non fit injuria*[822] doctrines among such other. This scenario is no stranger at international level having been employed in almost all of its liability regimes dealing with different subjects.

Derived from its predecessors, the CLC 69[823] & 92[824] and the HNSC[825], the Bunkers Convention has absorbed the defences on behalf of the shipowner as provided in Article 3.[826] These defences are subdivided into four sections, out of which three deals with total exoneration from liability as set out in Articles 3.3 (a), (b) and (c), while the fourth dealing with total or partial exoneration depending on the circumstance as described in Article 3.4[827]. The former categories deal with *force maejure* and *act of god*, although such expressions are found to be absent in the text, while the latter deals with malicious acts of a third party(s) and misconducts on the part of Coastal States respectively. The latter deals with malicious or negligent conduct of the victim under which, the component of exoneration firmly depends on the facts of the incident and subject to proof by the

---

821 See C. N. Brower, J. D. Brueschke, *The Iran-United States claims tribunal*, Kluwer, 1998, pp 466–471; L. T. Visscher, *Justifications and Excuses in the Economic Analysis of Tort Law*, RILE Working Paper No. 2008/03, February 21, 2008, published in *Economic Torts*, Hyderabad, (P. S. Khanum ed,), The Icfai University Press 2009, p. 22–57, available at SSRN: http://ssrn.com/abstract=1111547; N. S. Marsh, *The Ethiopian Law of Extra-Contractual Liability*, by G. Krzeczunowicz. [Addis Ababa: Faculty of Law. 1970. xv (preface and table of contents), 140 and 41 (Appendices)]. *International & Comparative Law Quarterly,* 20 , 1971, pp 788–789

822 See M. Lunney, K. Oliphant, *Tort Law Text and Materials*, 3rd ed., Oxford, pp 284–294; G. Kodilinye, *Commonwealth Caribbean tort law*, 3rd ed., Cavendish, 2003, pp 368–372; G. Stephenson, *supra* n. 688, pp 549–557; E. Stuhmcke, *Essential tort law*, 2nd ed., Cavendish, 2001, pp 72–76; H. J. H. W., *Tort—Volenti Non Fit Injuria—Knowledge of Risk*, The Cambridge Law Journal*,* 3 , 1927, pp 98–99; E. J. Weinrib, *supra* n. 687, p 317; J. Steele, *supra* n. 807, pp 279–287; Cohl, Karen, *Tort law*, I. P. I. Publishing, 1982, p 30; V. Harpwood, *supra* n. 651, pp 418–421

823 Articles III.2 (a), (b), (c) and III.3

824 Articles III.2 (a), (b), (c) and III.3

825 Articles 7.2 (a), (b), (c) and 7.3

826 Art. 3.3: "[N]o liability for pollution damage shall attach to the shipowner if the shipowner proves that: (a) the damage resulted from an act of war, hostilities, civil war, insurrection or a natural phenomenon of an exceptional, inevitable and irresistible character; or (b) the damage was wholly caused by an act or omission done with the intent to cause damage by a third party; or (c) the damage was wholly caused by the negligence or other wrongful act of any Government or other authority responsible for the maintenance of lights or other navigational aids in the exercise of that function"

827 Art. 3.4: "[I]f the shipowner proves that the pollution damage resulted wholly or partially either from an act or omission done with intent to cause damage by the person who suffered the damage or from the negligence of that person, the shipowner may be exonerated wholly or partially from liability to such person"

shipowner. Apart from these main factors, defences granted under Article 3.3 (a) pose a clear contrast from the rest due to the use of the words "resulted from" in its text. This implies that such causes would be considered as dominant or proximate causes even in the presence of negligence caused by personnel of the ship or contributory negligence of the respondents. For an extensive survey on how these defences could be brought into light by the shipowner in evading his liability under the Convention is highly inevitable. In order to clarify such situations it would be prudent to discuss the issues separately and more comprehensively.

The defence of 'act of war'

Article 3.3 (a) deals with a list of circumstances that could be brought in defence by the shipowner. A clear line of separation could here be drawn between natural disasters and man-made crisis. The latter has been granted foremost concern according to the proviso by citing act of war as of the main. Merchant ships have become utmost targets in situations of war, so that, much of the emphasis have directed towards them.[828] Evidently, many wars that were encountered up to now have imposed severe disasters and threats to commercial shipping.[829] This has turned a dark episode not only on the trade, but also on the volatile marine insurance market, depriving ships in obtaining any insurance cover for war risks.[830] Such refusal has resulted in denying compensation for victims by shipowners whose liabilities are to be borne by parties responsible for such warfare. This has ultimately resulted in negotiating international documents of this nature reser-

---

828 See G. G. Wilson, *Armed Merchant Vessels and Submarines*, The American Journal of International Law, vol. 24, no. 2 (Apr., 1930), pp. 337–339; E. Chadwick (2007), *The 'Impossibility' of Maritime Neutrality during World War 1.* Netherlands International Law Review, 54 , pp 337–360

829 See J. G. B. Hutchins, *The Effect of the Civil War and the Two World Wars on American Transportation*, The American Economic Review, vol. 42, no. 2, Papers and Proceedings of the Sixty-fourth Annual Meeting of the American Economic Association (May, 1952), pp. 626–638; D. H. Andersen, H. J. Voth, *The Grapes of War: Neutrality and Mediterranean Shipping under the Danish Flag, 1750–1807*, Uni. of Oxford, available at http://www.nuffield.ox.ac.uk/Economics/History/Paper18/18www.pdf

830 See G. Ceccarelli, *The Price for Risk-Taking: Marine Insurance and Probability Calculus in the Late Middle Ages*, Electronic Journ@l for History of Probability and Statistics, vol. 3, no. 1, June 2007, available at http://www.emis.de/journals/JEHPS/Juin2007/Ceccarelli_Risk.pdf; S. Hodges, *Law of Marine Insurance*, Cavendish, 1996, pp 337–338; H. T. Chester, *Round Table VI – Marine Insurance Law: Some Aspects of the Free of Capture and Seizure Clause*, 1943 A.B.A. Sec. Ins. Negl. & Comp. L. Proc. 240 (1943)

ving rights of defence for escaping liability. Nevertheless, one may not always underestimate situations of liability evasions in the light of war until such matters are justified on those grounds with the relevant degree of certainty as to rightful use of this defence by the shipowner. Sometimes, the margin of appreciation would turn out to be narrow in rejecting shipowners' right of defence on grounds of lack of proof.

In using the defence the shipowner must prove the existence of an act of war that resulted in causing pollution damage within the meaning of the Convention. Failing to provide justification with relevant proof would result in ultimate rejection of his defence and consequently imposing liability on him. There are many questions that surround the difficulty in ascertaining situations of war, starting from its definition up to the acceptance of such a defence in a legal forum as admissible. According to the Concise Oxford English Dictionary, 'war' is defined as "Hostile contention by means of armed forces, carried on between nations, states or rulers, or between parties in the same nation or state; the employment of armed forces against an opposing party in the state"[831]. Furthermore, the word "hostile" is defined as "of, or pertaining to, or characteristic of an enemy; pertaining to or engaged in actual hostilities"[832], while "hostility" being "The state of fact being hostile; hostile action exercised by one community, state, or power against another"[833].

In English jurisprudence,[834] the definition of 'war' seems to have taken a centre stage in a number of marine insurance cases. According to Halsbury's definition, which focus on the situation that deals particularly with the involvement of the United Kingdom, "At common law, there is no state of war between the United Kingdom and a foreign State, unless the Crown has declared war or hostilities have commenced by the Crown's authority".[835] The customary international law accepts the same situation as proclaimed by Article 1 of the Hague Convention 1907[836], which declares that States must "... recognise that hostilities between them must not commence without previous and unequivocal warning, which shall take the form either of a declaration of war, giving reasons, or of an ultimatum

831 C. Soanes, A. Stevenson eds., 11th ed. (revised), 2008, p 1627

832 *Id*, p. 689

833 *Id*

834 See W. J. Ronan, *English and American Courts and the Definition of War*, The American Journal of International Law, vol. 31, no. 4 (Oct., 1937), pp. 642–658

835 See Halisbury, *Laws of England*, Vol. 49, paragraph 101

836 Convention (IV) respecting the Laws and Customs of War on Land and its annex: Regulations concerning the Laws and Customs of War on Land. The Hague, 18 October 1907. Art. 1: "[T]he Contracting Powers shall issue instructions to their armed land forces which shall be in conformity with the Regulations respecting the laws and customs of war on land, annexed to the present Convention"

with a conditional declaration of war". Although the classical definition provides some useful criteria, state practices have not reflected its adherence. Evidently, the Korean and Vietnam wars in 1950s and 1960s never seem to have judged on this fact. Since the Second World War, States have not visibly commenced armed conflicts according to such theory. Throughout its practice, English courts have adopted the normal and popular meaning for the term 'war' in disputes concerning commercial contracts, although the courts are not particularly bound to follow such authorities in practice. There have been situations where certain deviations have occurred depending on the facts of some cases.[837] American courts on the other side are not bound by the fact that there exists of any absence of declarations as a necessity in accepting a state of war.[838] It was far clear from *Pan American World Airways Inc. v. The Aetna Casualty and Surety Co.*[839], that the insurance policies well excluded this implication from the definition of 'war' and 'warlike operations', whether there be a declaration of war or not.[840] In its judgment by the Circuit Court, Hays J. has well founded justifications for accepting a state of war, defining the term as meaning "... the employment of force between governments or entities essentially like governments at least *de facto*" and recognising such state of war in a course of hostility engaged in by such entities, in the present case the Popular Front for the Liberation of Palestine, that had at least significant attributes of sovereignty.[841]

In international law, the matter has gained tremendous concern of legal experts and jurists. The famous statement of W. E. Hall in his work on international law reads; "International law has no alternative but to accept war, independently of the justice of its origin, as a relation which the parties to it may set up if they choose, and to busy itself only in regulating the effects of the relation".[842] Although Article 2(4)[843] of the United Nations Charter prohibits its Members the use of force against any State, it accepts their right of self-defence by virtue of Article 51[844].

837 See M. D. Miller, *Marine War Risks*, 2nd ed., Lloyds 1994, p.42

838 See J. H. Mathias Jr,, J. D. Shugrue, T. A. Marrinson, *Insurance Coverage Disputes*, ALM Properties Inc., Law Journal Press, 2006, p 9; E. R. Anderson, J. S. Stanzler, L. S. Masters, *Insurance Coverage Litigation*, 2nd ed., Aspen, 2008

839 505 F.2d 989

840 [1974] 1 Lloyd's Rep. 207

841 See [1975] 1 Lloyd's Rep. 77

842 See Hall, *International Law*, 8th ed., p.82

843 "[A]ll Members shall refrain in their international relations from the threat or use of force against the territorial integrity or political independence of any state, or in any other manner inconsistent with the Purposes of the United Nations"

844 "[N]othing in the present Charter shall impair the inherent right of individual or collective self-defence if an armed attack occurs against a Member of the United Nations, until the Security Council has taken measures necessary to maintain international peace and security. Measures taken by Members in the exercise of this

Nevertheless, many States have turned down their pledges and have opposed the multilateral system in several occasions in pronouncing war against another. Most controversially, many of such incidents have occurred in complexity, upon being unable to identify the characteristic feature of whether a war exists of or not.[845] In addition the matter is sometimes more confusing due to uncertainties that gather around the issue of geographical extent of the conflict. The classic examples are the war that existed between United Kingdom and Argentina on Falkland Islands in 1982[846], the second Gulf War 1990–91[847] and the present insurgencies in Iraq and Afghanistan. In the first case, the two countries did not consider themselves as enemies of each other nor engaged in fierce battle, except certain incidents of exchange of fire and the British sinking of an Argentinean submarine within the so-called "exclusion zone". During such period, both countries operated their respective embassies and granted visas for travelling to each other's country and diplomats of both States visited each other. The only exception was that neither ships nor airlines of those two States were allowed to each other's ports.[848]

From the above discussion one may find several justifications that support defining the term 'war' under international law. Firstly, the matter may be dealt with customary law in ascertaining whether acts of belligerents have erupted between the States in conflict by viewing at the aspects such as breach of peace accord, breaking up of goodwill between high diplomats, termination of diplomatic services and the use of acts of force. Secondly, it would be proper to judge any violations of statutory undertakings and obligations such as the UN Charter in preventing from use of force or other manner inconsistent with them. Thirdly, the situation is better interpreted in taking its popular meaning and ascertaining whether the ingredients are well matched. Lastly, it should facilitate basic definitional requirements such as hostile contention, employment of armed forces,

right of self-defence shall be immediately reported to the Security Council and shall not in any way affect the authority and responsibility of the Security Council under the present Charter to take at any time such action as it deems necessary in order to maintain or restore international peace and security"

845 See K. N. Waltz, *Man, the State, and War: a theoretical analysis*, Columbia Uni., Press, 2001, pp 233–235; K. J. Holsti, *The State war, and the state of war*, Cambridge Studies in International Relations, Cambridge, 1996, pp 147–149; F. Wilmer, *The Social Construction of Man, the State, and War: identity, conflict, and violence in former Yugoslavia*, Routledge, 2002

846 See *The Falkland Island – A History of the 1982 Conflict: Preface to a Conflict*, available at http://www.raf.mod.uk/falklands/preface.html

847 See *The Gulf War 1990/1991*, available at http://www.rickard.karoo.net/articles/wars_gulf1990.html; P. J. Rowe, *The Gulf War 1990–91 in International and English Law*, Sweet & Maxwell, 1993; S. Schmidt, *Gulf War II (1990/91) – Iraq between United Nations' Diplomacy and United States Policy*, GRIN Verlag, 2006

848 See *supra* n. 846

comprising of acts by or against *de jure* or *de facto* governments[849] and so forth. An actual act of war is thereby confirmed to be somewhat more than a mere 'cold war'. In using the defence of act of war, the shipowner must prove the existence of relevant factors that might be necessary under the jurisdiction of the court of hearing due to the absence of a proper interpretation in the Convention and any uniform legal interpretation in international law. Therefore, the question is highly dependant on interpretation and its requirements as accepted by the court, although actual circumstances compel the shipowner in relying on such a defence. On the part of the victim, it would be necessary to judge this point correctly so that shipowners don't get any opportunity in escaping liability in the shadow of a war, when there has been actual fault or negligence for which he might be responsible.

The defence of 'act of civil war'

The Concise Oxford English Dictionary interpretation of 'war' includes 'civil war' in its meaning by way of inserting armed conflicts between opposing parties of one state, within such definition. This is an accepted theory in law where most of the experts find no exclusions, and supporting authorities are available in many jurisdictions. The situation was once clarified in the English judgment of *Pesquerias y Secaderos de Bacalao de Espana v. Beer* in which Lord Porter held that there was no essential difference.[850] Most importantly it is to be noted that civil war comprise of internal conflict between citizens of a State having necessary ingredients of *de jure* or *de facto* government statuses more than a mere riot. In general, international law treats civil war as purely internal affair that has no effect on the provisions dealing with use of force under the UN Charter.[851] The role of international law and international relations in situations of civil war is limited to certain parameters such as acknowledgement by third States on status of conflicting parties.[852]

849 See C. L. Cochran, *De Facto and De Jure Recognition: Is there a Difference?*, The American Journal of International Law, vol. 62, no. 2 (Apr., 1968), pp. 457–460; F. Coallier, R. Gerin-Lajoie, *Open Government Architecture: The evolution of De Jure Standards, Consortium Standards and Open Source Software*, Montreal, Feb., 2006, available at http://www.cirano.qc.ca/pdf/publication/2006RP-02.pdf

850 (1945) 79 Ll. L. Rep. 417; (1947) 80 Ll. L. Rep. 318; (1948–49) 82 Ll. L. Rep. 501; [1949] 1 All E. R. 845.

851 See Article 2(4) *supra* n. 843

852 See W. L. Walker, *Recognition of Belligerency and Grant of Belligerent Rights*, Transactions of the Grotius Society, Vol. 23, Problems of Peace and War, Papers Read before the Society in the Year 1937 (1937), pp. 177–210; B. A. Boczek, *supra*

Apart from the general interpretation, it is extremely necessary in the context of this work to identify important characteristics of civil wars in ascertaining the relevant degree of acceptance. According to the marine insurance dispute of *National Oil Company of Zimbabwe and Other v. Sturge*,[853] commonly known as the *Mozambique* case, where Seville J., applied the approach of providing ordinary meaning to the term 'civil war' that comprises of special characteristic of being civil i.e. being internal rather than external. The American approach seems to be somewhat situation oriented rather than relying on basic meanings of the terms. In *Somoza* ship case[854] the court took the view that it is a self evidentiary fact that civil war existed at the relevant time in Nicaragua. However, it accepts the requirement of being a conflict between governments which are *de jure* or at least *de facto* under the Pan Am case[855], where the Confederacy exercised certain degree of territorial sovereignty. Although identification difficulties are put to the minimum in clear cut cases where the conflict is between *de jure* or *de fact* governments, complexities could arise in certain instances where it is not always easily identifiable as to such statuses. In this sense, much can be relied on the tests developed in *Spinney's (1948) Ltd., and Others v. Royal Insurance*[856] that helps in ascertaining whether a state of civil war existed. Accordingly, it should be satisfied to the court that; (a) the conflict was between opposing sides having substantial community of aim coupled with use of force, (b) possessing objectives of seizing dominion over a whole or part of the State, change a government or government's policies or pursue a goal of achieving some other community of purposes by means of a war rather than a mere tumultuous upheaval, and (c) having certain degree of effect on the public life of the inhabitants with a considerable extent of scale of the conflict irrespective of the fact that such parties hold substantial proportion of territory.

The above mentioned clarification could be used in understanding the four corners of the 'civil war' context in paving way for allowing or disallowing shipowner's authorized defence under the Bunkers Convention. As in the case of war, the court exercising its jurisdiction on matters pertaining to oil pollution damage under the relevant provisions of the Convention, must employ effective means of ascertainment to the highest degree in order to combat mischievous attempts of the shipowner in escaping liability or rather transferring it to a State or other organization.

---

n. 528, pp 104–105; L. Moir, *The Law of Internal Armed Conflict*, Cambridge (2002 rep. 2003), pp 79–80

853 [1991] 2 Lloyd's Rep. 281

854 *O.P.E Shipping Ltd., and El Povernir Shipping Company Inc. v. Allstate Insurance Company Inc. and Others*, [1983] A.M.C. 22–30

855 *Supra*, n. 839 at p. 195

856 [1980] 1 Lloyd's Rep. 406

The defence of 'act of hostilities'

The term is highly associated with both 'war' and 'civil war' situations as well as with their respective nature. As described in the discussion concerning the literal meaning of 'war', act of hostility is connected to 'hostile' action exercised by one community, State or power against another.[857] Once again much of the authorities concerning the matter in private law can be found in marine insurance cases. In the English case of *Atlantic Insurance v. King*, Bailhache J. delivered his opinion on "hostilities" as "... means hostile acts by persons acting as the agents of sovereign powers, or such organized or considerable forces as are entitled to the dignified name of rebels as contrasted with mobs or rioters, and does not cover the act of a merely private individual acting entirely on his own initiative, however hostile his action may be".[858] The American courts have followed the same attitude as is evident from *Dairy Engineering Company of Asia v. American Home Assurance Company*, where Sweigert D.J., observed that "the term 'hostilities' and 'hostile act' ... has been defined as actual operations of war, either offensive, defensive or protective by a belligerent. It has also been held that the hostile act need not involve the overt use of a weapon which is, in itself, capable of inflicting harm; it can be an operation such as the extinguishments of a navigational light or the outfitting of ship – if done for a hostile purpose".[859] As can be seen, this view provides a much wider interpretation with regard to anticipated force that need not always be some kind of weapon. According to both literal and judicial definitions, it is an accepted fact that the alleged act must have been derived from a State relationship or that of a rebellion[860]. Even though the judicial view in both English and American courts support 'act of hostility' under the insured peril within the context of 'war' or 'civil war', there doesn't seems to be a binding force of such in other civil or commercial suits, such as claims under civil liability conventions. Hence, the literal meaning does not support such idea, which would in turn suggest that matters such as the Lockerbie incident[861], where the perpetrators were directly involved with the State of Libya even though there did not exist of any 'war', carry the relevant requisites.

However, it would not be easy to conclude the above mentioned fact under the civil liability conventions pertaining to oil pollution damage including the Bunkers Convention, as an act similar to the Lockerbie incident would more likely

857 See *supra* pp 193–194
858 See [1919] 1 K.B. 309
859 See [1971] A.M.C. 1001, paras. 14.22 – 14.24
860 For further discussion see *infra* p 202–203
861 Lockerbie aerial incident where Pan Am 103 Flight *Maid of the Seas* was bombed while in air over Lockerbie, Scotland in December 1988

be categorized as a malicious act that falls within the meaning of Article 3.3 (b). Therefore, it may be strongly contested that 'act of hostilities' used in Article 3.3 (a) must be interpreted within the context of 'war' or 'civil war' as in the manner in which they are all listed together in the same section of defences. Nevertheless, its prevalence in the absence of a 'war' or 'civil war' situation would not be exclusion for admission as a defence unless the relevant basic requirements haven't been met.

The defence of 'act of insurrection'

Since the evolution of political cultures, there is no other human activity than 'insurrection' that has ever come a long way.[862] History has proved many lessons in many jurisdictions what it really implies, even in the absence of a convincible definition. Literally it means: "The action of rising in arms or open resistance against established authority or governmental restraint; an armed rising, a revolt; an incipient or limited rebellion".[863] Even though, it seems to be a straight forward explanation, there are several complications and internally hidden relationships. Ironically, insurrection is connected to two other types of human unrest, namely revolution and rebellion. Therefore, it seems necessary to analyse important characteristics of those two in understanding the broad view of the main topic. Normally in marine insurance perils on war risks, all these three components are cited together in avoiding any confusions and misjudgements. But the CLC 69 & 92, HNSC and Bunkers Convention have no clear indication of the connected aspects, rather citing 'insurrection' as a defence.

According to the Concise Oxford English Dictionary, 'revolution' is defined as "A complete overthrow of the established government in any country or state by those who were previously subject to it; a forcible substitution of a new ruler or form of government"[864] and 'rebellion' as an "Organized armed resistance to the ruler or government of one's country; insurrection, revolt; open or determined defiance of, or resistance to, any authority or controlling power"[865]. From the out-

862 See J. Hogendorn, *Review: Blaming the British*, Reviewed work(s): *Britain and Nigeria: Exploitation or Development?* by Toyin Falola, The Journal of African History, vol. 29, no. 1, Special Issue in Honour of Roland Oliver (1988), pp. 133–134; J. P. Brennan , M. B. Gordillo, *Working class protest, popular revolt, and urban insurrection in Argentina: the 1969 Cordobazo*, Journal of Social History, vol. 27, 1994; R. Tomlinson, *French Peasants in Revolt: the Insurrection of 1851*, (1982) by Ted. W. Margadant. Princeton University Press, 1979, pp. xxiv + 379

863 *Supra* n. 831, pp 737–738

864 *Supra* n. 831, p 1232

865 *Supra* n. 831, p 1198

set of these definitions, it is clear that 'insurrection' plays the primary role to its progressive stages of development; 'rebellion' and 'revolution'.[866] Saville J. in the *Mozambique* case[867] held that "'Rebellion' and 'insurrection' have somewhat similar meanings to each other. To my mind, each means an organized and violent internal uprising in a country with, as a main purpose, the object of trying to overthrow or supplant the Government of that country, although insurrection denotes a lesser degree of organisation and size than 'rebellion'". The issue has not been so common in insurance cases, but in the *Pan Am* case[868] the court has made a further step in outlining the most important requirements, for the construction of 'insurrection' in legal context, as there need to be an existence of a violent uprising by a group or movement, which is acting for the specific purpose of overthrowing the constituted government and seizing its powers.

From the above analysis it could be concluded that both 'rebellion' and 'revolution' could not arise without its primary element of 'insurrection' which would have been the reason of inserting the latter type of civil unrest alone in the Convention, as taken from its predecessors, as a defence for liability. But however, it would have been more fruitful if such other related types were included, due to the fact that national courts which hear matters under this Conventional sphere would always pursue with its own domestic definitions rather than considering non-binding authorities of other jurisdictions in the absence of any unification of the laws concerning the interpretation of these terms, however popular they may be.

Specific shortcomings

The drafters of the Bunkers Convention have failed to improve or develop this specific area of war related defence rather than copying the relevant provisions from the prevailing CLC and HNSC regimes. Since the original draft that included these types of defences, namely the CLC 69, the area under discussion has come a long way having been subjected to various sensitive developments that

866 See A. V. Cicourel, Review: D'Antonio and Pike's "Rebellion, Revolution, and Reform", Reviewed work(s): Rebellion, Revolution, and Reform: New Forces for Change in Latin America by W. V. D'Antonio; F. B. Pike, Economic Development and Cultural Change, vol. 16, no. 2, Part 1 (Jan., 1968), pp. 315–327; J. H. Badgley, J. W. Lewis, *Peasant rebellion and Communist revolution in Asia*, Stanford, 1974, pp 273–276; J. C. D. Clark, *Revolution and Rebellion: state and society in England in the seventeenth and eighteenth centuries*, Cambridge (1986 rep. 1990)

867 See *supra* n. 853

868 See *supra* n. 839

took place in the modern history. Although international liability schemes of this nature have sometimes become insolvent in addressing such issues, marine insurance regimes tend to have coped competently with the relevant matters. One of the main issues is international terrorism[869], which has fallen as a thunder storm to the greatest concern of many States at present.

In early1980s the Market underwriters and Mutual War Risks Associations paid much concern on this issue that resulted in including 'terrorism' as a peril of war risk. Although, there existed of several incidents before such period that was considered within the meaning of piracy and consequently defined as a malicious act of a third party, the characteristic feature of this ever developing menace created much difficulty in ascertaining its actual assumption. This was mainly due to the fact that such perpetrators in most circumstances use ammunition and weapons employed in a war which then falls within the meaning of war risk policy. The same is true with the liability conventions of this nature, as acts of terrorism have more tendencies in falling within the provisions of war rather than as a malicious act of a private party. However, such conclusion seem to be unrealistic and unfound, taking into consideration of the literal meaning of Article 3.3 (b) where the "motive" places a greater role. In such context, act of terrorism could well be covered under that Article, as the particular act isn't done without the intention of causing damage that lead to loss or damage experienced by States and its interests of both public and private nature against whom such acts are aimed. But however, Article 3.3 (b) could be applied only in events where damage was wholly caused by such an act, and therefore it has no implications on damages that resulted from such acts, as in the case of war and related aspects in Article 3.3 (a). Furthermore, it is hard to include acts of 'terrorism' within the meaning of acts of 'hostilities', as far as international terrorism is concerned, due to the lack of a clear definition in law for 'hostilities' and wide range of attacks that are encountered against number of States rather than one single enemy. Therefore, non-inclusion of 'acts of terrorism' in Article 3.3 (a) is simply a huge shortcoming. Furthermore, there are conquering justifications that 'acts of terrorism' should have been included together with the defence of war and related aspects, starting from its definition and extending to its consequences. It would be prudent to discuss these as a matter of justifying the author's stand and to provide guidance for future reforms.

869 See B. Jenkins, *International Terrorism – a new mode of conflict*, NCJRS, 1975; Y. Sheffi, *Supply Chain Management under the Threat of International Terrorism*, Int'l Journal of Logistics Management, vol. 12, iss. 2, 2001, pp 1–11; D. Kanev, *Seaborne trade effects of international terrorism and effectiveness of the anti-terrorist policy*, Nikola Vaptsarov Naval Academy, Varna, Bulgaria, available at http://www.iamu-edu.org/generalassembly/aga6/pdf/s2-kanev.pdf; A. E. Evans, J. F. Murphy, *Legal Aspects of International Terrorism*, American Society of Int'l Law, 1978

A 'terrorist' is "a person who uses or favours violent and intimidating methods of coercing a government or country".[870] This definition clearly proves its characteristic feature of being specific enemy of a State that needs special attention under public law rather than classifying it as a third party private individual or group having basic criminal objectives such as 'pirates'.[871] It comprises of one or more individuals who create massive incidents of harm in an indiscriminate manner in order to achieve a recognisable and well defined public cause. Most often, their acts are not specifically aimed at creating a single harm, but expand to several, whether or not their motive could be achieved. Therefore, the intention behind a malicious act under Article 3.3 (b) could sometimes be an act of pressuring a State rather than causing damage that finally end up in a tragedy of oil spill from captivity. As it is necessary to prove that such act was the sole cause rather than a proximate or dominant one which surrounds the defences in Article 3.3 (a), the defence of a malicious act might easily be turned down in such extra ordinary circumstances.

On the other hand there are compelling reasons to include act of 'terrorism' as a specific inclusion in modern liability regimes especially since the '9/11 incident'. Even prior to such incident, more than twelve multilateral treaties and protocols were adopted to combat terrorism, one of which was aimed at shipping, namely the Convention for the Suppression of Unlawful Acts against the Safety of Maritime Navigation (SUA)[872], which recognise the gravity of the problem. In such a set-up it is hard to accept the failure of the drafters of the Bunkers Convention in disregarding the inclusion of 'terrorism' as a specific defence for the shipowner that could have brought in line with the other similar acts in Article 3.3 (a). Such an inclusion would not have been mere complementary, but a welcoming step in avoiding any misunderstandings and confusions as to its assertion within other types of defences.

---

870 *Supra* n. 831, p 1489

871 See P. Hardouin, R. Weichhardt, *Terrorist fund raising through criminal activities*, Journal of Money Laundering Control, Emerald Group Publishing, vol. 9, iss. 3, 2006, pp 303–308; D. J. Puchala, *Of Pirates and Terrorists: What Experience and History Teach*, Contemporary Security Policy, available at http://www.southchinasea.org/docs/Puchala,%20Pirates%20and%20Terrorists.pdf; D. R. Dillon, *Maritime Piracy: Defining the problem*, SAIS Review – Volume 25, Number 1, Winter-Spring 2005, pp. 155–165

872 Adopted on March 10, 1988 and entered into force on March, 1, 1992. See T. A. Mensah, *The Place of the ISPS Code in the International Regime – for the Security of International Shipping*, WMU Jour. of Maritime Affairs, vol. 3, no. 1, April 2004, pp 17–30; G. Plant, *The Convention for the Suppression of Unlawful Acts against the Safety of Maritime Navigation*, Int'l and Comparative Law Quarterly, 39, 1990, pp 27–56; E. S. Tzannatos, *A decision support system for the promotion of security in shipping*, Disaster Prevention and Management, vol. 12, iss. 3, MCB UP Ltd., 2003, pp 222–229

The defence of "a natural phenomenon"

The use of this expression is a clear deviation from the classical doctrine of 'act of God' that appears in general tort law and in most legislations imposing strict liability. Although, most of such regimes afford the tortfeasor a wide spectrum of application in relying on this defence, civil liability conventions of this nature seem to have restricted its scope to a narrow sense. Undoubtedly, a narrow scope is a clear evidence of the harmful effects of the consequences, which the drafters have addressed in their minds. But however, since the first approach, they have failed to clarify with detail the types of situations that would fall within the meaning of being "exceptional, inevitable and irresistible character"[873]. It provides unfavourable outcomes for both the shipowner and victim in certain instances depending on the jurisdiction. Nevertheless, one might find such a clause as a restriction of superfluous opportunities available for the shipowner in escaping his liabilities.

Basically, 'act of God' is a common defence for the liable party in almost every jurisdiction, as far as stricter forms of liability is concerned. The OPA 90 has applied this defence in clear form providing a complete exoneration from liability, in situations where the sole cause happened to be.[874] Similarly, United Kingdom, Germany, France and other European legal systems of both common and civil law recognise the defence of act of God as a valid excuse in their general laws of tort.[875] Although its primary recognition is well accepted likewise, most of these States require the tortfeasor to prove his duty of utmost care and due diligence applied in such circumstance. Therefore, such liable party is required to prove that the existence of damage was a result of matters falling beyond its competency, control, ability and efforts coupled with entirely natural process of occurrence.[876] The French law requires such an event to be unforeseeable and unpreventable having been externally occurred. For example, French courts consider that a storm is not an act of god "because ships, by their very nature, must be built to withstand force"[877]. Accordingly, conditions of heavy weather and storms are not considered as automatic events that exempt the shipowner from

873 See Art. 3.3(a)
874 See 33 USC 2703 Sec. 1003 (a) (1)
875 See *supra* pp 191–192
876 See Koch, *Contributory Conduct or Activity* in *Principles of European Tort Law*, Springer, 2005, pp 130–137; A. Calnan, *Anomalies in International Tort Law*, Tennessee Jour., of Law and Policy, vol.1, 2005, p 187; D. Kretzmer, *Transformation of Tort Liability in the Nineteenth Century: The Visible Hand*, Oxford Jour., of Legal Studies, vol. 4, iss. 1, 1984, pp 46–87
877 DMF, 1954, p.46

presuming liability, but reserve adjudication on predictability of consequences and the gravity of ultimate results based on each individual case.[878] The German law takes a similar view requiring the liable party to prove that the damage was caused by an unavoidable event.[879] But however, the application seems to have been made narrower by way of the introduction of the needs of due diligence and utmost care on the part of the liable party. Nevertheless, in Italy it is not sufficient to prove his utmost care, but also particular circumstances that resulted,[880] while the Dutch Civil Code[881] and corresponding articles in transport law[882] requires the liable party to prove that the event was preventable if it was known to his knowledge. In England, the situation purely depends on the area of application and type of law, where in certain legislations 'act of God' is not a defence.[883] Although, the Merchant Shipping Act 1995 does not specifically allow 'act of God' as a defence, it has clearly embodied the CLC expression. However, it accords heavy weather conditions as a least factor for evading liability thus accepting 'act of God' as a defence.[884] At common law strict liability, it is recognized as a major defence in view of *Rylands v. Fletcher*, but however elaborating its boundaries as extending only up to unforeseeable and unexpected occurrence that the human prudence would not recognize the possibility of existence of damage.[885]

As it is evident from State practices in view of enacting legislations with a reasonable degree of restrictions as to the defence of 'act of God', it is reasonable to suggest that international civil liability legislations of this nature have well understood the fact in narrowing the scope of defences in order to protect the victims from possible ill defined ways of escape for shipowners, while balancing the interests of the shipping industry. Furthermore, these legislations have overlooked the exceptions granted under the marine insurance laws[886] and laws of the Charter Parties[887] in terms of "perils of the sea", which both accord 'act of God' in plain meaning. Therefore, the implications of Article 3.3 (a) expression of the defence of 'a natural phenomenon of an exceptional, inevitable and irresistible character' is a quite welcome step in the Bunkers Convention, which is intended to be applied to all ships in general. From its narrow meaning, it is understood that only

---

878 See R. Rodiere, , E. du Pontavice, *Droit Maritime*, 11th ed., Precis Dalloz, Paris, 1991, p. 287

879 See Sec. 1 subs. 2 sent. 2 HPflG; 7 subs. 2 sent. 1 StVG; 26 subs. 1 AtG

880 Art. 2053 and 2054 of the Civil Code

881 Arts. 6: 175–177

882 Arts. 6:173–174

883 E.g. Animals Act 1971

884 See M. Grey, *Why heavy weather is not like Act of God*, Lloyd's List, March 2, 1989

885 See *Nichols v. Marsland* (1876) 2 Ex. D. 1

886 For example see, Rule 7, Schedule 1, Marine Insurance Act 1906, UK

887 See for example, Clause 41 of Texacotime 2

the natural occurrence or phenomenon as it is used, become the main subject. At first hand it may include earthquakes, cyclones, tornadoes, avalanches, severe hurricanes and the likes, but would obviously exclude bad weather conditions, strong rainfalls and winds, and of course any internal occurrences whether occurred due to natural process such as decays or not. But, acceptance or rejection of such a natural phenomenon would always depend on the satisfaction of the additional criteria i.e. being 'exceptional, inevitable and irresistible' in character.

The literal meanings of these terms would provide some basic food for thought. Especially in a situation where no legal authorities could be found with respect to the whole issue, it is profitable at least to make clarifications based on available tools. First of all, the character of being 'exceptional' is not so complicated to resolve, as it means "forming an exception; unusual; not typical"[888] in mode of occurrence. Basically, such an exceptional circumstance is considered to be somewhat beyond usual happenings. But the common play with the rest of the words, namely 'inevitable', which means "unavoidable"[889] and 'irresistible' meaning "too strong or convincing to be resisted"[890] pose much difficulty to the whole issue. In considering all these together, it is possible to draw certain prerequisites that would be helpful for the shipowner in determining his burden of proof. Collectively, it focuses on issues of "exceptional situations" where they were "unavoidable" even by employing "due care" and "foresight". Firstly, what are "exceptional situations"? This could be answered only on case by case basis depending on the type of the ship, due to the fact that similar ships possess same features and their ability to cope with a given situation remains the same, while a smaller ship might find such a situation to be 'exceptional' to itself. Therefore, one natural phenomenon which does not create much difficulty for the former might cause major problems to the latter of a different class. In such a circumstance the shipowner might have to rely on abundance of evidence ranging from the dynamics of his ship to available data of such other incidents where similar ships have been subjected to. Secondly, which situation is "unavoidable"? Again, it is a matter of fact based on the situation, mechanical ability of the ship and connected to the third issue; did the shipowner employ "due care" and "foresight"? Many authors seem to discuss this issue at length, and there are much case law that could be found in both common and civil law systems. Winfield stresses the point that "The criterion is not whether or not the event could reasonably be anticipated, but whether or not human foresight and prudence could reasonably recognise the possibility of such an event".[891] In exercising "due care" and "fore-

888 *Supra* n. 831, p 496
889 *Supra* n. 831, p 1566
890 *Supra* n. 831, p 751
891 See Winfield, *supra* n. 91, p 566

sight" the shipowner must prove that his ship was seaworthy[892] at the time it set out[893] and he possess all documentary evidence to that effect and furthermore the occurrence couldn't have been avoided using his reasonable knowledge.

Even though, the OPA 90, as stated above,[894] makes a straight forward application of 'act of God' as a defence, the Act specifies its definition in Section 1001 as "... an unanticipated grave natural disaster or other natural phenomenon of an exceptional, inevitable, and irresistible character the effects of which could not have been prevented or avoided by the exercise of due care or foresight". This clarification was not intended to be illustrated during the discussion beforehand by the author due to its relationship with the issues that were elaborated progressively. Convincingly, the OPA 90 interpretation well cope with the provision of the Bunkers Convention, and the other preceding conventions, though the former, tends to elaborate the essence in a much wider and clearer perspective. Therefore, one must bear in mind that the burden of proof in relying on this defence is extremely higher than one expects, but it would certainly depend on the requirements of substantive law of the jurisdiction in which, the case for oil pollution damage under the Convention is tried.

Narrowing the possibilities

Nowadays, the courts tend to deprive shipowners from relying on this defence due to the development of sophisticated technologies installed mostly in large ships and information provided by both coastal States and other media in taking precautions. Although, smaller ships are exempted from the obligation of being equipped with new technological aids, flag States are obliged to maintain standards introduced by ISG Code and governing rules of SOLAS Convention. Use of these techniques has considerably minimized incidents that occur due to

---

892 See A.H. Kassem, *The legal aspects of seaworthiness: current law and development*, 2006, Doctoral thesis, Swansea University, Unpublished, available at http://eprints.ucl.ac.uk/6988/1/6988.pdf; Cwinya-ai, Robert Ongom, *Implied Warranty of Seaworthiness Under United States Maritime Law: The Differences and Similarities to South African Marine Insurance Law* (December 21, 2008), available at SSRN: http://ssrn.com/abstract=1319064; M. D. Guner-Oezbeck, *supra* n. 126, pp 121–122; K. J. Rawson, *The carriage of bulk oil and chemicals at sea*, Chameleon, 1994, pp 51–55; S. Hodges, *supra* n. 217, pp 124–139; B. Soyer, *Warranties in marine insurance*, 2nd ed., Cavendish, 2006, pp 50–95; W. K. Tally, *Determinants of Ship Accident Seaworthiness*, Int'l Jour., of Maritime Economics, vol. 1, 1999, pp 1–14

893 See *The Marel* case [1994] 1 Lloyd's Rep. 624

894 *Supra*, p 206

natural phenomenon and have helped the conduct of modern maritime activities more incident-free. Therefore, the chances of shipowners to rely on this defence is becoming narrower due to the abilities of ship-borne equipments and constant information provided by coastal States on conditions of sea and atmospheric conditions, which in turn rescues the victim from possible less opportunities in obtaining compensation.

Malicious act of a third party

Article 3.3 (b)[895] stipulates the second set of defence that the shipowner could rely on in escaping liability under the Convention. Both the CLC regimes and HNSC provided the same defence in favour of the shipowner. The relevant provision is applicable if the shipowner could prove that "the damage was *wholly* caused by an act or omission done with the intent to cause damage by a third party" (emphasis added). Unlike in the first set of defences, malicious act of third party need to be proved as been the sole cause for damage that leads to draw a clear demarcation line between acts of this nature against wilful misconduct or contributory negligence on the part of the shipowner. Suppose that the act or omission of a third party was secondary to the primary cause of damage, the shipowner would be deprived of relying on this defence.

Although, the relevant provision does not anywhere use the term "malicious act", many authors and jurists refer these wordings as constituting such meaning in comparison with general tort law rules on defences.[896] The most important ingredient that must be available in such context is the "intention of such party to cause damage". Therefore, one must satisfy the existence of both *mens rea* and *actus reus* components on the part of the person who carried out such acts in order to invoke this defence. Hence, the shipowner must prove that such act was not foreseeable, and in anyway took precautions in preventing the commissioning of such act which became unpreventable. In proving *mens rea* and *actus reus* that lead to the attribution of 'malicious act' on the part of the third party, it is important to describe the nature of the act and the relationship of such party to the shipowner, who indeed should be an 'outsider'. As Lord Watson once described, 'malicious' intention, means and imply "an intention to do an act which is wrongful to the detriment of another".[897] Under English common law, the defence is

895 "[t]he damage was wholly caused by an act or omission done with the intent to cause damage by a third party"
896 See Lunney, *supra* n. 822, pp 692–693; G. Stephenson, *supra* n. 688, pp 401–402
897 *Aden v. Flood* (1891) A.C. 1

limited to the "mischievous, deliberate and conscious act of a stranger",[898] and it therefore possesses a characteristic difference from that of his acts of negligence. Once the intention is combined to the commissioning of the act, it completes the whole, which result in a prejudice towards the shipowner. Such prejudice in terms of civil liability conventions for pollution damage is normally aimed to create loss or damage to the ship, its owner and /or to other objects such as properties and interests, in broad sense, which could be considered as acts of sabotage.[899] But however, it does not limit to one type of acts of perpetrators within the meaning of sabotage rather extending to acts of terrorists, as discussed in the earlier discussion,[900] acts of pirates, thieves and many more potential third parties. As O'May describes, this includes persons acting maliciously or with a political motive and may sometimes include barratrous crew members who, out of spite, or for political reasons cause the ship to spill its oil by way of causing damage to such ship to the prejudice of the shipowner.[901]

Act of piracy as a defence

Even though 'act of piracy'[902] receives a greater attention in maritime activities especially with regard to trade and commercial maritime, it does not create much danger towards bunker oil spills unless the pirates had a political motive to cause pollution damage or the capsized ship was left to steam or sink by knowing that a damage of such nature would occur. Nevertheless, the subject gathers an important recognition in marine insurance law. In 1937 'piracy' was included as a

898 See *Perry v. Kendricks Transport Ltd.*, [1956] 1 W.L.R. at p. 87

899 See Marine Environment Law in Yearbook of International Environmental Law, vol. 4 (G, Handl et al., ed.), Lloyds of London Press Ltd., 1994, para 4.23

900 See *supra* pp 203–205

901 See O'May, *Marine Insurance*, Sweet & Maxwell, London, 1993, p.313

902 L. Orlando, *'Piracy' provisions under the Enforcement Directive and patent infringement*, Jour. of Intellectual Property Law & Practice Advance Access (originally published online on August 31, 2007), available in Jour. of Intellectual Property Law & Practice, vol. 2, no. 10, 2007, pp 642–643; B. H. Dubner, *The Law of International Sea Piracy: Developments in International Law*, vol.2, p 112; Martinus Nijhoff, 1980; L. Oppenheim, Sir. R. Roxburgh, International law: a treatise, vol. 1, 3rd ed., The Law Book Exchange, 2008, pp 434–440; A. P. Higgins, *War and Law.* The Cambridge Law Journal, vol. 1 1922, pp 177–184; P. Lehr, *Violence at Sea: Piracy in the age of Global Terrorism*, Taylor and Francis, 2007; African Programme and International Law Conference Report, *Piracy and Legal Issues: Reconciling Public and Private Interests*, Oct., 1, 2009, available at http://www.chathamhouse.org.uk/files/15221_011009piracy_law.pdf

'war peril' as a result of incidents that took place during the Spanish Civil War.[903] In modern days, it is no more considered within this aspect due to sophisticated means of weapons they encounter in launching attacks against commercial ships. Since 1982, many underwriters have been introducing new clauses for 'acts of piracy' as a marine peril[904] rather leaving it within 'war perils'. For example in the United Kingdom it is treated as a marine peril,[905] while still some countries and Protection and Indemnity Clubs include it under war risks.[906]

In international law, 'piracy' is generally regarded as an offence committed by persons with acts of violence on the high seas, outside the jurisdiction of any civilized State, and without recognized authority.[907] Many legal experts have criticized this view, finding several shortcomings that do not suite its real existence. In England some judge made interpretations could be found as in the case of *re Piracy Jure Gentium*[908], where the Privy Council viewed it as "any armed violence at sea which is not a lawful act of war".[909] Furthermore, Rule for Construction No. 8 of the Marine Insurance Act 1906, which does not define the term, states that "the term 'pirates' includes passengers who mutiny and rioters who attack the ship from the shore".[910] In *Banque Monetaca and Carystuiaka and another v. Motor Union Insurance Co. Ltd.*,[911] it was held that "piracy would be ruled out if there was any political motive". However, there are descending opinions of the court with regard to the place in which 'acts of piracy' could be committed and consequently defining such acts within that term. A classic example is the case *Republic of Bolivia v. Indemnity Mutual Marine Insurance Co. Ltd.*,[912]

903 See J. W. Boulton, *The Modern International Law of Piracy: Content and Contemporary Relevance*, Int'l Relations, vol. 7, No. 6, 1983, pp 2493–2511

904 See Z. Keyuan, *New Developments in the International Law of Piracy*, Chinese Jour. of Int'l Law, vol. 8, no. 2, 2009, pp 323–345

905 *Id.* See also Clyde &Co., *Shipping Update: Piracy*, Nov., 2008, available at http://www.clydeco.com/attachments/published/3001/Piracy%20(Nov%2008)%20(2).pdf

906 See H. Smith, *What risks dominate the shipping industry from a P&I perspective?*, 5th Int'l Marine Claims Conference, The Standard, Oct., 24, 2008, available at http://www.marineclaimsconference.com/2008/docs/HunterSmith.pdf; M. M. Kroll, G. Carpenter, O. Wyman, *Global Marine Practice: Charterparty piracy clause and maritime insurance*, available at http://www.marsh.co.uk/research/documents/CharterpartyPiracy_Technicalpaper_May09.pdf

907 See K. J. Goodacre, *Marine Insurance Claims*, 3rd ed., Witherby, 1996, p. 240

908 [1934] AC 586

909 See C. Fairman, *A Note on re Piracy Jure Gentium*, The American Journal of International Law, vol. 29, No. 3 (Jul., 1935), pp. 508–512

910 It was held in *Nesbitt v. Luslington* (1792) that rioters attacking from the shore are falling within the meaning of pirates

911 (1923)

912 [1909] 1 KB 785 at 799

where the court held that an act of piracy could not be committed by strangers to the ship within territorial waters or in navigable waters within the exclusive jurisdiction of a State(s), such acts being ascribed to "thieves", while Staughton J., in *Andreas Lemos* case[913] stated that "I see no reason to limit piracy to acts outside territorial waters. In the context of an insurance policy, if a ship is in the ordinary meaning of the phrase 'at sea', or if the attack upon her can be described as a 'maritime offence', then for the business purposes of a policy of insurance she is in my judgment in a place where piracy can be committed".

Act of barratry as a defence

In maritime activities, 'acts of barratry'[914] is not an uncommon thing. The term "barratry" is defined in Rule for Construction 11 in Marine Insurance Act 1906 as "includes every wrongful act wilfully committed by the master or crew to the prejudice of the owner[915], or as the case may be, the charterer". There doesn't seem to be any restriction in the Bunkers Convention in considering barratry as falling outside the ambit of Article 3.3 (b) unlike the OPA 90 which specifically stipulates that its defence on third party acts do not constitute act or omission of an employee or agent of the responsible party.[916] Therefore, act of barratry could well serve as a defence for the shipowner under the Convention.

Commonly, an act of barratry exists in activities of smuggling where an employee(s) of the ship is involved, while in some other instances there are evidences of malicious acts committed by the master or the crew against the ship and its owner for their own purposes. Nevertheless, how could one put an employee of the shipowner into the term of 'a third party'? The answer deals with the issue of type of relationship, as mentioned above,[917] as the decisive factor in such a circumstance. Normally, the shipowner and its employees or independent contractors possess a contractual relationship arising out of a written or verbal

913 *Athens Maritime Enterprises Corp., v. Hellenic Mutual War Risks Association (Bermuda)*, [1983] All ER 590

914 See S. Hodges, R. Carlile, *supra* n. 456, pp 509–534; H. W. Woolrych, *A practical treatise on the commercial and mercantile law of England*, p 135–137; S. Hodges, *Law of Marine Insurance*, *supra* n. 217, pp 292–306; A. Annesley, *A compendium of the law of marine insurance, bottomry, insurance on lives*, pp 82–88

915 In *Piermay Shipping Co. v. A. H. Chester* [1978] 1 WLR 411, it has been held that the words "to the prejudice of the owner" meant in effect, "without his consent or privity"

916 Section 1003 (a) (3) of 33 USC 2703 states "an act or omission of a third party, other than an employee or agent of the responsible party..."

917 supra p 210–211

contract of service or contract for service[918] respectively. In virtue of this relationship, the shipowner becomes vicariously liable for any wrongful act of the latter categories.[919] But however, when such a servant is acting outside his scope of employment, his relationship with the shipowner is no longer valid within master and the servant context. This has been confirmed in the English case of *Stevens v. Woodward*,[920] where such a party was considered as a stranger. Furthermore in *Joel v. Morison*,[921] it was held that such a member of crew who does a 'frolic of his own' through the commissioning of an act of barratry, becomes a third party. Therefore in such instances, the shipowner could rely on this defence upon proving that the master or crew acted in barratry and as a result constituted the important ingredients to be termed them as third parties.

Acts of strangers and guests

Certainly, acts of strangers fall within the meaning of 'third party' and therefore intentional acts of their own in causing damage could be included in this defence. The situation with regard to guests differ slightly, where lawful guests cannot be treated as 'third parties' while uninvited guests could be treated as strangers. In the context of shipping and maritime transport, it is always not possible to distinguish these two categories easily. It involves the 'control test'[922] that ascertains

918 See H. G. Beale, W. D. Bishop, M. P. Furmston, *Contract: Cases and Materials*, 5th ed., Oxford, 2008, p 740

919 See S. Baughan, *supra* 354, pp 261–264; M. M. Kroll et. al., *supra* n. 906; M. O. Tsavaris, *Calming Troubled Waters for Cruise Ship Owners and Their Passengers: Carlisle V. Carnival Corp*, The University of Miami Inter-American Law Review, vol. 35, no. 1, Symposium Edition: Forum Non Conveniens: Development and Issues over the Past Seven Years (Fall, 2003 – Winter, 2004), pp. 153–181; C. M. Davisl, Punitive Damages in Maritime Law: Exxon Shipping Co. v. Baker; Jun., 24, 2008 available at http://davismarine.com/articles/Exxon%20Shipping%20v%20Baker.pdf; A. M. Sheppard, (modern admiralty) *supra* n. 374, pp 563–568; J. P. Jones, *The Sky Has Not Yet Fallen on Punitive Damages in Admiralty*, Tulane Law Review, March 24, 2009, Forthcoming. Available at SSRN: http://ssrn.com/abstract=1367909; E. L. Johnson, *Government, Law and Courts in the Soviet Union and Eastern Europe*. General Editors, Vladimir Gsovski and Kazimerz Grzybowski, Stevens & Sons, Ltd. Mouton & Co., N.V. 1959. 2 Vols. xxxii and 2068, 1960, *International & Comparative Law Quarterly*, 9 , pp 716–720; E. J. Weinrib, *supra* n. 687, pp 665–694

920 (1881) 6 QBD 318

921 (1934) 6 C & P 501

922 See C. Y. C. Chu, Y. Qian, *Vicarious liability under a negligence rule*, Int'l Review of Law and Economics, vol. 15, iss. 3, Sep., 1995, pp 305–322; C. A. Hopkins. *Vicarious Liability—Negligence on the Way to Work.* The Cambridge Law Journal,

whether the shipowner exercised his control over such personnel. Therefore, it is not the conduct of the person concerned which is important, but the shipowner's ability and competency to control his acts.[923] However, in relying on the defence of malicious act of a third party, the shipowner anyway has to prove much more things as stated above,[924] such as foreseeability factor, his precautionary steps and unavoidability of the incident.

The gravity of the onus

Even though the conventional regime allows the shipowner to invoke this defence upon the satisfaction of the requirements as to malicious intention and the commissioning of the act or omission, it does not go further in prescribing any relevant criteria. But in actual practice of States, the shipowner may be required by the court to prove his case beyond mere allegations. In other words, he would have to 'show cause' his due care and actions that were taken in response to avoid any wrongdoing by such a third party. Firstly, it is vital that he proves that such an act was not foreseeing. In England, it is important to prove that the act was not foreseeing, even in situations, where the defendant had a defence under the strict liability theory.[925] Many European legal systems bear the same implication, but they differentiate such acts on grounds of intentional and unintentional commissioning. In the United States, the opinion is divided, where courts in Alaska impose strict liability on the defendant without providing any opportunities for defences in cases of dangerous goods,[926] while courts in other states are reluctant to impose strict liability on the defendant in such situations. But the OPA 90 has evidently applied the foreseeability requirement through its Section 2703 (3) (A) and (B)[927] together with the requirements on exercising due care and taking pre-

48, 1989, pp 368–370; R. H. Stern, Vicarious liability for infringement, *IEEE Micro*, vol. 24, no. 5, Sep./Oct. 2004, pp. 6, 79–82; Hall, M. Isabel, *After Waterhouse: Vicarious Liability and the Tort of Institutional Abuse*, Journal of Social Welfare and Family Law, vol. 22, no. 2, 2000, pp. 159–173, available at SSRN: http://ssrn.com/abstract=894115; An Efficiency Analysis of Vicarious Liability under the Law of Agency, The Yale Law Journal, vol. 91, no. 1, Nov., 1981, pp. 168–206

923 See *Perry v. Kendricks Transport Ltd., supra*, n. 898, at 90 for a discussion on this matter under general law of tort

924 Supra 210–211

925 See *Northwestern Utilities Ltd., v. London Guarantee and Accidents Ltd*, [1936] A.C. 108

926 See T. V. van Flein, *Allocation of Fault nad Products Liability: A comment on safety products and human error*, 19 Alaska Law Rev., 141

927 "[a]n act or omission of a third party, other than an employee or agent of the responsible party or a third party whose act or omission occurs in connection with any

cautionary actions by the shipowner, in preventing malicious acts of third parties. In essence it requires the shipowner to establish by a preponderance of evidence that he;

"(A) exercised due care with respect to the oil concerned, taking…
(B) took precautions against foreseeable acts or omissions of any such third party and the foreseeable consequences of those acts or omissions;"

Generally, these requirements are essential ingredients of 'act of third party' defence in most jurisdictions' strict liability theories. For example, the French law requires the defendant to prove that the alleged act was unforeseeable and unpreventable, while in Germany unavoidable events are thus restricted to acts having external influence. Therefore, it would have been much beneficial if the Convention dealt with this aspect in including the type of care and commitment that is expected from the shipowner rather than allowing him to rely on much basic evidence in the wake of oil spill incidents. Such an approach would certainly create a vigil eye on him to take precautionary actions before some incident exists due to one or more of the above mentioned facts; i.e. act of piracy, barratry, a stranger and the likes.

State liability

Although the topic acquires a broad meaning, it is not the intention here to discuss or analyse the whole issue with regard to liabilities of States in varying contexts such as liabilities for wanton destruction of the marine environment or environmental resources during a war situation, but to confine the present scope of discussion to "*negligence and wrongful act of any government or responsible authority for the maintenance of lights and other navigational aids*"[928] which

contractual relationship with the responsible party (except where the sole contractual arrangement arises in connection with carriage by a common carrier by rail), if the responsible party establishes, by a preponderance of the evidence, that the responsible party-- (A) exercised due care with respect to the oil concerned, taking into consideration the characteristics of the oil and in light of all relevant facts and circumstances; and (B) took precautions against foreseeable acts or omissions of any such third party and the foreseeable consequences of those acts or omissions;"

928 See N. M. Hunnings and Research Officers of the British Institute of International and Comparative Law (1970), *Pollution of the Seas*. International & Comparative Law Quarterly, 19 , pp 341–343; S. Bloodworth, *Death on the High Seas: the demise of TOVALOP and CRISTAL*, Florida State Univ., Jour., of Land Use & Environmental Law, 1998

could be invoked by the shipowner as a defence.[929] According to Article 3.3 (c), the shipowner is exempted from liability if he proves that:

"the damage was wholly caused by the negligence or other wrongful act of any Government or other authority responsible for the maintenance of lights or other navigational aids in the exercise of that function".

Again the primary necessity rest on the fact that the cause being the sole cause for 'pollution damage'. Secondly, it must consist of an act of negligence or wrongfulness on the part of the coastal State. In order to establish the responsibility of the coastal State in maintaining navigational aids including lights, one must look into the origination of such an obligation on the part of such State by virtue of international public law. Historically, States have respected the idea of providing navigational aids for the safety of navigation taking place in its own respective waters.[930] The customary law as well as both the Territorial Sea Convention[931] and the Law of the Sea Convention lay this obligation on coastal States requiring them to "give appropriate publicity to any dangers to navigation, of which it has knowledge within its territorial sea".[932] This duty has been well accepted in *Corfu Channel* case[933] and has become binding on coastal States under IMO Conventions requiring them to provide basic navigational services such as lighthouses, hydrographical and meteorological information, warning and ship routing systems and rescue facilities.

Within the context of 'navigational aids' coastal States are expected to provide several types of facilities for the safety of navigation. This obligation is clearly stipulated in the Convention for the Safety of Life at Sea (SOLAS), 1974, another important milestone legal regime[934] that came up under the auspices of the IMO. According to its Chapter V, Regulation 14:

929 Though "State liability" falls as an important issue in situations where the shipowner is exonerated from liability for which a State(s) is liable, the main aim of the Bunkers Convention is to implement liability against the shipowner in circumstances where he is liable and exonerate him in such absence, and therefore it is inappropriate and irrelevant to pay much emphasis.

930 See Capt. T. Hughes (2009). *When is a VTS not a VTS?*. Journal of Navigation, 62 , pp 439–442; NILOS, *International Organizations and the Law of the Sea: Documentary Yearbook*, Kluwer, 1999, p 817

931 Convention on the Territorial Sea and the Contiguous Zone, done at Geneva, April 29, 1958 entered into force Sep., 10, 1964. Full text can be found in UN Treaty Series, vol. 516, p 205. Also available at http://untreaty.un.org/ilc/texts/instruments/english/conventions/8_1_1958_territorial_sea.pdf

932 TSC Art. 15(2) and LOSC Art. 24 (2)

933 *United Kingdom v. Albania*, I.C.J. Rep., 1949, p 4

934 See *supra* p 7

"The Contracting Governments undertake to arrange for the establishment and maintenance of such aids to navigation as, in their opinion, the volume of traffic justifies and the degree of risk required, and to arrange for information relating to these aids to be made available to all concerned".

These stipulated aids to navigation consist of varying equipments and services used for the proper conduct of ships towards an end result of safe and efficient marine activities. Such aid to navigation are two folded, namely ship-borne facilities such as equipments and installations and facilities provided from outside the ship such as lights and other amenities of coastal States' and information disseminated from other ships navigating in the same area. As the topic compels, it is the idea of this discussion to focus on the issue of navigational aids provided by the coastal States in viewing possible circumstances in which the shipowner could rely on the provided type of defence.

The obligation

As mentioned earlier, the obligation to provide navigational aid by coastal States have officially been proclaimed under the UNCLOS and SOLAS Conventions. Article 60(3) of the UNCLOS dealing with islands, installations and structures in the EEZ states that "due notice must be given of the construction of such artificial islands, installations or structures, and permanent means of giving warning of their presence must be maintained". Article 24(2) requires coastal States to give "appropriate publicity to any danger to navigation of which it has knowledge within its territorial sea". Furthermore, States bordering straits "should by agreement cooperate in the establishment and maintenance in a strait of necessary navigational and safety aids or other improvements in aid of international navigation".[935] In addition to Regulation 14 of Chapter V of SOLAS Convention, it imposes an obligation on coastal States to disseminate information relating to such aids to all concerned. In fulfilling these obligations, coastal States are required to pay much attention to coastal areas and natural access routes, wrecks and other types of danger that pose severe threats, and to sea routes within its waters. Many national laws have inserted these obligations in positive sense understanding the necessity and ultimate dangers to their own coastal areas.[936] The importance of providing navigational aids has been felt even before the adoption of internatio-

935 Art. 43

936 See A. N. Ince and E. Topuz, *Modeling and Simulation for Safe and Efficient Navigation in Narrow Waterways*. The Journal of Navigation, 57 , 2004, pp 53–71; L. Williams, *Navigational Aids*, Cavendish, 2008

nal rules, by many historic seafaring nations. For example, the French law has considered the signalling obligation as a fact of positive law in way back in 1885, where it has been held by the Administrative Supreme Court "Conseil d'Etat" in the case of *Le Borgne* that the administration is liable for inadequate maintenance of public structures in events of maritime accidents.[937] The United Kingdom has somewhat similar obligations as evident from its historic case law.[938] In present context, Section 195(1) of the Merchant Shipping Act 1995 provides that, "subject to the subsequent provisions of Part VIII of the Act and the rights of any local lighthouse authority, the general lighthouse authorities are responsible for the superintendence and management of all lighthouses, buoys and beacons within their respective areas". Furthermore in its law of harbours, the United Kingdom regards that the navigational safety functions of harbour authorities comprise dredging to maintain navigational channels, the provision of lights, buoys and beacons to mark the channels and give warning of dangers, and the removal of wrecks and other obstructions to navigation, as highly inevitable.[939]

Nowadays, the whole subject acquires a broad view due to the modern developments of technological aids that are been used by both ships and coastal States, and further been highly recommended by relevant international authorities such as IMO and regional arrangements such as the EU. These comprise not only lights and lighthouses but also radio aids as mentioned expressly in Regulation V/14 of the SOLAS Convention, meteorological aids as laid down in Article 4 of SOLAS Convention, ship weather routeing procedures and hydrographical service aids provided by IMO and IHO[940] in issuing charts and nautical publications including maritime safety information and disseminating danger messages and navigational warnings.[941]

Nature of proof

The expected onus on the part of the shipowner in relying on the defence granted by the Bunkers Convention is that, such government or governmental authority

---

937 See, G. J. de Matons, *La responsabilité de l'Etat pour le fonctionnement de certains services maritimes*

938 See for example, 17 QBD 795 (1886)

939 For a comprehensive discussion on this see, Douglas and Green on *The Law of Harbours, Coasts and Pilotage*, 5th ed., Informa, 1997

940 International Hydrographic Organization. See Home Page, http://www.iho-ohi.net/english/home/

941 See J. Kasum, *Updating Sea Charts and Navigational Publications*, The Journal of Navigation, vol. 56, 2003, pp 497–505

which was responsible for providing such navigational aids acted in negligence or wrongfulness in causing the damage. A recent litigation concerning oil spill which applied this exception was the case of *Tsesis*[942] in 1977. The Supreme Court of Sweden held that the shipowner is at a better position in applying this defence in an event where there have been defects and faults in hydrographical charts provided by the coastal State.[943] In this case the court has held that 'a marine chart' falls within the meaning of navigation aid under the CLC Article III.2(c). However, the decision was not unanimous as three judges dissented citing the term 'maintenance' as meaning physical and technical up-keeping rather than a mere failure to up-date marine charts. Since this controversial judgement, there have been numerous views and debates on this issue attempting to search positive acceptances and rejections within the meanings of 'maintenance' and 'navigational aids'.

One must always bear in mind that this defence could only be applied by courts when it can be proved that there existed of a fault on the part of the responsible authority in the exercise of the maintenance of lights or other navigational aids. Therefore, a mere absence of navigational aids would not pave way for the shipowner in relying on this defence.[944] Most importantly it must be proved that maintenance to prevailing navigational aids has not been met. But of course the difficult question rests on whether the failure to modify charts is to be regarded as lack of maintenance within this meaning. On the other hand, one could argue that charts are not navigational aids which require constant physical maintenance and upkeep such as lights. But however, coastal States have a duty to update charts and other navigational aids that are important for ships navigating in their respective waters, so that the failure on the part of such States should not deprive shipowners from transferring liability. There have been several occasions where courts have considered the issue in ascertaining types of navigational aids that fall within the provided meaning. In *Jose Marti* case[945], the Swedish Supreme Court has dismissed 'pilotage services' as navigational aid[946] while in *Nissos Amorgos* case[947], the court in Venezuela accepted the system of bulletins to ships informing of the current depth of a dredged channel as navigational aids[948], within the

942 *Tsesis* grounded on October 26, 1977 in the Stockholm archipelago

943 For the facts of case see, Colin de la Rue, *supra*, n. 316 pp. 89–90, and for an analysis of this decision see, H. Tiberg, *Oil Pollution of the Sea and the Swedish Tsesis decision*, 1984 LMCLQ, pp. 218–226

944 See *The Irish Stardust* case [1977] 1 Lloyd's Rep. 195

945 1981

946 See IOPC Fund Annual Report, 1987, p 16

947 1997

948 See, 71 FUND/EXC.54/INF.3 and 71 FUND/EXC.54/10, para 3.1

meaning of the CLC provisions, which is the same as in the Bunkers Convention. Therefore, it is clear that the interpretation seems to depend somewhat on the understanding of the court of relevant services, and would certainly be based upon respective national law in circumstances where such provisions prevail. Nevertheless, it would be intelligent to suggest that a court must consider all types of aids that were feasible to the coastal State at the time of the incident in question and determine whether proper improvements have been carried out by such State within the broad meaning of 'maintenance', notwithstanding the common literal meaning of this term alone. In such a context, it would have been much beneficial for the outcome of the Conventional goals, if the drafters inserted the term 'maintenance and improvement' rather than leaving the present effect in force.

Proportionate Liability

As mentioned at the beginning of this Chapter, the shipowner gains a fourth set of defences, based on conduct of the victim. His component of exoneration varies from zero to one hundred percent depending on the proof of his case. This particular defence acquires a complete separate Article which deals with the subject that recognises its thorough importance in the field of marine pollution damage. According to Article 3.4;

"If the shipowner proves that the pollution damage resulted wholly or partially either from an act or omission done with intent to cause damage by the person who suffered the damage or from the negligent of that person, the shipowner may be exonerated wholly or partially from liability to such person".

Here again, the word 'resulted' appears in the context implying the 'causation' to be 'dominant or proximate' rather than being the 'sole'. Furthermore, it *prima facie* suggests that the shipowner would still be able to use this defence in failure to achieve an outcome in favour of him in cases where State liability is involved under Article 3.3 (c). In other words, he will be able to support an argument based on negligence or wrongfulness of the coastal State that is responsible to carry out its required obligations that were discussed under that particular Article, where such an act or omission was not a 'sole cause' for damage.[949] On the other hand, the shipowner would still be able to use this defence in failing to use Article 3.3 (c) on the ground that there existed some negligent act of his own or his employee(s) such as the master, in avoiding damage, which has been further suppor-

949 *Supra* pp 217–219

ted by negligent or wrongful act of such State.[950] In the present context, the exoneration would deeply depend on the contribution of the victim in ascertaining proper components of liability between the liable party and the victim. There are two types of conduct, on the part of the victim, involved inside the meaning of this Article. They are in essence, malicious conduct and contributory negligence committed by the victim. These two grounds are discussed in length in the following paragraphs in order to have a broad idea of their scope of application.

Malicious conduct of the Victim

The successful use of this particular defence in achieving allowed exoneration needs to prove "an act or omission done with the intent to cause damage by the person who suffered the damage". This indeed means malicious conduct of the victim that resulted in creating pollution damage. As illustrated earlier with regard to 'malicious acts of third parties' in the foregoing Chapter,[951] it is worthless to repeat the discussion about important characteristics that form the basis for such acts or omissions. Most importantly the shipowner must prove the existence of both mental and physical components on the part of the victim in carrying out the ultimate outcome. However, his physical appearance at the incident is irrelevant as such acts could well be coordinated upon delegation. Anyway, these types of acts or omissions are very remote, unlike third party involvements. But however, it would not be always possible and prudent to turn down such arguments as even in rare circumstances, such incidents may still appear.

In any event, malicious conduct of the victim forms a clear defence for the liable party in most respected jurisdictions. Apportionment of liability components in such circumstance depend basically on respective national law, either according to specific related regime or general tort law regime. Furthermore, the characteristic feature of the ultimate result and seriousness of the damage sometimes fall as basis or decisive factors for such apportionment in most of those jurisdictions. Likewise in the case of contributory negligence, the victim would be considered as a co-offender[952] and tried under the general rules of tort liability while holding convictions to voluntary assumption of risk to himself or leading it to *volenti non fit injuria.*

950 For further discussion see, Colin de la Rue, *supra*, n. 316, pp. 89–90
951 *Supra*, p 210–211
952 See *infra*, p 227

Contributory negligence

The doctrine of 'contributory negligence' appears in almost all civil and common law systems, internationally.[953] In English Common Law the term is traditionally used as 'contributory negligence' while in continental European systems it is known as 'contributory fault'.[954] Provisions containing this aspect especially on apportionment of liability can be found in either statues such as civil codes or in case law. For example, the French Code Civil does not contain any provisions dealing with the subject, but many law reports of the Cour de Cassation prove the availability of this doctrine.[955] The German law on the other hand does include explicit provisions with regard to the matter as evident from its BGB Section 254 paragraph 1 which states that;

> "If a fault of the injured party has contributed to the occurrence of the damage, the duty to compensate and the extent of the compensation to be made depend upon the circumstances, especially upon how far the injury has been caused predominantly by one of the other party".

The use of the term "*Mitverschulden*" in the German text is quite ambiguous that it contains number of tortfeasors who are jointly liable for the damage. Therefore, international lawyers seem to be unsatisfied in using the German interpretation, and favours the English term of 'contributory negligence' in first place. But however, the English common law did not posses provisions for proportionate fault in respect of contributory negligence in actions of tort, allowing it to be a complete defence for the tortfeasor. As Winfield and Jolowicz state, at common law, the victim can recover nothing, if it could be proved that his own negligence contributed partly towards the damage.[956] Nevertheless, traditional maritime law in the UK allowed the apportionment of damage[957] by virtue of Maritime Con-

953 R. C. McCullough, *Doctrine of Contributory Negligence and Comparative Negligence*, South Carolina Trial Lawyer Transcript, 1989; C. S. Sangco, *Philippine Law on Torts and Damages, Volume 1*, 1993, pp 37–47; *Contributory Negligence. Doctrine of Last Clear Chance Retained by Comparative Negligence Statute*, Harvard Law Review, vol. 52, no. 7 (May, 1939), pp 1187–1188; J. C. Macintosh (1922). Contributory Negligence. The Cambridge Law Journal, 1 , pp 185–191

954 For a comprehensive study see, W. van Gerven , *Cases, Materials and Text on National, Supranational and International Tort Law*, Hart, 2000

955 See, *Bardeche v. Jonier* (Bull. Civ. 1987. II. 86) D 1988. Jur 32, in which it was held, that a keeper of a thing under Art. 1384 (1) C. civ., (other than a vehicle) shall be exonerated in part if he or she proves that the victim's fault contributed to the occurrence of the damage caused by that thing.

956 See Winfield, *supra* n. 91, p. 247

957 See Case Law Analysis, *Waste and Liability in Environmental Law*, vol. 14, iss. 1, Oxford, 2002, pp 75–84

ventions Act 1911.[958] The English law was later modified in this respect via The Law Reform (Contributory Negligence) Act 1945[959], which replaced its status as a complete defence to a defence that reduces the damages to the extent of the contribution of the victim. Its Section 1(1) states that;

"Where any person suffers damage as the result partly of his own fault and partly of the fault of any other person or persons, a claim in respect of that damage shall not be defeated by reason of the fault of the person suffering the damage, but the damages recoverable in respect thereof shall be reduced to such extent as the court thinks just and equitable having regard to the claimant's share in the responsibility for the damage".

The existence

With regard to its application under the relevant provision of the Bunkers Convention, the shipowner can allege contributory negligence on the victim for two basic types of occasions, namely his contribution to the initial damage and failure to prevent or limit further damage being occurred. This is a situation which corresponds to an inter-play of 'contributory negligence' and 'personal misconduct'. In both these cases, the victim is treated as a second offender. Circumstances may even lead to the formation of 'multiple' contributory negligence on the part of the victim, depending on the facts of the case. Once the defendant pleads contributory negligence on the part of the victim as a defence, he must prove such existence to the credibility of his arguments.

Proving contributory negligence

As Brazier[960] put forward the requirements for the establishment of contributory negligence on the part of the victim under English Law, it is obviously common in any circumstance that the tortfeasor must prove that; (a) the negligence of the victim exposed himself to the injury, (b) such negligence contributed to his injury and (c) there was fault or negligence on the part of the victim.[961] Although it provides a basic thought for judgement, a further elaboration of important elements such as causation, violation of a duty of care and determination of its standard are indeed necessary.

958 Under Section 12, it authorized the court in collision cases to apportion the loss based on the degree of fault of each party. See also *Admiralty Commissioners v. SS Volute* [1922] 1 AC 129 HL

959 For text see, http://www.opsi.gov.uk/RevisedStatutes/Acts/ukpga/1945/cukpga_19450028_en_1

960 J. Murphy, H. Street, *supra* n. 117

961 *Id.*, p. 238

Causation

The most important element in this respect is not the contributory act towards the occurrence, but the victim's conduct that contributed to his damage. Therefore, there should exist of a positive link between the contributory act of the victim and the resulted damage upon which he relies on the issue of compensation. In establishing such causation, one should refer to the requirements of causation under the specific legal regime, which was discussed in the earlier Chapter.[962]

Duty of care

One may wonder why it is necessary to consider this issue, as the victim does not bare any duty of care on the part of the tortfeasor. It is extremely correct to think in that direction. But most importantly, what is necessary in this respect is that the victim has failed to take reasonable care of his own safety in avoiding damage. This opinion was convincingly upheld in a famous case in England by Viscount Simon,[963] and is applied as a rule against able persons.[964] In most jurisdictions this would result in a reduction of compensation, due to the application of 'the mirror principle' that places the victim on the opposite side and treat him as, being injured by an act of, someone else who appears in the mirror image.[965]

Standard of care

Depending on the circumstance and the law pertaining to the relevant event, the expected standard of care that rests on the conduct of the victim varies.[966] As Winfield and Jolowicz point out rightly "the law certainly does not require the claimant to proceed on his way like a timorous fugitive constantly looking over his shoulder for threats from others".[967] Lord Denning once said that "a person is guilty of contributory negligence if he ought reasonably to have foreseen that, if he did not act as a reasonable, prudent man, he might be hurt himself and in his

962 *Supra* pp 146–158

963 *Nance v. British Columbia Electric Ry Co., Ltd.*, [1951] AC 601 at 611, per Viscount Simon, [1951] 2 All ER 448

964 In cases of children, disables and old persons the rule is not applied in its entirety. There are certain exempted circumstances where it is not used depending on the ability of the victim. But however, it is not important to discuss that in detail as such incidents are very rare in oil pollution damage cases

965 See D. Chabas, *1973 Chron.*, pp 207–210

966 See V. Harpwood, *supra* n. 651, p 461; M. Lunney, K. Oliphant, *supra* n. 822, pp 190–191

967 *Supra* n. 91, p. 254

reckonings he must take into account the possibility of others being careless".[968] In other words the victim is expected to execute due diligence in avoiding damage caused by his own acts and further avail himself from any occurrence that helps enlarges the damage.

Apportionment of damage

Article 3.4 of the Bunkers Convention does not deal with the term 'apportionment' of liability, but rather 'exoneration' of the shipowner from liability. Hence, it does not specifically address the matter of division of such exoneration leaving it to be decided by national courts. Therefore, in such a context, the respective national courts would always employ its general rules applied in tort law remedies for apportionment of liability in cases of contributory negligence. As mentioned earlier, apportionment would most often depend on the seriousness and gravity of the damage and factual contribution of the victim's negligent act or omission. In many jurisdictions each party would bear half of the liability, in circumstances where each other's contribution cannot be determined due to insufficient evidence.[969] In clear cases, courts are not reluctant to apportion liability based on the facts and evidence ranging from least values to higher numbers.[970]

Interestingly, The Law Reform (Contributory Negligence) Act 1945 of the United Kingdom directs courts to assess the reduction of the award of damages in such events on grounds of 'just and equitable having regard to the claimant's share in responsibility for the damage'.[971] What is 'just and equitable' is left for the determination of the court, which would most certainly rely on principles of common law and decisions of the judiciary. In *Stapley v. Gypsum Mines Ltd.*,[972] Lord Reid has drawn a method for pursuing with the notion rather than providing a concrete interpretation. He goes onto state that "... what is just and equitable, must have regard to the blameworthiness of each party, but the claimant's share in the responsibility for the damage cannot, I think, be assessed without considering the relative importance of the acts in causing the damage apart his blameworthiness". In essence, he leaves the issue on the act of contributory negligence of the victim and the standard of care expected from him. In some other cases, judges

968 See *Jones v. Livox Quarries Ltd.*, [1952] 2 Q.B. 608 at 615

969 For example see Section 1304 of ABGB for Germany and Articles 2056, 1227 and 2055 III of the Civil Code of Italy for Italy.

970 Different countries adopt different modes, like Germany for instance uses percentage system while France and Belgium use fractions.

971 Sec. 1(1)

972 [1953] AC 663, at 681

have simply apportioned reductions in numbers rather than attempting to provide justifications based on the notion of 'just and equitable'.[973]

In *Aegean Sea* litigation it was found that the state-employed pilot of the coastal State, Spain, was at fault by way of contributory negligence and the ship-owner who was indeed liable was able to obtain an exoneration amounting to 50% of the established claim.[974] Furthermore, in *Patmos*[975] litigation reduction was made for damage to the marine environment on the ground that the Italian Messina Port Authority acted in contributory negligence over improper use of dispersants in preventing pollution damage.[976] This proved that the defence is not only associated with the initial damage but also with consequential damage that occur afterwards and negligence of authorities in taking preventive measures or failing to take such measures with due care. One must make sure that in the above cases the IOPC Fund always played a major role in determining the apportionment of such nature irrespective of specific clauses or figures, but under the Bunkers Convention, it would solely be dependant on respective national regime.

Specific problems

Although, almost all jurisdictions allow apportionment of liability in cases of contributory negligence even under strict liability regimes, it is unclear whether courts would turn down any reduction on the basis of proved intentional misbehaviour of the tortfeasor committing the tortuous act in other words where it could be proved 'actual fault or privity' on his part even in the presence of contributory

973 For example, in *Froom v. Butcher* [(1976) QB 286], Lord Denning has ordered a reduction of 25% for the failure to wear seat belt and consequently failing to prevent injury plus another 15% for failing to reduce the severity of injuries as a result of such, while in *Gregory v. Kelly* [(1978) RTR 426], 40% reduction was made for the failure to wear seat belts coupled with the knowledge of the faulty footbrake

974 Decision of June 18, 1997. *Sed quaere*

975 Incident occurred on March 21, 1985, when the *Patmos* and the *Castillo de Montearagon* collided in the Straits of Messina off Sicily in Italy. The Patmos was carrying 828,300 barrels of Kirkuk crude oil while the Castillo de Montearagon was in ballast. See NOAA/ Hazardous Materials Response and Assessment Division, Seattle, Washington, *Oil Spill Case Histories: Summaries of Significant US and International Spills*, available at http://response.restoration.noaa.gov/book_shelf/26_spilldb.pdf, pp 145–146; D. Wilkinson, *Moving the boundaries of compensable environmental damage caused by marine oil spills: The effect of two new international protocols*, Jour., of Environmental Law, vol. 5, iss. 1, 1993, pp 71–90

976 See Decision of the Court of Appeal, January 1994, and see also IOPC Fund Annual Report 1994, pp. 36 and 38. See also P. Sands, Principles of International Environmental Law, 2nd ed., Cambridge, 2003, p 919

negligence still on the part of the victim. Especially in a regime such as the Bunkers Convention where the shipowner is strictly liable without any opportunity for excuse other than the provided defences, this question is open for discussion. In the proper opinion, the shipowner should not be provided with any reduction, except in circumstances where the intentional oil discharge has taken place for the safety of the ship or in such other distress and to the minimum justifiable limit. However, the shipowner would still have an opportunity to rely on the defence under Article 3.4 on circumstances where a coastal State in particular has failed to grant assistance to the ship at the time of need or in an emergency and furthermore such a State has failed to limit the extent of damage by way of applying proper preventive measures. In addition, the wording of the relevant Article which states that "…, the shipowner *may* be exonerated …" does provide a wider margin of appreciation for the court in deciding this issue.

Capter 7: Limitation of Liability

Since the introduction of strict liability in maritime legislations at both domestic and international levels, legislators have opted for implementing limitation regimes[977] separating the concept from the doctrine of absolute liability[978] and affording the liable parties, the concessions of limited liability with a defined extent. There are many reasons for justification of this idea that can be outlined dating since the adoption of classical rules on liability. Especially in the field of maritime law, there were divulging propositions based on many aspects which have been addressed by law-makers ranging from the idea of protecting private maritime interests, insurance concerns to reasonableness of remedies. But however, limitation of liability is neither a bar to fair justice nor a sword for the shipowner. Lord Denning once described his attitude in the case of *Bramley Moore*[979] that, "limitation of liability is not a matter of justice. It is a rule of public policy which has its origin in history and its justification".

Classical Approaches

The most difficult issue in solving limitation figures and adopting a certain formula for such calculation became the central point on which many States in the past paid much attention. The most classical doctrine that was developed based the limits on the value of the ship after casualty, upon which the question of limitation of liability actually arose. Based on this aspect, States adopted their

977 For detail discussion see A. Rein, *International Variations on Concepts of Limitation of Liability*, 53 Tul. L. Rev., 1259 (1978–1979); A.H.E. Popp, *Limitation of Liability in Maritime Law – An Assessment of Its Viability from a Canadian Perspective,* 24 J. Mar. L. & Com. 335 (1993); D. Dion, *Canadian Approach to Limiting the Liability of pleasure craft owners*, 24 J. Mar. L. & Com. 561 (1993); G. Gauci, *supra* n. 130

978 See E. C. German, *Seller Beware – Strict Liability not Absolute Liability*, 37 Ins. Counsel J. 44 (1970); P. Salvador-Coderch, N. Garoupa, C. Gomez-Liguerre, *Scope of Liability: The vanishing distinction between negligence and strict liability*, European Jour., of Law and Economics, Springer, vol. 28, no. 3, Dec., 2009, pp 257–287; P. Zweifel, J. R. Tyran, *Environment Impairment Liability as an Instrument of Environmental Policy*, Ecological Economics, vol. 11, iss. 1, Sep., 1994, pp 43–56

979 [1964] 2 Lloyds' Rep. p. 200

respective limitation regimes in their own domestic legal structures. In general, there existed of two main systems that were used by two main groups of States, namely, the 'execution system' practiced by Germany and Scandinavian States and the 'abandonment system' practiced by France and the United States. Under the former, claims against the incident were thus restricted to the value of the ship and the freight after a casualty without having any force on other personal assets of the liable party. In the meantime, the liable party in turn possessed the priority right of recovery for any maritime liens[980] he owes against such value. Under the latter, the liable party was personally liable for limitable claims subject to the abandonment of the ship and freight to the claimant in order to recover the damages. In addition, both these systems allowed for the distribution of amounts based on the system of priority rules for maritime liens[981] and in accordance with the ratio to the aggregated amount calculated up to the institution of proceedings leading to limitation.

In contrary, England had its own system for limitation of liability that was developed way back during the 18th and 19th century. Even though, the English law recognized a system of limitation based on the value of the ship and freight, such value was considered to be the value before the casualty. Its legislations adopted between 1850–60 expressly indicated such values[982] that turned out to be a modern innovation of legislations dealing with the subject based on fixed monetary figures, which later became an acceptable platform for the adoption of international rules. In England, this development was later incorporated into its classical maritime legislation, namely, the Merchant Shipping Act 1894.

Concerns over the issue in the Legal Committee

During the initial drafting period of the Bunkers Convention, many delegations spoke on resolving the issues concerning the provisions for limitation of liability

980 See N. B. Richards, *Maritime Liens in Tort, General Average, and Salvage*, 47 Tul. L. Rev., 569 (1972–1973)

981 See *Priorities of Maritime Liens*, Harvard Law Rev., vol. 69, no. 3 (Jan., 1956), pp 525–528; I. d'Almeida Pires-Filho, *Priority of Maritime Liens in the Western Hemisphere: How Secure Is Your Claim?*, The University of Miami Inter-American Law Review, Vol. 16, No. 3 (Spring, 1985), pp. 505–537; D. Yang, *A Comparative Analysis of Maritime Lien Priority under United States and Chinese Maritime Law*, 23 Tul. Mar. L. J. 465 (1998–1999); F. G. Harmon, *Discharge and Waiver of Maritime Liens*, 47 Tul. L. Rev., 786, Apr., 1973

982 These values were figured out as 8 Sterling Pounds per ton for normal ships, and 15 Sterling Pounds per ton for passenger ships out of which 7 Sterling Pounds per ton to be awarded for personal claims.

with the view of inserting the best possibility. Throughout the development of the field of commercial maritime law, the subject has gained tremendous momentum from time to time and from one specific field to the other. Especially, in similar preceded conventions to the Bunkers Convention, such as the CLC and HNSC, [983] specific limitation regimes have been introduced without referring to the prevailing common international regimes on limitation of maritime claims.[984] At present stage of drafting, the delegations were not convincingly of the view of introducing such an innovation due to various problems that might encounter[985] the smooth prevalence and growth of this particular regime and in particular, the problems concerning the universal acceptance. Therefore, the proposed draft that was tabled initially did not carry such complications. Article 5 which dealt with limitation of liability was extremely short in context stating that;

> "the shipowner shall be entitled to limit liability in accordance with the applicable international convention or the national law of the State the courts of which have jurisdiction in accordance with article 8, paragraph 5".[986]

The connected Article 8.5 stated that;

> "After a limitation fund has been constituted by the shipowner or other person providing security for the shipowner's liability, the courts of the State in which such fund is constituted shall have exclusive jurisdiction to determine all matters relating to the apportionment and distribution of the fund".[987]

In its entirety, Article 5 did not clearly point out the relevant international regimes it has referred to. Basically, the international regime on limitation of liability covers number of conventions that have come into existence since the adoption of the most classical version in 1924[988]. Furthermore, various States have become

---

983 See Article V of CLC 69/92 and Article 9 of HNSC

984 See *CLC, CLC 1992, 1992 Oil Pollution Convention, 1992 Civil Liability Convention* in *International Documents on Environmental Liability* (H. Descamps et. al., ed.), Springer, 2008, pp 41–53; T. J. Schoenbaum (1998), *Oil Pollution at Sea.* By Ganci Potthard [Chichester: John Wiley 1997, xxx + 313 pp. ISBN 0–471–97066–2. No price given], International & Comparative Law Quarterly, 47 , pp 966–966; K. J. Rawson, *The carriage of bulk oil and chemicals at sea*, Inst.., of Chemical Engineers, 1994, pp 28–31; J. Wren, *Overview of the Compensation and Liability Regimes under the International Oil Pollution Compensation Fund (IOPC)*, Spill Science & Technology Bulletin, vol.6, iss. 1, Feb., 2000, pp 45–58

985 See Wu Chao, *Liability and Compensation for Bunker Pollution*, 33 J. Mar. L. & Com. 553 (2002)

986 *Supra*, n. 51

987 *Id.*, at p. 6

988 Int'l Convention for the Unification of Certain Rules relating to the Limitation of Liability of the Owners of Sea-going Vessels, signed at Brussels on August 25, 1924

parties to different international conventions[989] that have come up in relation to limitation of liability, and have designed their respective national regimes in accordance with such international regime to which such State is a party. Therefore, the situation would have become much complicated if the proposed draft was left for adoption without clarifications and modifications, for which most States could not have consented. In response, the briefing of the Comite Maritime International (CMI), the institution responsible for the drafting of the first ever limitation convention in 1924, made a significant contribution by addressing important areas of concern over the issue of relevant laws and other aspects.[990] In virtue of such concerns, relevant issues needed further observation and clarification to identify the best possible solution.

1957 Brussels Limitation Convention[991]

Due to the failure of the 1924 Convention, which was acceded by only 15 countries, the revision of the international regime was greatly welcome by the international community of maritime States. During 1960–70s, as many as 40 States including most European States, Australia, Canada, Japan and India became State Parties to the new Convention and inserted these provisions in their respective national regimes. However, United States, Italy and most of the South American States abstained from entering into the new regime, while the Soviet Union and Brazil, which have inserted 1924 Convention provisions in their respective domestic regimes, remained outside the application of this Convention. Although, the new Convention made significant changes and adopted important rules concerning limitation, it was not a complete success.[992] As far as bunker spills and the relevant provisions of the then proposed Draft Bunkers Convention are concerned, the 1957 Convention had many shortcomings. Apart from the failure to obtain universal recognition, the Convention encounter many problems

989 For example, Int'l Con., relating to the Limitation of Liability of Owners of Seagoing Ships, done at Brussels on Oct., 10, 1957, Conv., on Limitation of Liability for Maritime Claims, done at London on Nov., 19, 1976, which replaced and abrogated the 1924 and 1957 Conventions and 1996 Protocol to Amend the London Limitation Convention 1976, done at London on May 02, 1996

990 *Supra*, n. 142, p. 3

991 1957 Convention Relating to the Limitation of the Liability of Owners of Seagoing Ships

992 See W. W. Eyer, *Shipowners' Limitation of Liability. New Directions for an Old Doctrine*, Stanford Law Review, vol. 16, no. 2 (Mar., 1964), pp. 370–393; P. J. S. Griggs, *Obstacles to Uniformity of Maritime Law*, The Nicholas J. Healy Lecture, available at http://www.comitemaritime.org/year/pdfiles/obst_uniform.pdf

including one of the major problems of failing to specify in clear language, types of damages that could be subjected to limitation. This resulted in different States interpreting it in different ways under their respective national regimes, that sometimes led to confusions between the meanings of 'damages' and 'debts' as in the case of the United Kingdom, where the shipowner's right to limit did not extend to damages beyond stipulated amounts, under the then prevailed amended Merchant Shipping Act of 1894.[993] Leaving such confusion would leave the shipowner without the right to limit for claims arising out of clean-up and preventive measures, especially in a regime intended to deal with bunker spills.

Furthermore, issues arising out of economic loss, pure economic loss and environmental damage would enlarge the problem to much greater extent, suppose that the 1957 Convention is to be applied for limitation of liability claims under the Bunkers Convention. In addition, obtaining insurance under the requirement of compulsory insurance would become a hard obstacle for the shipowners in events of some damages or rather recurring and extended damages and losses are not so covered under a limitation regime. Finally, the need for a new regime seem to have been driven forward by concerns over the emergence of super-tankers in the late 1960s coupled with potential environmental consequences of casualties created by such ships for which adequate level of compensation was not in presence.

Insurance as a Basic Element

As far as the shipping trade is concerned, the shipowner is normally required to undertake the risks for the dangerous activity to which he steps-in. The risk factor is generally laid on the hands of the insurance provider,[994] without whom, the shipowner is unable to carry out his performances in normal circumstances due to his own convictions and as matter of satisfying statutory requirements. As Wet-

993 See Section 505. See also *The Stonedale (No.1)* [1956] AC 1

994 See G. Ceccarelli, *supra* n. 830; M. G. Kohn, *Risk Instruments in the Medieval and Early Modern Economy*, Dartmouth College, Department of Economics Working Paper No. 99–07, February 1999., available at SSRN: http://ssrn.com/abstract=151871; P. Bennett, *Environmental governance and private actors: enrolling insurers in international maritime regulation*, Political Geography, vol. 19, iss. 7, Sep., 2000, pp 875–899; T. Nast, *Weathering the Storm: How the Lessons of History will enable the Marine Insurance Industry to survive the current economic downturn*, available at http://works.bepress.com/cgi/viewcontent.cgi?article=1000&context=timothy_nast; P. Bennett, *Mutuality at a distance ? Risk and regulation in marine insurance clubs*, Environment & Planning A 2000, vol. 32, pp 147–163, available at http://www.envplan.com/epa/fulltext/a32/a3215.pdf

terstein points out, the wide spread use of insurance and particularly third party insurance substantially reduces any possible risk for shipowners.[995] Therefore, insurance has become an integral part of maritime activities, especially with regard to channelling of liability of the shipowner. Most importantly, liability rules must afford the shipowner with the right to limit, which could be passed on to the insurance provider. In the present day of a volatile insurance market, no insurer would ever be willing to grant coverage for liabilities to which no such limitation exists. It is correct to say that the insurance industry relies on the principles of limitation of liability.[996] As Abecasis describe, no insurer would ever be willing to give unlimited coverage especially for oil pollution damage.[997] In addition, the insurer expects to enjoy the same rights as afforded towards the assured, in avoiding any less-convenient position into which the insurer could fall.

Instruments such as CLC, HNSC and Bunkers Convention, which enacts a system based on strict liability, need to afford a favourable regime on limitation in order to attract insurance providers towards the supply of proper policies. Reluctance of such insurers would result in a total failure of the entire efforts. Satisfaction of these ends have enabled shipowners in obtaining insurance cover, and in modern days, P & I Clubs have come forward in facilitating such objectives, especially covering most of the areas of risks involved in shipping including ship bunkers, due to the development of the concepts of limitation that are afforded to liability insurers as well. They are nowadays, afforded with most of the rights and privileges enjoyed by shipowners together with the rights of defence[998], offsetting and limitation of liability[999]. This is an important element that the liability insurer is required to possess in facing litigations based on direct actions[1000] against insurers.

995 *Damage from International Disasters in the Light of Tort and Insurance Law*, 1990, p 100

996 See L. Burglass, *Limitation of Liability from a Marine Insurance Viewpoint*, Tul. Law Preview, vol. 53, 1979, p. 1364

997 See *supra*, n. 92, p. 507

998 See Art. 7.10 of the Bunkers Conv., which states that, "... [I]n such a case the defendant may invoke the defences (other than bankruptcy or winding up of the shipowner) which the shipowner would have been entitled to invoke, including limitation pursuant to article 6."

999 See Art. 7.10 of the Bunkers Conv., which states that, "...[F]urthermore, even if the shipowner is not entitled to limitation of liability according to article 6, the defendant may limit liability to an amount equal to the amount of the insurance or other financial security required to be maintained in accordance with paragraph 1."

1000 See Art. 7.10 of the Bunkers Conv., which states that, "...[A]ny claim for compensation for pollution damage may be brought directly against the insurer or other person providing financial security for the registered owner's liability for pollution damage."

Apart from the idea of affording right to limit as a matter of protecting the trade as well as to obtain insurance under the Bunkers Convention, the Legal Committee was expected to arrive at a certain conclusion on setting up amounts of limitation for pollution damage caused by bunker spills. The matter was surrounded by many issues such as the need for compulsory insurance for ships over a certain threshold,[1001] uncertainty as to the values of such insurance due to the absence of agreed limitation amounts and overlapping limitation funds that could result in incidents involving bunker oil spill from a ship carrying HNS and at the same time releasing of its cargo including oil from the particular ship, as observed by the Australian delegation[1002]. As pointed out at the Legal Committee meeting by way of a note by the International Chamber of Shipping, the latter situation would complicate the whole process due to the fact that such an incident would end up in constituting three limitation funds, i.e. for Bunkers Convention, CLC and HNSC, which would be an extreme 'potentially expensive exercise and an inefficient use of insurance capacity'.[1003] These remarks not only deprived the enactment of the requirement of another separate set of limitation figures under the Bunkers Convention, but also turned down the demands of delegations who spoke in favour of such an enactment.

The Final Settlement

The Legal Committee at its 77th Session arrived at a general agreement that the limitation amounts must be tied to the global limitation regime, namely the Convention on Limitation of Liability for Maritime Claims (LLMC) 1976,[1004] and drops the idea of setting up free standing limits for the proposed Bunkers Convention as in the case of the CLC and HNSC.[1005] Thereafter, the attention was directed towards another aspect due to the proposal made by the delegation of the Government of Australia, namely, whether to apply the limitation provisions

1001 What later became Art. 7.1 stating, "[T]he registered owner of a ship having a gross tonnage greater than 1000 registered in a State Party shall be required to maintain insurance or other financial security, such as the guarantee of a bank or similar financial institution,..."

1002 *Supra* n. 49

1003 *Supra* n. 58, p. 4

1004 That states in Art. 6, "[i]nternational regime, such as the Convention on Limitation of Liability for Maritime Claims, 1976,..." as well as in Art. 7.1, "[i]nternational limitation regime, but in all cases, not exceeding an amount calculated in accordance with the Convention on Limitation of Liability for Maritime Claims, 1976,..."

1005 *Supra* n. 139, p. 22

of the LLMC 1976 as amended by the LLMC Protocol 1996 or inserting the limits of the LLMC Protocol 1996 in the provisions of the Bunkers Convention.[1006] Although, the new limits under the '96 Protocol would work well in the Bunkers Convention, most delegations were unwilling to include such instrument due to the fact that it was not in force at the particular time.[1007] Therefore, many delegations favoured the position of retaining LLMC 1976, as amended, together with the supplement of invoking respective national legislation with regard to amounts of limitation as may be applicable.[1008] As the '96 Protocol would serve the purposes of the Bunkers Convention, especially with regard to higher liability limits, seven delegations including the United Kingdom introduced a draft conference resolution[1009] urging the States to become Party to the said newly amended instrument of 1996, and thereby denounce the Conventions of 1924, 1957 and 1976, as appropriate, giving effect to the LLMC Protocol 1996 to come into force.[1010] Thereafter the Conference adopted the Resolution on Limitation of Liability, one of the three Resolutions it adopted under the entire work towards the adoption of the Final Act, and made part and parcel of the Bunkers Convention.[1011]

The Effect and Application of the Relevant Provision

As it stands at present under the Convention, Article 6 expressly states that;

"Nothing in this Convention shall affect the right of the shipowner and the persons providing insurance or other financial security to limit liability under *any applicable national or international regime*, such as *the Convention on Limitation of Liability for Maritime Claims, 1976, as amended*". (Emphasis added).

Accordingly, the Convention affords the liable party to limit liability based on the relevant law that is applied in the jurisdiction in which proceedings take place. In such context, one may identify five different types of regimes that are prevailing universally. Firstly, there may be States that apply the LLMC 1976 Convention[1012] with its pure effects, secondly, States which apply the said Convention with certain modifications and reservations prescribed by the law of such State to the extent they don't contradict with the Convention, thirdly, States that apply LLMC

1006 *Supra*, n. 380, p. 17 at para 116
1007 LLMC Protocol 1996 came into force on the 13th May 2004
1008 *Supra*, n. 185, p.17 at para 121
1009 See LEG 81/4/3, 21st February 2000
1010 *Supra* n. 407, p. 5 at para 18
1011 See LEG 83/11, 6th September 2001, p. 2, para 7
1012 As at Nov., 30, 2009, 52 States have ratified the Convention having a percentage of 49.94% of the total global tonnage

1976 provisions modified by LLMC 1996 limits, fourthly, States which apply LLMC 1996 limits alone[1013] and lastly, States which are not parties to either of these limitation Conventions and purely apply applicable national legislations.

From the outset of Article 6 of the Bunkers Convention, it is clear that the shipowner, liability insurer and financial security provider possess the right to limit in any proceedings instituted against such persons or entities under the relevant provisions of the Bunkers Convention. Furthermore, they have the option of utilizing this right under the applicable national law or the international law, which is in force in a given jurisdiction, in which they are eligible to invoke such right, under the relevant law.

Applicable Law

Suppose that the country, in which the limitation is invoked according to the requirements of the particular law, is not a party to the LLMC 1976, then the limitation proceedings will be based on the applicable law of that State to the extent it affords any limitation to such actions. Although many States apply the *lex fori* to such limitation proceedings, one might encounter situations where a choice of law still exists. In such circumstance, the courts will afford the defendant to invoke either the *lex fori* or the applicable law of the State in which he holds nationality pursuant to the acceptance of private international law rules and to the extent such conflict of laws is respected under such jurisdiction, pursuant to the Rome Convention[1014]. Suppose that such State does not respect the application of foreign law in such context, *lex fori* would then prevail as the applicable national law.[1015]

Where such country is already a Party to the Limitation Convention, as described, its limits will have the necessary force of law. The relevant wording of

1013 As at Nov., 30, 2008, 37 States have ratified the 1996 Protocol having a percentage of 42.12% of the total global tonnage

1014 1980; in force from April 1, 1991

1015 See T. M. Bos (2007). P. van der Grinten; T. Heukels, eds., *Crossing Borders: Essays in European and Private International Law in Honour of Frans van der Velden*, Kluwer, Deventer 2006, xvi + 262 pp., ISBN 90–13-03040-8.. Netherlands International Law Review, 54 , pp 168–173; M. R. Badykov, *The Russian Civil Code and the Rome Convention: Applicable Law in the Absence of Choice by the Parties*, Jour., of Private Int'l Law, vol. 1, no. 2, Hart, Oct., 2005, pp 269–289; A. V. Y. Ong, *Issues in the Application of Depecage in Chinese Private International Law*, Chinese Jour., of Int'l Law, vol. 8, no. 3, 2009, Oxford, pp 637–655; J. O'Brian, R. Smith, *Conflict of Laws*, 2nd ed., Cavendish, 1999, pp 351–360; W. Tetley, *New Development in Private International Law: Tolofson v. Jensen and Ganon v. Lucas*, The American Jour., of Comparative Law, vol. 44, no. 4, Autumn 1996, pp 647–668

the ongoing provision would pave way for the application of the limits under the LLMC Protocol 1996, to any proceedings that take place in relation to limitation, once such Protocol comes into force in the relevant jurisdiction, which has so ratified.

In essence, it would be up to the court in which the limitation is invoked, to apply the relevant law which applies to that jurisdiction, either, the national limits or the international limits, and thereafter distribute the amounts of compensation accordingly. As it is more advantageous for a State to fix higher limits based on its own national law or international regime, if it is a Party or has not denounced the Convention after been ratified, States must utilize the best suitable option.[1016] However, having higher limits would make the circumstances much difficult with regard to the shipping interests of such State, especially with regard to obtaining and maintaining insurance, and in particular may jeopardise its whole financial structure. Therefore, as urged by the Conference itself, and for the smooth prevalence of the maritime liability system in the world, it would be prudent on the part of all States to come under one international limitation regime, such as the LLMC Protocol 1996, as early as practicable.[1017] This would enhance and promote a unified and harmonized system of limitation of liability that would suit and encourage all parties concerned, such as the victims, shipowners, insurers and financial security providers.

Limitation of Liability under LLMC 1976

The 1976 Convention replaced the 1957 Convention as a whole by introducing several new provisions together with higher amounts of liability limits expressed in terms of units of accounts[1018], replacing the then-prevailed system based on Poincaré gold francs. Based on this regime, many States Parties have inserted the relevant provisions by giving entirely the same effects in their respective national legislations[1019] while some have made reservations as allowed by the Convention

1016 See R. Price, *The Carriage of Hazardous Cargo by Sea: A UAE Law Perspective*, Arab Law Quarterly, vol. 10, no. 4, 1995, pp. 310–330; A. J. Rodriguez, P. A. C. Jaffe, *The Oil Pollution Act of 1990*, 15 Tul. Mar. L. J. (1990–1991)

1017 For further illustration see L. de la Fayette, *Achievements in 2001*, Environmental Policy and Law, vol. 32, no. 3–4, IOS Press, 2002, pp 146–151

1018 System based on Special Drawing Rights (SDR) as defined by the International Monetary Fund (IMF)

1019 For example, see H. Karan, *Revising Liability Limits in International Maritime Conventions: A Turkish Perspective*, 34 J. Mar. L. & Com. 615; H. Kim, *Korean Maritime Law Update: 2007 – Focused on the Revised Maritime Law Section in the Korean Commercial Code*, Jour. of Maritime Law and Commerce, vol. 39, no.3,

itself and certain extra modifications. Therefore, it is important to refer to the national law on limitation with regard to the subject, before invoking the right of limitation under such jurisdiction. The ongoing discussion would prove such necessity and analyse such situations under various headings as it may warrant. In the following, it is the idea to discuss important aspects that need to be unveiled under the topic of the application of the LLMC 1976 in order to ascertain the true sense of applying that instrument within the meaning of limitation of liability as prescribed under the Bunkers Convention. It is therefore important to note that only the matters arising out of the Bunkers Convention are taken into discussion in realising the scope of such applicable law on limitation of liability together with special emphasis on prevailing situations of several States which have not acceded to any international Conventions on limitation, as mentioned above.

Scope of Application I

According to Chapter IV, Article 15;

> "This Convention shall apply whenever *any person referred to in Article 1* seeks to limit his liability before the Court of a State Party or seeks to procure the release of a ship or other property or the discharge of any security given within the jurisdiction of any such State. ..." (emphasis added)

Under the relevant provisions, it is obvious that only the persons who are referred to in Article 1 are eligible to invoke the right to limitation within such application. Therefore, the foremost attention is drawn towards analysing and identifying such eligible persons who fall within such category.

Persons Entitled to Limit Liability

(a) Shipowners

Article 1 of Chapter I categorically express the eligibility of persons and entities, although the following discussion only focuses on its relevance to the Bunkers Convention, as the main Convention in question, for which such application is intended to be applied. According to Article 1.1;

2008, Available at SSRN: http://ssrn.com/abstract=1284676; J. Hare, *Limitation of Liability – A Nigerian Perspective*, available at http://web.uct.ac.za/depts/shiplaw/fulltext/harepapers/limliab-nigeria.pdf

"Shipowners and ..., as hereinafter defined, may limit their liability in accordance with the rules of this Convention ..."[1020]

Article 1.2 states that;

"The term "shipowner" shall mean the owner, charterer, manager and operator of a sea-going ship".

The right to invoke limitation is clearly vested on the shipowner as required under the meaning of the Bunkers Convention.[1021] Furthermore, the latter's definition suits the list of persons described in Article 1.1 of this Convention in a much easier and straight forward manner.[1022] However, the meanings of the relevant persons and entities are undefined and not further illustrated in clear sense. Therefore, one may use the meanings provided by the Bunkers Convention and further illustrations supplied in the discussion with regard to such meanings in the relevant context, once limitation proceedings are instituted with reference to bunker spills and resulting pollution damage.[1023] The only difference is that the LLMC 1976 refer to a much bold and broader term of 'charterer' unlike the term 'bareboat charter' used in the Bunkers Convention. This would include all categories of charterers as specified in the earlier discussion concerning the application of 'responsible party'[1024] within the meaning of the Bunkers Convention, but it is far clear that 'bareboat charterers' are indeed entitled to limit liability, as they fall within such meaning, and furthermore the issue here relates to right to limit concerning liable parties who fall within the ambit of the Bunkers Convention. As the said Convention clearly stipulates that "no claim for compensation for pollution damage shall be made against the shipowner otherwise than in accordance with this Convention"[1025], and that "the shipowner at the time of an incident shall be liable for pollution damage caused by bunker oil on board or originating from the ship..."[1026] referring the 'shipowner' as the "owner, including the registered owner, bareboat charterer, manager and operator of the ship"[1027], the width of the conventional meaning, affords a fair opportunity under the LLMC 1976 for all such responsible parties concerned. In short, 'shipowners', as defined, who may

1020 States in its entirety that "Shipowners and salvors, as hereinafter defined, may limit their liability in accordance with the rules of this Convention for claims set out in Article 2"

1021 See *supra* pp 84–85

1022 See Art. 1.3 of the Bunkers Convention

1023 See *supra* pp 85–101

1024 See *supra* pp 94–97

1025 Art. 3.5

1026 Art. 3.1

1027 Art. 1.3

become liable under the Bunkers Convention are fully entitled to invoke right to limitation under the LLMC 1976.

Most jurisdictions that have embodied the LLMC 1976 have expressly included the above mentioned persons in their respective domestic law on limitation for marine claims, but in some States, further clarifications and additions to the prevailing texts can be identified. For example, in Japan, it is expressly stated that such right is extended to 'members with unlimited liability, in case of a corporation'.[1028] The Dutch Commercial Code uses a different approach by stating the terms 'hirers and any users of a ship' in addition to owners and charterers, [1029] which undoubtedly include bareboat charterers and managers. Spain as a State Party to the LLMC 1976 interprets the 'shipowner' as 'operator and disponent owner', which include a quite broad perspective. However, it lacks a clear meaning for 'disponent owner' in its Ports and Merchant Marine Act 1992[1030], which deals with the subject of limitation, and therefore has to depend on interpretations available in case law, which thus mean to include managing agents and charterers.[1031]

As with regard to States which are not Parties to the LLMC 1976, and therefore apply respective national laws on limitation, the situation is quite interesting to analyse. In the United States, the OPA 1990 accords the right to limit as a general rule for the 'responsible party' described in Section 2704.[1032] Under the Russian Merchant Shipping Code 1968, the 'shipowner' is entitled to such right[1033], as the 'shipowner' means "a person who runs in his own name as owner or on any other legal basis"[1034], thus enabling the qualification of being the proprietor of the ship, persons who are actually in possession of the ship on the basis of business ownership or operational management, a time-charterer or bareboat charterer under a contractual obligation, manager on trust and etc.[1035] However, as the Russian Federation has now acceded to the Protocol '96, the situation would not in anyway contradict. The South African Merchant Shipping Act, 1951, which is modelled on the English Act defines the 'shipowner' within the meaning embodied by the Bunkers Convention,[1036] and does not pose much difficulty.

1028 Art. 2 para 1 sub para 2 and 3, Art. 3 para 1 and Art. 98 of the Law concerning Limitation of Liability of Shipowners

1029 See Sec. 740

1030 Came into force on 24th November 1992

1031 See Green, M. Fernando on Spain in *Limitation of Liability for Maritime Claims* (P. Griggs, R. Williams eds.), 3rd ed., LLP, 1998, pp. 311–312

1032 See US Code Collection, Title 33, Ch. 40, Sub-Ch. 1

1033 See Art. 274

1034 See Art. 10

1035 See A. Makovskai on Russia in *Limitation of Liability for Maritime Claims* (P. Griggs and R. Williams eds.), 3rd ed., LLP, 1998, p. 281

1036 See Sec. 263(2)

Apart from the general rule concerning the characteristic feature of the construction of the term 'shipowner' who may limit accordingly under the Convention, it also enables such person to invoke rights on limitation in circumstances where actions are brought *in rem*.[1037] Under its Article 1.5, such right is so conferred by its provision which states that;

"In this Convention the liability of a shipowner shall include liability in an action brought against the ship itself".

Therefore, it corresponds to actions that could be brought under both *in personam* and *in rem* procedures for claims for compensation, for damage caused by bunker oil spills.

(b) Liability Insurers and Financial Security Providers

As provided by Article 6 of the Bunkers Convention, the right to limit extends to two further categories, namely 'the person or persons providing insurance or other financial security'.[1038] Such right is duly granted by the LLMC 1976 as well to such liability insurers.[1039] Even though, it does not mention anything concerning its application towards financial security providers, there are convincible reasons to uphold their inclusion within this meaning. According to Article 7.1 of the Bunkers Convention the requirements of having compulsory insurance is delegated towards the fulfilment of maintaining insurance to that effect or prove such competency by way of a financial security obtained in way of a guarantee issued by a bank or similar financial institution.[1040] Furthermore, it has also mentioned such value of the liability cover to be an amount equal to the limits of liability accepted under the applicable national law or the LLMC 1976, as amended. The entire pro-

1037 See S. C. Derrington, J. M. Turner, *The Law and Practice of Admiralty Matters*, Oxford, Ch. 10, available at www.gbv.de/dms/spk/sbb/recht/toc/527702471.pdf; A. Dandal, P. Gillies, Characterizing the Action in rem in Australia and the Implications on International Commercial Arbitration, 40 J. Mar. L. & Com. 271, Apr., 2009; S. Sime, A Practical Approach to Civil Procedure, 10th ed., Oxford, 2007, p 148; G. Gauci, supra, n. 130; D. M. Collins, Comments on the American Rule of in rem Liability, 10 Tul. Mar. L. J. 71, Spring 1985

1038 See *supra* pp 93–95

1039 See Art. 1.6 which states that "[A]n insurer of liability for claims subject to limitation in accordance with the rules of this Convention shall be entitled to the benefits of this Convention to the same extent as the assured himself."

1040 Stating "...[s]uch as the guarantee of a bank or similar financial institution, to cover the liability of the registered owner for pollution damage in an amount equal to the limits of liability..."

vision leads to the fact that the financial security provider is a substitute for the liability insurer in case where insurance could not be obtained by the shipowner for some or other reasons, and thereby posses the same rights and obligations of such liability insurer due to such absence. It is understood that the financial security provider act on behalf of the obligations that normally rest on the hands of the liability insurer, and therefore, should be vested with the same rights that are enjoyed by the insurer. On the other hand the CLC 92 affords the same rights to both the insurer and financial security provider in relation to rights of limitation.[1041] Most importantly, one must treat financial security providers within the meaning of liability insurers and to the effect that they are merely substitutes for such liability insurers who do not appear in a given circumstance in any financial undertaking for associated risks. In other words, the financial security provider undertakes the same risks that are expected to be covered by the liability insurer and under that context he can be considered as the insurer, although it lacks certain legal implications, such as important characteristics of insurance contracts and so forth, which leaves the former in a much different position. Nevertheless, it would be prudent on the part of all States to make necessary amendments in their respective national legislations on limitation of liability in order to overcome possible confusions, by including the term 'financial security providers' together with 'liability insurers'.

The LLMC 1976 extends the right of limitation to liability insurers by its Article 1.6, which states;

"An insurer of liability … shall be entitled to the benefits of this Convention to the same extent as the assured himself".[1042]

Being a quite new provision which did not appear in the former version of 1957, this enables the insurer to limit his liability in direct actions instituted against him. Most importantly, his right extends only to the 'same extent' as the assured himself. It places him in no worse situation than the assured, and adversely, looses it in events where the assured is deprived of the invocation of such right. Almost all States provide this opportunity to liability insurers in order to proceed successfully with claims in direct actions.

(c) Exempted Persons

Article 15.1 of the LLMC 1976 further provides an option on States Parties to wholly or partially exclude shipowners who do not have their habitual residence or

1041 See Art. V.11
1042 See *supra* n. 1039

principal place of business in a State Party,[1043] at the time when the conventional rules on limitation are invoked. In essence, it gives the States Parties the power to legislate such provisions in their respective national laws so as to give force of law for the application or non-application against non-States Parties to LLMC 1976. Most importantly, it would be necessary to check the relevant law of a State Party in which the proceedings take place, whether or not such exemptions have been made.

In exercising such option, many States Parties have indeed included provisions with respect to such matters in their respective national laws. For example, Section 352.1[1044] of the Indian Merchant Shipping Act 1958, expressly include foreign shipowners, irrespective of their origin, within the meaning of 'the owner of a seagoing vessel', if such ship lies within India, including its territorial waters.[1045] Therefore, India has not excluded the application of the rules of the Convention to shipowners of non-States Parties, unless such ship rests outside its territorial sovereignty. On the other hand, the English Merchant Shipping Act 1995 has not made any reservations based on this provision, and therefore, provide the right of limitation to persons falling within the ambit of Article 15.1 of the LLMC. Likewise, many States Parties have abstained from making such reservations such as Japan[1046] in particular.[1047]

Scope of Application II

The right of limitation afforded to shipowners, within its definition, under the LLMC 1976 could only be enjoyed by 'shipowner' of a 'seagoing ship'.[1048] There-

---

1043 Which states that "...each State Party may exclude wholly or partially from the application of this Convention any person referred to in Article 1 who at the time when the rules of this Convention are invoked before the Courts of that State does not have his habitual residence in a State Party or does not have his principal place of business in a State Party or any ship in relation to which the right of limitation is invoked or whose release is sought and which does not at the time specified above fly the flag of a State Party."

1044 Which states that "Limitation of liability of owner for damage.- (*1*) The owner of a ship, whether an Indian ship or not, shall not, if any loss of life or personal injury to any person, or any loss of or damage to any property or rights of any kind, whether movable or immovable is caused without his actual fault or privity,-..."

1045 No significant change has been effected by the amended Merchant Shipping (Amendment) Act 2007 relevant to the prevailing provision

1046 Acceded on 04.06.1982, and entered into force on 01.12.1986

1047 For the list of States Parties to the 1976 Convention, their accession and entering into force see http://www.ecolex.org/ecolex/ledge/view/RecordDetails;jsessionid=813C42F0F5707D9A343DF010BBEE4A1E?id=TRE-000117&index=treaties

1048 See Art. 1.2

fore, the application is restricted to a certain category of ships instead of a general and wide application.

Category of Ships – Inclusions and Exclusions

Here again, the Convention has failed to provide any specific definition for the term 'seagoing ship' despite providing exclusions of some ships from its application. The Convention has expressly excluded (a) air-cushion vehicles and (b) floating platforms constructed for the purpose of exploring or exploiting the natural resources of the sea-bed or the subsoil, by its Article 15.5[1049], and ships which are falling within the meaning of the CLC[1050] together with nuclear ships[1051] for which different regimes apply. In addition, Article 15.4 excludes ships which are constructed for, or adopted to, and engaged in drilling: when there exists of a higher limit of liability, than that provided for in Article 6 of the Convention, under respective national law of such State Party or when such State Party has entered into other international convention regulating the system of liability for such category of ships.[1052]

Furthermore, it is vested on the States Parties with the option of regulating the provisions of the Convention, through respective national law, for ships intended for navigation in inland waterways[1053] and for ships which are less than 300 tons of weight[1054]. In addition, a State Party may also exclude wholly or partially, any ship which does not fly the flag of a State Party to the LLMC 1976, at the time such limitation rules are invoked,[1055] likewise in the case of shipowners, which was discussed earlier.[1056] Therefore, with such exclusions, the scope of application has turned out to be limited, but a careful observation would prove that the definition of 'ship' under the Bunkers Convention complies with the relevant category as expressed, with the exception of seagoing ships of less than 300 tons, hovercrafts and other vessels used for drilling and any other purposes as specified, to the extent they fall within the ambit of the Bunkers Convention, which indeed qualify within the meaning of 'any seagoing vessel and seaborne craft, of any type whatsoever' as provided by the Bunkers Convention.[1057]

1049 "[T]his Convention shall not apply to: (a) air-cushion vehicles; (b) floating platforms constructed for the purpose of exploring or exploiting the natural resources of the sea-bed or the subsoil thereof"
1050 See Art. 3 (b)
1051 See Art. 3 (c) and (d)
1052 See Arts. 15.4 (a) and (b)
1053 See Art. 15.2 (a)
1054 See Art 15.2 (b)
1055 See Art. 15.1
1056 *Supra* pp 249–250
1057 See *supra*, pp. 39–44

Most importantly, one must bear in mind that the Bunkers Convention applies to ships of such defined category which use bunker oil for its operation or propulsion, and such category therefore need to have been included in the LLMC 1976 for invoking limitation of liability under that particular instrument in legal proceedings commenced under the Bunkers Convention. Suppose that a certain type of ships does not fall within the scope of application of the LLMC 1976 then it is necessary to check for the relevant applicable national or international law[1058], if any, dealing with limitation for such particular ship.

Using the conventional ruling regarding this matter, many States have legislated in their respective national laws different types of coverage under which conventional system of limitation could be applied. For example, the Canadian Shipping Act by virtue of Section 576(3) extends such right to ships whether they are seagoing or not, while the Greek Code of Private Maritime Law[1059] excludes ships that navigate in inland waters[1060] as seagoing, but recognises floating rigs over certain capacity, which are used for exploration and exploitation of natural oil and gas, as seagoing ships[1061]. On the other hand many States Parties have made special provisions allowing the application of right of limitation under their respective national laws. In Canada, Section 578 of the Canadian Shipping Act provides a domestic limitation regime for ships of under 300 tons, while in the Peoples Republic of China there seem to be a different regime for ships between 20–300 tons, whereas the French law does not distinguish between different types of vessels, but applies a different regime for vessels engaged in drilling. The interesting point is that, platforms are expressly excluded from the scope of application of the LLMC 1976, and there doesn't seem to exist of any applicable rules to the limitation of liability for such crafts, as long as they are unqualified to be regarded as ships. [1062] The situation is somewhat similar in the United Kingdom, where the definition of 'ship' under the Merchant Shipping Act 1995 does not include 'floating platforms'[1063], thereby leading to the conclusion that the liability is unlimited. However, the subject of hovercraft is covered by the

---

1058 Suppose that a certain State is a party to another Convention, such as dealing with ships engaged in drilling as stated in the LLMC 1976, and thereby rely on limitation regime described in such instrument

1059 See T. Karatzas, "*The Greek Code of Private Maritime Law*" Kluwer, 1982. See also S. N. Vlachos, "*Greece: New Code of Private Maritime Law*", The American Journal of Comparative Law, vol. 7, No. 4 (Autumn 1958), pp 599–601; D. P. Christodoulou, "*Introduction to the Greek Legal System*" available at http://jurist.law.pitt.edu/world/greececor2.htm

1060 Article 1(1) stipulates that a ship is a vessel of not less than a registered net tonnage of 10 destined to navigate at sea by own means.

1061 See Article 1 of Legislative Decree 117/1974, Govt. Gazette A310/23.10.1974

1062 See J. Rohart on France in P. Griggs and R. Williams, "*Limitation of Liability for Maritime Claims*", 3rd Edition, LLP, 1998, p. 177

1063 See Paragraph 12 of Schedule 7, Part II

Hovercraft (Civil Liability) Order 1986 that makes specific provisions on limitation of liability.[1064] On the other hand, the OPA 1990 provides a wide coverage within its meaning of 'vessel' as described in Chapter 2 of this work.

Scope of Application III

In order to apply conventional regime on limitation of liability in the court of a State Party it is necessary to identify the types of claims covered under its rules. In doing so, one must employ a careful observation on Article 2 which deals with "claims subject to limitation" together with exempted claims that are described in Article 3 and conduct of the liable party that might turn out to be a bar for invoking the right to limitation as described in Article 4.

Claims Subject to Limitation

With respect to loss and damage that can be caused by bunker oil spills[1065] for which claims for compensation against the shipowner could lie in a court having proper jurisdiction[1066], it would be possible to categorise such relevant claims as applicable under the provided scope of the LLMC 1976, as follows;

(a) claims in respect of loss of life or personal injury or loss or damage to property, occurring in direct connection with the operation of the ship, and consequential loss resulting therefrom;[1067]
(b) claims in respect of loss resulting from infringement of rights other than contractual rights, occurring in direct connection with the operation of the ship or salvage operations;[1068]
(c) claims in respect of the raising, removal, destruction or the rendering harmless of a ship which is sunk, wrecked, stranded or abandoned;[1069] or
(d) claims of a person other than the person liable in respect of measures taken in order to avert or minimize loss for which the person liable may limit his liability in accordance with this Convention, and further loss caused by such measures.[1070]

1064 Later amended by Ord. 1987
1065 See *supra*, pp. 54–62
1066 See *supra*, Ch. 4
1067 See Art. 2.1 (a)
1068 See Art. 2.1 (c)
1069 See Art. 2.1 (d)
1070 See Art. 2.1 (f)

In addition, claims which are subjected to limitation do not depend on the nature of liability as it is clearly stated in Article 2.1 by way of the phrase "whatever the basis of liability may be"[1071]. This infers all types of liabilities including strict liability, which is the base of the Bunkers Convention.

Most importantly, loss or damage under the Bunkers Convention must first satisfy the relevant requirements as stipulated within the term 'pollution damage' in order to qualify for claims of compensation. As discussed in Chapter 3 with regard to its meaning and types of damage that would fall within such meaning as discussed in Chapter 6, are extremely relevant and form part and parcel to the present discussion. In order to ascertain the types of loss and damage covered under the on-going provision of the LLMC 1976 for which the shipowner could limit his liability, an in-depth study is highly necessary. Therefore, it is the intention at present stage to analyse the 'types of claims covered' in order to achieve clarity in context.

(A) Claims under Article 2.1(a)

According to the relevant provision, the shipowner is entitled to limit liability with respect to four main types of claims, namely;

(1) loss of life;
(2) personal injury;
(3) loss of damage to property; and
(4) consequential loss resulting from above three categories.

All these types of claims could arise with respect to bunker oil spills, but the first is considerably remote, while the second possess a certain probability, but is of course remains very low. Majority of the claims would fall under the third category, while the fourth would have a considerable degree of occurrence, especially connected to the third category of claims. According to the continuation of the on-going provision, the first three types of claims must have occurred as a result of its direct connection with the operation of the ship. Therefore, loss or damage occurred due to any subsequent measures aftermath an incident is clearly excluded from its ambit. This leads to the fact that the fourth category relates only to consequential loss which occur due to the existence of the former three that have direct connection to the ship's operation.[1072]

1071 It states "[S]ubject to Articles 3 and 4 the following claims, whatever the basis of liability may be, shall be subject to limitation of liability:"

1072 Some States in their national law dealing with limitation of liability refer to the phrase "in the navigation or management" instead of using the LLMC 1976 word-

All claims that would be forwarded with respect to the above categories of loss or damage occurred due to the main requirement of been contaminated with bunker oil, is therefore well covered under the relevant provision. Even though the wordings are straight forward, one may encounter several unclear and sensitive situations upon an in-depth observation. Importantly, one has to identify what types of proprietary rights fall within the meaning of 'property' and whether national law of respective jurisdiction would respect the idea of 'marine environment' to be within such meaning.[1073] Further and much broader confusions would constitute around the fourth category of claims, which does not provide a clear meaning. Although it refers to "consequential loss resulting therefrom", the idea is somewhat ambiguous. Does it means any consequential physical loss or covers a wide spectrum including that of economic and much debated pure economic loss? Furthermore, to what extent would it expand or rather what type of connection should there be between the principal loss or damage and the consequential loss, especially concerning the remoteness aspect?

In pure meaning, the bold term of 'consequential loss' extends to physical as well as economical loss resulting as a matter of loss or damage caused to the principal object. These would naturally include, costs of repairs, replacement costs and loss of revenue with regard to property damages, while loss of income, financial loss occurring from inability to be engaged in one's employment and loss sustained by the dependants in cases of loss of life or personal injury. With respect to the extent of such loss, it must not encompass the doctrine of 'just and equitable' as prescribed in most international and national legislations. As far as damage to marine environment is concerned, the Bunkers Convention impose a limit to such extent as 'reasonable measures of reinstatement actually undertaken or to be undertaken', as a condition for claims for consequential loss. Therefore, the interpretation of consequential loss would always depend on the decisions of the court in respective jurisdiction.

While some jurisdictions expressly allow pure economic loss claims and other associated costs such as litigation costs, interests on expenditure and the like to be legitimate claims within the meaning of consequential loss, the courts must always bear in mind that such claims must be directly connected to the operation of the ship and to the relevant loss or damage caused to one or more of the above

---

ing. This is used mainly by LLMC 1976 non-ratified States. Until Singapore acceded to the Convention its position remained as outlined before, due to the influence of the classical Merchant Shipping Act, 1894 which had the same phrase. In substance, there doesn't exists of any differences with regard to the application of the Bunkers Convention, as such meaning includes also acts or omissions connected with breach of contract.

1073 See *supra* pp 140–141

first three categories, and especially that the principal damage or loss is a result of contamination in qualifying for compensation under the Bunkers Convention, and only such claims are subject to limitation as provided by Article 2.1(a) of the LLMC 1976 or according to the requirements laid down by the applicable law of that State. For example, the Merchant Shipping Act 1995 in particular has made specific provisions with regard to non-tankers for bunker spills, referring liability for damage only to damage to property for which limitation could be sought under Article 2.1(a) of the LLMC 1976.[1074] This however covers consequential damage connected to such damage to property within that meaning, which would include pure economic damage to the extent it could be recovered under the English Law. On the other hand, civil law countries which do not distinguish between consequential loss and pure economic loss would certainly allow limitation as far as the claims of such nature could fall within the meaning of Article 2.1(a).

(B) Claims under Article 2.1(c)

The expression of Article 2.1(c) establishes few important elements connected with the types of claims mentioned therein. It refers to claims with regard to infringement of rights occurring in direct connection with the operation of the ship as in the earlier case, but expressly excludes infringement of rights connected with losses described in the preceding Articles, namely Articles 2.1 (a) and (b), and loss resulting from infringement of contractual rights. Here contractual rights mean to include rights arising out of a contract with the liable party, namely the shipowner. Although it does not specify types of rights which are so referred, i.e. whether statutory, natural or all such types, it obviously means to include rights of persons or entities having substantial interest to exercise such rights.

Claims by fishermen, hoteliers and others depending on the use of natural resources and tourism industry would certainly fall into this category. Eventhough, loss or damage to their proprietary interests, sometimes if any, would become subjects to claims mentioned in the preceding Articles while pure economic loss issues which certainly arise due to infringement of rights would not become subjects, unless within the meaning of this particular Article. Therefore, the residual claims that might be forwarded under the Bunkers Convention will certainly fit into this meaning enabling the shipowner to limit liability for such claims. However, different State practices might bring about some problems concerning the interpretation and application of this provision. Especially in common law countries, pure economic loss is relatively less recoverable, and rights are more re-

1074 See Sec. 168

ferred to as rights granted by Statutes or upon licences.[1075] Such countries might therefore confine the application to that scope, while civil law countries certainly interpret this provision to mean to include economic loss, as these countries are reluctant to distinguish consequential loss and pure economic loss, as stated before.[1076] Another example is that some countries practice the system of issuing licence to commercial fishermen and hoteliers, and therefore refer only to such permit holders within the meaning of infringements of rights, while in many jurisdictions, such system of licence and permits do not exist, and do not confine this to such narrow meaning. Nevertheless, it is important to identify the types of rights covered under the interpretation provided in the jurisdiction in question, if LLMC 1976 is applied as the applicable law for limitation proceedings in that jurisdiction. In particular, country like South Africa, which is not a party to the LLMC 1976, refer to infringements of any kind of rights.[1077]

(C) Claims under Articles 2.1(d) and (f)

The reason behind addressing these two Articles under one heading is mainly because of their relevance to claims concerning 'clean-up and preventive measures'. Article 2.1(d) includes claims concerning clean-up expenses, while Article 2.1(f)[1078] deals with claims in respect of preventive measures. The former gives way for limiting liability in respect of claims stemming from costs incurred towards the raising, removal, destruction or rendering harmless of a ship which is sunk, wrecked, stranded or abandoned. It has covered a wide range of activities associated with clean-up measures, which are normally undertaken by salvors under the guidance of both the shipowner and national authority, or the latter alone in the absence of the owner. This type of claim is more often referred to as a debt rather than costs, which the shipowner is expected to repay, in common law jurisdictions as shown in *Stonedale No. 1*.[1079] Nevertheless, the relevant provision could be used to limit claims for clean-up expenses once a ship is wrecked. But, under the Bunkers Convention, clean-up costs will not necessarily be related to

1075 See Griggs and Williams, "*Limitation of Liability for Maritime Claims*", 2ed, 1991, p.21

1076 *Supra*, p. 255–256

1077 see Section 261(1) of Merchant Shipping Act 57 of 1951 of South Africa. Prior to the accession by India, the situation remained same as evident Section 352. 1 of (Indian) Merchant Shipping Act, 1958

1078 "[c]laims of a person other than the person liable in respect of measures taken in order to avert or minimize loss for which the person liable may limit his liability in accordance with this Convention, and further loss caused by such measures"

1079 [1956] A.C. 1 (H.L.)

wrecks, as pointed out by the CMI during the meetings of delegations.[1080] Accordingly, there may arise of situations where bunkers are lost from double bottoms as a result of grounding, without the ship being wrecked. The primary goal in case of response to bunker spills is to avoid and minimize pollution damage been caused rather than just remove un-threatening wrecks.[1081] In such circumstance the clean-up costs will not be associated with the relevant provision of Article 2.1(d) under which liability could be limited, but it would be possible to make use of Article 2.1(f) for such limitation. Hence, the LLMC 1976 provides States Parties the possibility to make reservations as to the application of Article 2.1(d), which some State Parties have exercised to preserve a policy of unlimited liability for wreck removal.[1082]

For example, the United Kingdom has made such reservation and maintains a policy of unlimited liability for wreck removal.[1083] Eventough, Spain has not made any reservations against the LLMC 1976, the Spanish Ports and Merchant Marine Act[1084] which imposes liability on the shipowner for all costs caused by sinking of a ship, grants powers to the Port Authority to recover all such expenses from the shipowner without any limitation being implemented.[1085] But due to the effect of the Spanish Constitution, which gives the LLMC 1976 superiority over the said Act,[1086] leads to the conclusion that claims of such nature will anyway be subjected to limitation.

As the Bunkers Convention does not define the nature of clean-up measures for which compensation is granted, thus referring to a bold phrase of 'costs of pre-

---

1080 See *supra*, n. 142, p.5

1081 See Ships' Bunkers Convention Conference proposed for 2000–2001, LEG/80, Oct., 11–15, 1999 available at http://www.imo.org/Newsroom/mainframe.asp?topic_id=280&doc_id=752. See also J. Michel et al, *Potentially Polluting Wrecks in Marine Waters*, An Issue Paper Prepared for the 2005 Int'l Oil Spill Conference, available at http://www.iosc.org/docs/iosc_issue_2005.pdf; A. Findley, *Practical Considerations in the Recovery of Oil from Sunken and Abandoned Vessels*, available at http://www.oilspillresponse.com/pdf/Articles%20From%20IOSC/Practical%20Considerations%20in%20the%20Recovery%20of%20Oil%20from%20Sunken%20and%20Abandoned%20Vessels.pdf; Wreck Oil Removal Programme (WORP), *2009 Demonstration Project Overview*, available at http://www.mlaus.org/content/ItemContent/10431Events.pdf

1082 However, this area is now covered under the IMO adopted Nairobi International Convention on the Removal of Wrecks, 2007 and therefore forms part of a distinct regime at present. Text available at http://www.simsl.com/Articles/Wreck%20Removal%20Convention.pdf

1083 See paragraph 3 of Schedule 7, Part II and section 185 of Merchant Shipping Act 1995

1084 Law of 24 November 1992

1085 See Art. 107

1086 See Art. 96 of the Constitution

ventive measures', national authorities may well claim wreck removal costs. Such action could further be justified by such authorities under the meaning of 'costs of reasonable measures of reinstatement' of the environment, and thereby presenting claims for wreck removal. In any circumstance, the shipowner will be able to limit his liability for such claims under the relevant provision of the LLMC 1976, but most importantly it would be necessary to check whether the jurisdiction in question has made any reservations permitted by the Convention.

As stated before, clean-up costs except costs for wreck removal could however be subjected to limitation under Article 2.1(f), which makes specific provisions concerning preventive measures taken to avert or minimise loss and further loss caused by such measures. Doubtlessly, claims for costs of preventive measures and further loss caused by such measures that could be entertained under the Bunkers Convention fall within the relevant meaning of Article 2.1(f) and shall be subjected to limitation accordingly. But however, it clearly states that claims of the person liable shall not be subjected to limitation, with respect to costs of measures taken by a third party by virtue of a contract with the shipowner that fall outside the ambit of limitable claims. Furthermore, he would only be able to limit claims for costs of preventive measures undertaken by others pursuant to the wording "a person other than the person liable".

All these limitable claims mentioned in Article 2.1 are further supplemented by Article 2.2[1087], which allows the shipowner to limit liability even in actions brought against him by a third party such as an insurer or financial security provider by way of recourse or for indemnity under a contract or otherwise. However, it restricts the shipowner to limit liability for claims set out in Articles 2.1(d) and (f), as is relevant to the Bunkers Convention, to the extent they relate to remuneration under a contract between him and a third party[1088], such as a national authority carrying out salvage, wreck removal or clean-up and preventive measures.

In summing up, it is possible to conclude that most potential claims that would give rise under the Bunkers Convention are compatible with the relevant limitable claims mentioned in Article 2 of the LLMC 1976.[1089] But however, one must

1087 "[C]laims set out in paragraph 1 shall be subject to limitation of liability even if brought by way of recourse or for indemnity under a contract or otherwise. However, claims set out under paragraph 1(d), (e) and (f) shall not be subject to limitation of liability to the extent that they relate to remuneration under a contract with the person liable."

1088 See W. Poor, *A Shipowner's Right to Limit Liability in Cases of Personal Contracts*, The Yale Law Journal, vol. 31, no. 5 (Mar., 1922), pp. 505–513

1089 See E. Røsæg, H. Ringbom, *Liability and Compensation with Regard to Places of Refuge*, Final Report, Oct., 2004, Scandinavian Institute of Maritime Law, Uni. of Oslo, pp 20–22, available at http://www.emsa.europa.eu/Docs/other/places%20of%20refuge%20study%20def.pdf

bear in mind that it is important to identify the applicable law of the jurisdiction in which the actions are brought under the Bunkers Convention and acquire an understanding about the extent and scope of interpretation of different aspects as to loss, damage and costs in realising which claims are really limitable in that selected jurisdiction.

Claims Excepted from Limitation

The LLMC 1976 expressly excludes the following claims from the scope of its application under Article 3;

(a) claims for salvage or contribution in general average;
(b) claims for oil pollution damage within the meaning of CLC 1969 or of any amendment or protocol thereto in force;
(c) claims subject to any international convention or national legislation pertaining to limitation of liability for nuclear damage; and
(d) claims against the shipowner of a nuclear ship for nuclear damage;
(e) claims by servants of the shipowner or salvor whose duties are connected with the ship or salvage operations, including that of their heirs, if limitation of liability is not allowed pursuant to such contract entered between the parties or if permitted limitation exceeds amount provided by Article 6.

Accordingly, none of these claims are connected with the liability or prospective claims that could arise under the Bunkers Convention. As the CLC regime has its own limitation provisions inside the Convention, it is well understood of its exclusion from the scope of the LLMC 1976. Furthermore, bunker spills of CLC conventional ships are so covered within the meaning of that Convention, a special emphasis is not needed. Therefore, it can be concluded that neither of the claims which may arise under the Bunkers Convention is thereby excepted from the scope of application in respect to claims subject to limitation.

Conduct Barring Limitation

The shipowner's right to limitation is restricted in situations where his 'actual fault or privity' could be proved.[1090] Although, the 1957 Convention used this ex-

1090 See A. M. Tettenborn, Reviewed work(s): *Limitation of Shipowners' Liability: The New Law* by Nicholas Gaskell, The Cambridge Law Journal, vol. 46, no. 3 (Nov.,

pression, the LLMC 1976 provides a much illustrated provision instead, in order to avoid different interpretations of the phrase in different jurisdictions. Nevertheless, several States seem to use this principle in their respective national legislations even at present.[1091] Evidently, these different jurisdictions have embodied different meanings giving rise to many legal debates. In order to obtain a clear understanding about its meaning with respect to its application in States where such expression still earns much recognition in their domestic legislations dealing with right to limitation, it would be worthwhile to have a general observation.

Concept of Actual Fault or Privity

During the time it appeared in Merchant Shipping Acts of the United Kingdom, the courts were vested with intense obligation to provide suitable meanings. Therefore at present, one may find many decisions which have dealt with the common issue of defining and illustrating the components of this doctrine. In most jurisdictions, the problem connected with the meaning of the 'owner' is settled by the development of the concept of the "alter ego".[1092] The term 'actual fault' is connected with act or omission of the liable party as shown in the case of *The Marion*, where the ship's managers were held to be at fault upon failing to ensure that the ship was equipped with adequate up-to-date charts on board.[1093] As a result such act or omission would lead to make even the shipowners liable for their 'actual fault' with respect to such failure. The meaning of 'privity' was dealt in the famous case of *The Eurysthenes* within the context of the Marine Insurance Act 1906, in which it was held that shipowner is liable if the ship has been sent to sea with the knowledge that the ship was unseaworthy, and this constitute 'privity' on the part of such shipowner under Section 39(5) of the Act.[1094] The Court of Appeal further held in this context, that 'privity' means and includes 'knowledge and consent'.

1987), pp. 535–536; K. S. Matison, *Comparison of Shipowner's Limitation of Liability Schemes*, Lloyd's Maritime Training Programme, Lane Powell Spears Lubersky LLP, available at http://www.lanepowell.com/wp-content/uploads/2009/04/matisonk_002.pdf

1091 See J. M. Joseph, *Limitation of Shipowners' Liability in Malaysia*, available at http://www.jbc.com.my/resource/LIMITATION%20OF%20SHIPOWNERS.pdf; R. Wolfson, *The English and French Carriage of Goods by Sea Enactments*, Int'l & Comparative Law Quarterly, vol. 4, 1955, pp 508–532

1092 See *supra*, pp. 90–92

1093 [1984] 2 Lloyd's Rep. 1

1094 [1976] 2 Lloyd's Rep. 171

Though Singapore applies this expression,[1095] there are no reported decisions outlining its meaning, but however decisions of *Lennard's Carrying Co. Ltd.*,[1096] *The Marion* and similar cases are of persuasive value. In India the situation remains unchanged, as well as in South Africa. In the latter State, the actual fault or privity of the owner in case of a company refers to the person or persons with whom the chief management of the company's business resides, although such a person need not necessarily be a director. Therefore, for example, the head of the company's subsidiary which handles shipping matters is the person whose action is the very action of such company.[1097] In the United States the phrase 'privity and knowledge' is said to exists where the owner has actual or constructive knowledge of the fault or condition, involving the owner's keen interest in obtaining necessary information by way of reasonable inquiry and inspection.[1098] In such circumstance it is measured on grounds of perusal rather than mere knowledge.[1099] Nevertheless, right to limitation would be granted if the liable party found to have acted in a reasonable manner. With respect to corporate bodies, only certain officials are chargeable with corporate knowledge, especially persons to whom corporate management is delegated, and within whose scope of authority actual fault or privity has existed. In such context various types of personalities fall within, such as managing agents, officers, and supervising employees and etc.

Restriction as to Article 4

The LLMC 1976 has deviated from its predecessor with respect to this aspect, rather addressing the issue much comprehensively without using the concept of "actual fault or privity". According to its Article 4;

"A *person liable* shall not be entitled to limit his liability *if it is proved* that the *loss* resulted from *his personal act or omission*, committed *with the intent to cause such loss*, or *recklessly and with knowledge that such loss would probably result*". (emphasis added)

There are several important characteristics that need further observation and clarification in the above mentioned provision. For the convenience of understanding these issues, they would be addressed in detail in the following manner;

1095 See Sec. 136 of the Merchant Shipping Act

1096 *Supra*, n. 444

1097 See Prof. Staniland, Hilton, on South Africa in P. Griggs and R. Williams, *Limitation of Liability for Maritime Claims*, 3rd ed., LLP, 1998, pp. 299–300

1098 See *Spencer Kellogg & Sons, Inc. v. Hicks* (*The Linseed King*), 285 U.S. 502 (1932)

1099 See *States S.S. Co., v. United States*, 259 F.2d 458, 466, 1957 AMC 1181 (9th Cir. 1957)

The Person Liable

Naturally, it refers to persons described in Article 1[1100] as persons entitled to limit liability, including the shipowner, charterer, manager, operator[1101] or the liability insurer[1102] or the financial security provider of the first four categories, as it applies to the Bunkers Convention. As the right to limitation exists once the liability is established, the liable person or persons could be one or more of the above parties. As far as right to limitation is concerned, it is only the person(s) who are not barred could qualify. This would mean that some are eligible for limitation, while some are not with respect to a ship involving one incident. On the other hand the liability insurer or the financial security provider could limit liability only when the assured is not barred from limitation.

Normally, the shipowner, manager, charterer and operator are entered together in P & I insurance in respect of the same ship, and therefore, there could arise of situations where the liability insurer is able to limit liability on behalf of an assured who is not barred, but however, looses such right on behalf of another assured whose been barred under Article 4[1103].

His Personal Act or Omission

Accordingly, it is the personal act or omission of the person liable that construe the imposition of bar to limitation. With regard to natural persons, it should mean personal to himself, the so liable party, and in respect of a company the concept of *alter ego* should be applied. Furthermore, the definition for 'company' provided in the ISM Code 2002 would be of some value.[1104] The physical component of the misconduct, which obstructs the right to limit must, consists of either an act or omission that was executed.

1100 Art. 1.1, which states "[S]hipowners and salvors, as hereinafter defined, may limit their liability in accordance with the rules of this Convention..."

1101 See Art. 1.2

1102 See Art. 1.6

1103 "[A] person liable shall not be entitled to limit his liability if it is proved that the loss resulted from his personal act or omission, committed with the intent to cause such loss, or recklessly and with knowledge that such loss would probably result."

1104 According to its sec. 1.1.2, „Company" means the owner of the ship or any other organization or person such as the manager, or the bareboat charterer, who has assumed the responsibility for operation of the ship from the shipowner and who, on assuming such responsibility, has agreed to take over all duties and responsibility imposed by the Code.

Loss and Such Loss

As the LLMC 1976 deals with various types of loss and damage in Article 2[1105], it is not entirely clear what is meant as 'loss' in this context. Obviously, it would not have been the intention of the drafters to include something outside the scope of the Convention, and therefore it could be concluded that 'loss' means to include all types of losses and damages including costs and other expenses described in Article 2, as discussed earlier[1106].

Consequently, 'such loss' as expressed in this provision would be nothing more than anticipated loss and damage that would probably result upon an incident and such probable results that were known to the liable party.

Intent to Cause such loss

This constitutes the proper element of '*mens rea*' required to commit such act or omission. Importantly, it should be proved that the shipowner had the intention to cause the loss, if bar to limitation is to be based on this issue.

Recklessly and with Knowledge that such loss would probably result

It is clearly stated in Article 4, that the conduct barring limitation should have arose due to the liable party's act or omission that occurred as a result of his intention to cause such loss or his recklessness, and knowledge that such loss would probably result. English jurisprudence provides some explanation to the term 'recklessly' or 'recklessness' in *R. v. Caldwell*[1107] and *R. v. Lawrence Stephen*[1108], which connotes either carelessness or utter heedlessness of consequence that has led to a result which the wrongdoer has neither considered the probability nor possibility of its likelihood.

Upon the fulfilment of or in other words proving the necessary elements of the requirements in this provision, will lead to a situation where the shipowner being deprived of the right to limitation in claims for compensation for bunker oil pollution damage against him in jurisdictions where both the Bunkers Convention as well as the LLMC 1976 applies. Nevertheless, States which have not ratified the LLMC 1976 would still have such situations in their applicable law in depriving

1105 See *supra* pp 253–261
1106 *Id*
1107 [1982] A. C. at p. 354
1108 [1982] A. C. at p. 520

limitation, such as to the effect of the 'actual fault or privity' clause. In any case one must thereafter identify the person on whom the onus of proof lies in such a situation.

Onus of Proof

Unlike the 1957 Convention which laid the burden of proving shipowner's actual fault or privity on the determination of the court in accordance with the *lex fori*,[1109] the LLMC 1976 has prevented itself from any specific expression. It could be found from the *travaux préparatoires* of the 1976 Convention that the drafters had no intention in including such a provision, eventhough it was suggested by the Working Group on Basic Issues relating to the Limitation System to lay the burden of proof in a similar manner as of its predecessor.[1110] The present text affords the right to limit pursuant to the fulfilment of Article 2 requirements thus subjecting it to Article 4, which automatically vest the burden on the claimant. This is reflected by the use of the words "subject to Article 3 and 4 the following claims... *shall* be subject to..." in Article 2 and "A person shall not be entitled to limit... if it is proved..." in Article 4. While Article 2 allows such claims to be subjected, Article 4 disallows it on grounds of misconduct of the shipowner proved with evidence. Therefore, it is obviously clear that the party challenging the right to limitation is required to prove conduct barring limitation in terms of Article 4.

In the recent judgment in United Kingdom, *The Capitan San Luis*, Clarke J., pointed out the effect of Article 4 by stating that "the shipowner merely has to establish that the claim falls within Article 2 of the Convention. Once he establishes that, he is entitled to a decree limiting his liability, unless the claimant proves the facts required by Article 4".[1111] States which apply principles of the old regime determine this issue in terms of *lex fori* thus imposing the onus on the shipowner.[1112] But in the United States, the situation is two folded as the shipowner bears the burden of proving lack of privity or knowledge, while the initial burden of proving negligence or unseaworthiness rests on the claimant.[1113]

Russia on the other hand maintains a completely distinct and separate approach without having any provision dealing with conduct barring limitation. In

1109 See Art. 1(6)
1110 See O.R. p. 134
1111 See [1994] 1 All ER 1016
1112 For India, Singapore and South Africa see S. Thacker, H. Ajaib and Prof. H. Staniland respectively in *Limitation of Liability for Maritime Claims* (P. Griggs and R. Williams eds.), 3rd ed., LLP, 1998, p. 193, 287 and 299 respectively
1113 See *Farrell Lines Inc. v. Jones*, 530 F.2d 7 (5th Cir. 1976)

other words, the Russian legislation does not deprive the shipowner of his right to limit on grounds of his fault. This enables the shipowner to exercise right to limit in any circumstance whatsoever.[1114]

Limits of Liability

The modern system of limitation figures is based on the vessel's tonnage.[1115] The situation is basically the same with regard to international law as well as national regimes dealing with limits of liability. Both LLMC 1976[1116] and its Protocol 1996[1117] provide a scale of figures depending on the tonnage of the vessel in dispute. Limits of liability that should apply to claims under the Bunkers Convention fall within the scope of the general limits described in Article 6 of LLMC 1976 as amended by Article 3 of Protocol 1996. Furthermore, both these regimes stipulate a minimum fixed level for considerably 'small ships'. According to Article 6.1 (a) (i) of LLMC 1976, such fixed value is applicable to ships with a tonnage not exceeding 500 tons, while Protocol 1996 has increased such fixed value to ships not exceeding 2,000 tons. Although, the LLMC 1976 excludes 'small ships' of less than 300 tons from its scope of application under Article 15.2 (b)[1118] these relevant figures could be taken into account in circumstances where States Parties intend to regulate or has already included such ships within the system of limitation of liability via specific provisions of national law in pursuant to the option vested by the relevant provision of the Convention. Nevertheless, many State Parties have excluded the application of this figure to ships less than 300 tons, leaving them outside the scope of application of LLMC 1976. However, some States have even separated the application of this provision; to ships less than 300 tons of tonnage and ships between 300 – 500 tons of tonnage, rather than leaving a common figure for all ships of less than the minimum tonnage expressed in Article 6.1 (a) (i)[1119]. For example, France has fixed half of the limitation figure applicable to ships less than 500 tons, to ships which are less than 300 tons of tonnage.[1120] In certain instances, one may see that some States Parties have excluded the application of

1114 See A. Makovskia on Russia in *Limitation of Liability for Maritime Claims* (P. Griggs and R. Williams eds.), 3rd ed., LLP, 1998, p. 282

1115 See Ch. 2 of LLMC 1976

1116 See Art. 6: The General Limits and Art. 7: The Limit for Passenger Claims

1117 See Art. 3 and Art. 4

1118 "A State Party may regulate by specific provisions of national law the system of limitation of liability to be applied to vessels which are:…(b) ships of less than 300 tons."

1119 "333,000 Units of Account for a ship with a tonnage not exceeding 500 tons,"

1120 See the Law of 25 December 1986

these figures to ships of less than 300 tons due to their non-application within the purview of the LLMC 1976 while fixing the minimum level as for ships between 301 -500 tons of tonnage. The Chinese law is an example to this effect.[1121]

Evidently, the respective limits are divided into two main categories depending on the nature of the claims, namely claims in respect of loss of life or personal injury and claims in respect of any other type, allowed by virtue of Article 2. The respective figures are expressed in terms of "units of account", which are referred to as Special Drawing Rights (SDR) in Article 8[1122]. The relevant SDR is derived from the system adopted by the International Monetary Fund (IMF)[1123], and the procedure involved in the conversion of specific figures pertaining to limits of liability, is well described in the said Article 8[1124]. The situation with respect to States Parties which are not Parties to the IMF is somewhat complicated. Article 8 provides such States Parties to the LLMC regimes to determine the value of its national currency in terms of the SDR in accordance with a method adopted by that State Party.[1125] Nevertheless, Article 8.2 provides optional criteria for State Parties,[1126] which are not members of the IMF, and States Parties whose law does not permit the application of Article 8.1. Accordingly, States Parties may *at the time of signature without reservation as to ratification, acceptance or approval or at the time of ratification, acceptance or accession or at any time thereafter*, declare itself to be bound by the method described in Article 8.2. The relevant system is based on "monetary units", unlike "units of account", and the figures have been further amended by Protocol 1996[1127]. The definition of these "mon-

1121 See G. Sunlai on China in *Limitation of Liability for Maritime Claims* (P. Griggs and R. Williams eds.), 3rd ed., LLP, 1998, p. 150

1122 "The Unit of Account referred to in Articles 6 and 7 is the Special Drawing Right as defined by the International Monetary Fund. The amounts mentioned in Articles 6 and 7 shall be converted into the national currency of the State in which limitation is sought, according to the value of that currency at the date the limitation fund shall have been constituted, payment is made, or security is given which under the law of that State is equivalent to such payment. The value of a national currency in terms of the Special Drawing Right, of a State Party which is a member of the International Monetary Fund, shall be calculated in accordance with the method of valuation applied by the International Monetary Fund in effect at the date in question for its operations and transactions. The value of a national currency in terms of the Special Drawing Right, of a State Party which is not a member of the International Monetary Fund, shall be calculated in a manner determined by that State Party."

1123 The changing values of SDR into national currencies can be found daily in the currency page of the *Financial Times*

1124 See Art. 8.3

1125 See Art. 8.4

1126 Arts. 8.2(a)(i)-(ii), 8.2(b)(i)-(ii) and 8.2(c)

1127 See Art. 5

etary units" is given in Article 8.3, which is based on the value of gold at the particular time.

However, States which are not Parties to the LLMC regimes maintain systems based on respective applicable national law. For example, India measures the units in terms of gold francs, as was in force under the 1957 Convention, and use a conversion method to its currency in a manner stipulated in Section 352-B of the (Indian) Merchant Shipping Act 1958. South Africa on the other hand applies Section 11 of its Shipping General Amendment Act 23 of 1997 according to which, the limits would be fixed. But however, it uses SDR as the units of account in converting the value into its own currency. Russia being a Party to the LLMC Protocol 1996 with respect to the limits of liability applies the limitation figures in accordance with the provisions of that Protocol alone without changing substantive law that is in force under 1968 Merchant Shipping Code.

Priorities of Claims

Under the LLMC 1976, separate limitation figures are provided for two categories of claims as mentioned before, namely with respect to loss of life or personal injury and other claims including property claims. This implies that in case of claims arising out of these two categories there need to exist of two different sets of limitation amounts, commonly describe as funds[1128]. According to Article 6.2[1129], the priority for payment lies with the first category, and claims in excess of the availability of funds in such category then moves into the second to rank rateably with the claims therein.[1130] With respect to the claims that would give rise under the Bunkers Convention, the first category will remain remote, while most claims fall under the second. Here, one must understand that the personal injuries that might occur during the process of clean-up and preventive measures fall within

1128 See Waltons & Morse LLP, *Limitation of Liability for Maritime Claims: New Limits Come into Force*, available at http://www.waltonsandmorse.com/Bulletins2.jsp?bulletinID=10 (last visited 01.01.2010); C. J. Giaschi, *Limitation of Liability; Introduction / Case Summaries*, Giaschi & Margolis, available at AdmiraltyLaw.com, http://www.admiraltylaw.com/limitation.htm

1129 "Where the amount calculated in accordance with paragraph 1(a) is insufficient to pay the claims mentioned therein in full, the amount calculated in accordance with paragraph 1(b) shall be available for payment of the unpaid balance of claims under paragraph 1(a) and such unpaid balance shall rank rateably with claims mentioned under paragraph 1(b)."

1130 See CMI, *Singapore 1, Implementation and Interpretation of 1976 LLMC Convention: Draft Report*, available at http://www.comitemaritime.org/singapore/implellmc/implemen_draft.html (last visited 01.01.2010)

the meaning of Article 2.1 (f), and not under Article 2.1 (a), which would not therefore gain priority over the second category within the meaning of Article 6.2. In such context, those injuries should be treated equally with the rest of the claims in the said second category. Suppose that there doesn't exist of claims falling within the first category, then there would lie only one single fund. In such circumstance, all claims rank rateably with other claims of the same category; to the extent such other claims do exist.

The relevant idea seems to prevail even in non-LLMC States. For instance, it prevailed in India before it became a party while in South Africa[1131] it prevails at present and Singapore applied the so-called *pari passu* system of shifting the unsatisfied first category claims to the second one while using a *pro rata* basis for division of the amounts in such second category fund. Eventhough the OPA 1990 does not distinguish various claims, but rather fix a common amount;[1132] the procedural law evidently allows personal injury claims to rank in prior.

Apart from allowing priority for first category claims against the second, the LLMC 1976 also grants the option for the States Parties to legislate at their will in respective national laws that claims in respect of damage to harbour works, basins and waterways and aids to navigation shall have such priority over other claims that fall within the purview of the second category.[1133] In such circumstance, these claims would certainly rank higher than any loss of life and personal injury claims if they shift to the second category due to its non-settlement under the first fund. While some States Parties have given force to this aspect in their applicable law, some have indeed prevented themselves allowing equal opportunities for all claims.

Finally, matters arising with respect to the calculation of a ship's tonnage, which indeed the decisive measurement of the limitation figures, has been settled by way of referring it to the scope of application of the International Convention on Tonnage Measurement of Ships, 1969[1134]. Accordingly, the ship's tonnage shall be the gross tonnage calculated in accordance with the tonnage measurement rules contained in Annex 1 of the said Convention.[1135] Regulation 3 of the Annex 1 specifies the method through which the Gross Tonnage (GT) should be determined. It provides a certain formula[1136] based on the moulded volume of the

1131 See Sec. 261 (1) (c) of the Merchant Shipping Act 1951, as amended by Act of 1997

1132 See 33 USC Sec. 2704 (a) (2)

1133 See Art. 6.3

1134 Adopted in London on June 23, 1969 and came into force on July 18, 1982. full text can be found at http://www.admiraltylawguide.com/conven/tonnage1969.html

1135 See Art. 6.5

1136 $GT = K_1V$, where V indicates total volume of all enclosed spaces of the ship in cubic meters, while $K_1 = 0.2 + 0.02\log_{10}V$

entire ship including its hull, deck structures and all enclosed spaces without any deductions, exceptions or special allowances. After the coming into force of this instrument, Contracting States re-measured ships registered in their respective jurisdictions or flying its flag in order to maintain a uniform system as to the measurements of ships engaged in international maritime business, especially sea-going ships of certain capacity. Normally, ships of Contracting Parties should carry International Tonnage Certificates[1137] issued by their relevant government authorities and these certificates are accepted as valid instruments by governments of other Contracting States. Therefore, ascertaining the gross tonnage of a ship is not a major problem in this regard, but some disputes might occur with respect to States which do not apply this method of calculation. However, most ports nowadays require tonnage certificates consistent with international rules and regulations, as it is considered as the basis for the calculation of port dues, and therefore, this requirement would enable to maintain a harmonized situation worldwide.

The Procedural Aspect of Limitation

Commencement of an action or the invocation for limitation is the first basic step involved in the process of limitation of liability. It may happen at any stage as required by the applicable law of the State in which limitation is invoked. As the LLMC 1976 clearly points out, in Article 1.7, *the act of invoking limitation of liability shall not constitute an admission of liability*, and therefore, the shipowner is able to commence limitation proceedings at any stage by way of instituting such action subject to the requirements of the law of that State. In the present case concerning the Bunkers Convention, neither it nor the LLMC 1976 stipulate the exact time at which such proceedings should take place. But however, it is somewhat clear from Article 15 that such action may be commenced whenever the 'shipowner' seeks to limit his liability before the Court of a State Party or seeks to procure the release of a ship or other property or the discharge of any security given within the jurisdiction of any such State. More often, national laws require shipowners to constitute a limitation fund once an incident of damage has occur-

1137 An example of an extract of an International Tonnage Certificate issued by the United States Coast Guard in compliance with the 1969 Convention can be found at http://www.usmma.edu/waterfront/kingspointer/General%20Documents/Int%20Tonnage%20Certificate.pdf and examples of extracts of such a certificate issued by Ireland and Republic of Liberia for instance can be found respectively at http://www.asl.ie/fleet/tonnage_pdf/t_rover.pdf and http://www.handytankers.com/handy/certificates/fileShow.asp?intFileId=3552

red or an action for compensation is brought. Prior to its accession, the practice in Singapore as a non-LLMC State was somewhat different. In a limitation action brought in its competent court, the shipowner was required to prove his case in order to obtain a decree of limitation, and it is a precondition for the grant of such decree that the shipowner wholly or partly admits liability.[1138] This evidently contradicts with the above mentioned Article 1.7 of the Convention.

As far as the LLMC 1976 is concerned, it has made provisions dealing with different procedures involved in respect of situations where the shipowner seeks to commence limitation proceedings against limitable claims.

Invocation without the constitution of a Limitation Fund

Article 10 allows the shipowner to invoke proceedings without constituting a limitation fund.[1139] However, it has curtailed such possibility by way of granting an option on States Parties to legislate in its respective national law, the necessity of the constitution of a fund before invoking the right to limitation.[1140] Furthermore, the Convention has made provisions for the determination of questions of procedure to be solely dependent on the national law of such State Party, in which action is brought.[1141]

Some States Parties seem to have given no force to the reservation allowed in Article 10.1, but the exact situation in some of those are certainly not that clear. For example, Article 59–60 of the Decree of 1967 in France expressly requires the shipowner to file an application specifying circumstances of the incident, in the Commercial Court, if he intends to seek limitation, and further specify the maximum amount of the limitation fund rather than making any provision with regard to any necessity of a constitution of such a fund. The only authoritative means could be found in the Court of Appeal judgement in the *Rennes*,[1142] where it has been held that a person may benefit from limitation even if there has been no fund constituted. Other States Parties expressly upheld the necessity of the

1138 See Ord. 70 (Admiralty Proceedings) of the Rules of Court 1996 and Section 137 of the MSA

1139 Art. 10.1, which states that "Limitation of liability may be invoked notwithstanding that a limitation fund as mentioned in Article 11 has not been constituted…."

1140 "…However, a State Party may provide in its national law that, where an action is brought in its Courts to enforce a claim subject to limitation, a person liable may only invoke the right to limit liability if a limitation fund has been constituted in accordance with the provisions of this Convention or is constituted when the right to limit liability is invoked."

1141 See Art. 10.3

1142 See DMF 1983, 739, 742f

constitution of a fund as a pre-requisite to invocation of limitation. Sometimes, the language used or intention expressed in relation to such a necessity might create lack of clarity. In Japan for example, Article 26 of the applicable law states that; "the proceedings to limit liability become effective when a decision to commence the limitation proceedings is rendered", while Article 25(3) provides that the court may not render a decision to commence limitation proceedings *if the applicant does not comply with the decision to form a limitation fund.*

While the situation takes either of the forms in non-LLMC States, Russia in particular do not provide provisions for the possibility of creating a limitation fund by the shipowner, under its applicable law, namely the Merchant Shipping Code.[1143]

Invocation upon the constitution of a Limitation Fund

Chapter III of the LLMC 1976 specifically deals with the provisions concerning "The Limitation Fund". According to Article 11.1;

> "Any person alleged to be liable may constitute a fund with the Court or other competent authority in any State Party in which legal proceedings are instituted in respect of claims subject to limitation..."

In respect of claims filed in accordance with the provisions of the Bunkers Convention, the limitation proceedings should thereby be instituted with that particular court. This provision affords the claimant, a clear margin of selection as to the jurisdiction in which action for compensation should be instituted. In other words, it provides some sought of opportunity for the claimant to make use of forum shopping. But however, such a possibility is restricted to a certain degree due to the requirements established under Article 9 of the Bunkers Convention[1144], concerning jurisdictional matters. Nevertheless, due to that fact that the limitation figures vary in States which apply LLMC 1976 and Protocol 1996, the claimant would be able to choose out of the specified jurisdictions, the one which still applies the low limits of LLMC 1976. Hence, the claimant is in a position to select such State that narrowly interprets claims subject to limitation, and thereby succeed in obtaining compensation in full. Therefore, the shipowner will not be in a position, in limitation proceedings, to force a claimant to a jurisdiction of his own choice.

1143 See *supra*, n. 1098, p. 283
1144 See *supra* p 117

As in the case of commencement of limitation proceedings the Convention nowhere defined the exact stage at which such a fund may be constituted. The matter is therefore seems to have left for the determination by the national law or procedural rules of the competent court.[1145] Likewise, many jurisdictions enable the shipowner to constitute the fund at any stage even before an action is brought by the claimant, LLMC 1976 allow the constitution of such fund at any stage, as there does not exist of a compulsory requirement of a constitution of the fund under the Convention, and furthermore, the act of constituting the fund at the initial stage does not constitute an admission of liability.[1146] But however, the situation might sometimes be different in other States.

The volume of the Fund and the competent forum

The relevant Article stipulates that such Fund be constituted with the Court or other competent authority in a sum of such amounts set out in Article 6 together with the interest thereon from the date of the occurrence[1147] of the incident that caused the loss or damage. Furthermore, the particular fund may be constituted adequately by way of monetary deposit or financial guarantee acceptable by the relevant forum according to the law of the State.[1148] However, the Convention is silent on the issue of the rate of interest that it has referred to under the relevant provision, leaving such issues for the determination of the State via its national law.

With regard to these issues State Parties have adopted different methods. For example, the Merchant Shipping Act of UK provides that the Secretary of State may prescribe the rate of interest.[1149] Further, the Admiralty Court is vested with the powers of entertaining the constitution of the fund in case of England and Wales, while the Court of Session in Scotland, and it is necessary to establish such fund by way of cash deposit. The situation in France links to Chapter VII of the Decree of 1967, which requires an application be made to the Commercial Court of the port of registration for French vessels, while in respect of non-French vessels the competent court shall be the Commercial Court of the place where the

1145 See *Seismic Shipping Inc and another v Total E & P UK plc (The Western Regent)*: Limitation of liability – 1976 Convention – jurisdiction to grant limitation decree – limitation fund – foreign litigation of limitable claim – anti-suit injunction, JIML 11 [2005] 4, 263

1146 Read in conjunction, Arts. 11 and 1

1147 See Art. 11.1

1148 See Art. 11.2

1149 See Para. 8 of Sch. 7 of Part II

incident occurred, first French port of call after the incident where it was arrested or security was posted. While the competent court differs in such a way from one jurisdiction to the other, LLMC States would need to make special emphasis in their respective laws to enable such courts to entertain claims under the Bunkers Convention, once it is acceded by such State, in order to harmonise the situation required by Article 11 of the LLMC 1976, or at least make provisions concerning transfer of actions and the fund so constituted to courts which are thus competent to hear claims made with respect to so-ratified Bunkers Convention.

Provisions relating to the constitution of the fund in such circumstance can be found even in non-LLMC States. In most instances the issue is addressed specifically in the applicable law of such States and in most cases, stipulating the required procedural steps involved. In India for example, the situation is best described in the Merchant Shipping Act, allowing the constitution of the fund with the High Court by way of cash or security.[1150] Sometimes, some States might even require the shipowner to constitute more funds for one distinct incident, such as in the case of United States, where certain state laws require supplementary funds in addition to the main funds required under the OPA 1990.[1151] In effect, the right of states to impose such additional requirements is not pre-empted by the OPA 1990, which clearly states in Section 1018 (a) (1) that neither it nor the limitation Act of 1851 shall "affect or be construed or interpreted as pre-empting, the authority of any state or political subdivision thereof from imposing any additional liability..."

The LLMC 1976 further provides under Article 11.3, the possibility of constituting the fund by any person alleged to be liable, and who indeed is entitled to benefit from limitation prescribed by the Convention. Furthermore, in situations where anyone out of such persons has constituted the relevant fund, it shall be regarded as constituted by all persons so liable. This facility also extends to the liability insurer of such persons mentioned therein, but only up to the extent that such insurer represents the assured for the liabilities of such assured. This reflects the idea that the insurer in no way qualify to constitute a fund for himself, to exercise right to limitation for his own liabilities, if any.[1152] Such a case would in no way arise under the meaning of the Bunkers Convention as it deals only with regard to liability against the 'shipowner' as defined, and under the relevant Convention, and furthermore, it is clear that this facility will be afforded to persons mentioned within such definition of 'shipowner', and their liability insurers including financial security providers, as the case may be, in direct actions.

1150 See Sections 352-C (1) – (3)
1151 Not dependent only on the federal law
1152 It applies only to persons mentioned in Article 9

Distribution of the Fund and safeguards

Article 12 of the LLMC 1976 deals with the subject of distribution of the fund among the claimants upon the judgement of the court, and in accordance with the relevant procedure involved in the relevant law of the State Party. Hence, Article 13 imposes restriction on exercising any right in respect of such claim against any other assets of a person by or on behalf of whom the fund has been constituted. Therefore, the constitution of a fund would not only enable the shipowner to set the maximum payable compensation for a distinct incident but also to safeguard all his interests including the ship and other properties belonging to him from seizure. This highlights the importance of constituting a fund rather than commencing limitation proceedings without it.[1153]

Governing law

According to Article 14, the rules concerning the constitution and distribution of the limitation fund and all procedural rules connected thereto are to be governed by *lex fori* in which the fund is constituted.[1154] Therefore, the Convention provides the law of the State Party, the supremacy, to deal with such procedural aspects. As seen during the entire discussion, this is not the only instance where such force of the national law was lifted to its best. Therefore, many authors tend to reflect the idea that the Convention itself is procedural in nature rather than substantive.[1155]

1153 See J. T. Biezup, T. J. Abeel, *The Limitation Fund and its Distribution*, 1979, 53 Tul. L. Rev. 1185

1154 "Subject to the provisions of this Chapter the rules relating to the constitution and distribution of a limitation fund, and all rules of procedure in connexion therewith, shall be governed by the law of the State Party in which the fund is constituted."

1155 See also the judgment of *Caltex v. BP* [1996] 1 Lloyds Rep. 286 at p. 294

Conclusion

The International Convention on Civil Liability for Bunker Oil Pollution Damage, 2001, was a creation to fulfill the objectives of filling the gap that existed in the entire field of marine oil pollution liability and compensation.[1156] Evidently, the preamble to the said Convention expressly declares such objectives as a matter of imposing liability on the shipowner with the view of compensating the victims in a manner associated[1157] with the norms of natural justice. Imposing such liability alone would not suffice the attainment of the intended end results that need to be coupled with several important elements. Therefore, a truly effective application and enforcement is highly relevant.[1158] Certainly, this end is highly dependant on matters of proper effectual drafting and the force of law which it imposes[1159]. Throughout the entire discussion based on liability issues arising out of this Convention, the author emphasized important elements outlining its effectual operation as a liability regime intended to provide compensation to possible victims. Although, there exists of several shortcomings[1160] in this regime as whole, it can be recognized as a successful draft. Nevertheless, such shortcomings are numerous, and indeed subject to criticism, but however, it would be intelligent to draw

1156 See L. de la Fayette, Achievements in 2001, Environmental Policy and Law, vol. 32, no. 3–4, 2002

1157 See *supra* p 30

1158 See E. Louka, *International Environmental Law: Fairness, Effectiveness and World Order*, Cambridge, 2006, p 465

1159 See for example, the discussion paper No. CB(1)1344/08-09(03) of Transport and Housing Bureau of Hong Kong, *Legislative Council Panel on Economic Development Bunker Oil Pollution (Liability and Compensation) Bill*, April 27, 2009 available at http://www.legco.gov.hk/yr08-09/english/panels/edev/papers/edev0427cb1-1344-3-e.pdf; Flagship, *Bunker Pollution Convention may fill the last compensation gap*, iss. no. 4, April 2001, available at http://www.liscr.com/liscr/LinkClick.aspx?fileticket=D9Q0EhltD1w=&tabid=154&mid=538 (last visited on 07.01.2010); Directive 2004/35/CE of the European Parliament and of the Council of 21 April 2004 on Environmental Liability with regard to the Prevention and Remedying of Environmental Damage, Notice to Members No. 8, 2009–2010, July 2009, available at http://www.westpandi.com/WestPandI/NoticesToMembers/2009-2010/2009-08+2.htm

1160 See Yearbook of International Cooperation on Environment and Development 2003–2004 (O. S. Stokke, O. B. Thommessen, Int'l Environmental Governance Set, Oct. 2009

attention to ways and means of overcoming these negative points for the sake of both the victim as well as the shipowner.

Saving the Responder

The Convention is badly drafted on issues pertaining to responder immunity[1161]. Unlike the prevailing similar regimes such as the CLC and HNSC, the Bunkers Convention does not pay much attention to resolve this issue, due to its efforts concentrated centrally on imposing liability on the 'shipowner' alone[1162]. In other words, the Convention is aimed at imposing all liabilities arising out of the pollution incident on the shipowner rather than addressing connected areas of dispute. In such context, the victim would fall into many difficulties in obtaining compensation for which the shipowner might sometimes be able to escape liability. In such circumstance, the application of the Convention becomes obsolete, thus implying the victim to invoke remedies under prevailing national laws.

Nevertheless, the Convention has left a huge margin of appreciation on the court seized with such matter to adjudicate on the relevance of measures executed by persons taking preventive measures[1163]. Courts of different jurisdictions would thereby be able to interpret in their own legal language as to such relevance, which would indeed generate complex theories. However, the prevailing deficiency is somewhat difficult to overcome in a regime such as the Bunkers Convention which focuses primarily on issues concerning liabilities of shipowners rather than compensating pollution damage itself. The best alternative would be either to interpret the Convention in a way that it includes responder liabilities within the ones of the shipowner while providing a right of recourse to such shipowner against such responder, as is intended by the Convention, or make special emphasis with regard to the relevant subject as falling outside the its scope. In the present text, neither is expressed in clear language. The ambiguous meaning would tend to de-stabilize the exact objective of the drafters, who indeed tried to maintain a compromise between different interests.

1161 See *supra* pp 105–110
1162 See *supra* pp 84–85
1163 See *supra* pp 65–66

Avoiding "forum shopping"

Although, the jurisdictional aspect is well founded in the Convention[1164] with the view of minimizing opportunities of forum shopping, it thus tends to reflect some ways of penetration from certain loop-holes. The best alternative in overcoming such situations would have been to confer jurisdictional competency to an international tribunal, such as the International Tribunal for Law of the Sea (ITLOS)[1165] or make reference to ad-hoc conciliation system[1166]. With respect to such a persuasion, one may find no corresponding provisions under the CLC or HNSC. Therefore, it could be argued that such a requirement would leave the Bunkers Convention in an unacceptable situation. This would not be always a true justification, as other similar regimes possess supplementary funds[1167], which this Convention really lacks, and therefore, it would be quite an acceptable move in isolation. Nevertheless, the ITLOS would not be the best solution considering its characteristic feature of attending to public law issues[1168], irrespective of its somewhat-less workload at present.

Establishing an ad-hoc tribunal seems to be somewhat workable and attainable solution. In this respect, one could consider the provisions of the Space Convention, 1972[1169], under which a system of *Claims Commission*[1170] exists. Having an international forum of such nature would minimize associated problems and avoid opportunities of forum shopping concerning maximized claim-grants and minimized limitation-grants, but the absence of set of rules concerning substantive laws would generate further problems. The only way to avoid such complications would deeply depend on the adoption of uniform rules of private international law.

1164 See *supra* pp 115–116

1165 See Home Page, http://www.itlos.org/

1166 Such as arbitration or other means of dispute resolution stipulated in the UNCLOS, UN Charter and other similar international multilateral instruments

1167 Such as IOPC Fund for CLC and HNS Fund for HNSC

1168 N. Klein (2008), *The Rules of the International Tribunal for the Law of the Sea: A Commentary* by P Chandrasekhara Rao and Ph. Gautier (eds) (Martinus Nijhoff Publishers, 2006; ISBN-13: 978-9004152403, 521 pp, hb)]. *International and Comparative Law Quarterly,* 57 , pp 999–1004; M. Ravin, *ITLOS and Dispute Settlement Mechanisms of the United Nations Convention on the Law of the Sea*, 2005, available at http://www.un.org/Depts/los/nippon/unnff_programme_home/fellows_pages/fellows_papers/mom_0506_cambodia_itlos.pdf

1169 The Convention on International Liability for Damage Caused by Space Objects, 1972. full text of the Convention can be found at http://fletcher.tufts.edu/multi/texts/BH595.txt

1170 See Arts. XIV–XX

Limitation v. Adequate Compensation

It is far from clear that the main objective of the Convention is to provide adequate compensation to the victims.[1171] The Convention indeed expressly use the term 'adequate' in several places, but it has boldly dealt with the manner in which it could be attained. Obviously, one may question the real meaning of being 'adequate', whether it denotes a term thus used for mere expression as for the sake of using or it is an underestimation for de-valuation. Referring to the entire scope, adequacy is short of achievement. The deficiency lies not on issues of adjudication or proof of loss or damage, but on the issue of limitation, which is further complimented by the absence of a supplementary fund.

Even though, strict liability concept operates as the principle vein in this regime[1172], its subjection to limitation undervalues the whole system. Obviously, such a move is not a deficiency in its entirety, but the failure in arriving at a separate limitation regime within the Convention, as is stipulated in the CLC and HNSC, makes it a paralyzed piece of work. Although, there existed of divulging reasons to arrive at this solution, leaving the matter on the hands of the LLMC or applicable national regime[1173] would not in anyway fulfill the requirement of adequately compensating the victims, especially considering the gravity of the main issue, the 'pollution damage'. Progressive development of the subject and increasing claims in the future would prove that attaining adequate compensation within the prevailing context is somewhat unachievable.

The absence of a supplementary Fund

The other problem associated with obtaining adequate compensation is the absence of a tier fund such as that exists in CLC and HNSC regimes via Fund Convention and HNS Fund respectively. Obviously, there is nothing much to do about this matter rather leaving it on the respective authority of State Parties. The main reason for such an absence is formulated by the fact that, unlike other similar regimes, there are no stakeholders from the receiving end. In the case of both CLC and HNSC, the shipowner carries out his business for the interests of the cargo owners at the other end, and therefore, it was possible to impose certain burdens on both the oil and HNS industry, as the case may be, to contribute towards the respective funds established under those regimes. In such context, the respective

1171 See *supra* pp 15–17
1172 See *supra* pp 19–20
1173 See *supra* pp 241–243

contributory funds made up by the oil or HNS importers, as it may, have been able to provide extra compensation that would be needed to fill the gap.

In the case of the Bunkers Convention, such an approach would not yield any success, as it is only the shipowner who will have to bear the entire fees in case of such a contribution. This would in turn leave the shipowner in an extremely uncomfortable situation. But however, one could suggest a fund consisting of tax contributions of the persons and trade involved with environmentally hazardous activities such as that exists in Canada[1174] and in some states of the United States. In my opinion, it would be better to constitute so-called common environmental funds made up of not only such industries' or businesses' but also the tax contributions of the general public, who ultimately are the objects which enjoy the benefits of the maritime trade, especially concerning activities such as fishing. In such a circumstance, a State should separate a considerable amount of public tax for the establishment and maintenance of a fund of that nature. Although, this would not be possible at common international level, such an approach would be highly welcome by the shipping industry as well as interested parties. The absence of internationality character would however create unfavourable consequences, if such funds are constituted nationally, but bilateral and multilateral arrangements could minimize such associated problems. Furthermore, a strategy of this nature would promote the up-keep of social welfare and enhance social gains in the context of just and reasonableness. However, the introduction and implementation of such a system would not always be an easy task due to various social and economic constraints at national level, but its prevalence would certainly bring along tremendous benefits.

Concluding Remarks

Drawing the attention to the collectivity and entirety of the provisions of the Bunkers Convention, it could be concluded that its creation is a much of a success. This regime has been able to fill the existing gap that prevails in the whole system of marine liability and compensation for oil pollution damage while addressing many aspects of the problems connected with the present day commercial maritime activities. Fortunately, the Convention came into force much earlier than anticipated, and the author is confident that the international maritime community, especially States which have not become parties to it, would accede to this

1174 See Transport Canada, Canada's Government takes action to protect the environment with changes to the Marine Liability Act, June 25, 2009, news report available at http://www.tc.gc.ca/eng/mediaroom/releases-nat-2009-09-h096e-3293.htm (last visited 05.01.2010)

Convention in accommodating a much important issue that need to be resolved in filling the said gap in the field of international commercial maritime law, and especially to benefit from afforded advantages of the particular regime. Certainly, this would comfort both the victims of oil pollution damage as well as the shipping industry to a very great extent.

Bibliography

Books

A. Aarnio, *The rational as reasonable: treatise on legal justification*, Kluwer, 1987

A. Makovskai on Russia in *Limitation of Liability for Maritime Claims* (P. Griggs and R. Williams eds.), 3rd ed., LLP, 1998

A. Annesley, *A compendium of the law of marine insurance, bottomry, insurance on lives*

A. Aust, *Handbook of International Law*, Cambridge, 2005

A. Beever, *Rediscovering The Law of Negligence*, Hart, 2009; G. Chakrabati, *Law of Negligence*, R. Cambray & Co. Ltd., 2007

Abecassis and Jarashow, *Oil Pollution from Ships,* Steven & Sons, 2 Ed, 1985

A. Berger, *Encyclopedic dictionary of Roman Law*, The American Philosophical Society, 2004

A. Bonomi, P. Volken, *Yearbook of Private International Law: Vol. X, 2008*, Sellier, 2009

A. Carette, *Herstal van en vergoeding voor aantastingen aan niettoegeeigende milieubestanddelen (1997)*

A. E. Branch, *Elements of shipping*, 8th ed., Routledge, 2007

A. E. Chircop, O. Linden, *Places of refuge for ships: emerging environmental concerns of a maritime custom*, Maritnus Nijhoff, 2006

A. E. Evans, J. F. Murphy, *Legal Aspects of International Terrorism*, American Society of Int'l Law, 1978

A. Khee-Jin Tan, *Vessel Source Marine Pollution: the Law and Politics of International Regulation*, Cambridge Univ. Press, 2006

A. Mandaraka-Sheppard, *Modern Admiralty Law: with risk management aspects*, Cavendish, 2001

A. M. Sheppard, *Modern maritime law and risk management*, Routledge-Cavendish, (2001, 2nd ed., 2007)

A. M. Sheppard, *Modern Maritime Law and Risk Management*, 2nd ed., Informa Maritime and Transport, 2009

A. Odeke, *Bareboat charter (ship) registration*, Kluwer, 1998

A. Porat, A. Stein, *Tort liability under certainty*, Oxford, 2001

Archbold (2005)

A. S. Bell, *Forum shopping and venue in transnational litigation*, Oxford, 2003
A. Watts, *The International Law Commission: 1948–1998*, 1999
B. A. Boczek, *The Law of Treaties* in *International law: a dictionary*, Ch. 6, Scarecrow, 2005
B. H. Dubner, *The Law of International Sea Piracy*: *Developments in International Law*, vol.2, Martinus Nijhoff, 1980
B. Kwiatkowska, *The 200 mile Exclusive Economic Zone in the new Law of the Sea*, Springer, Dordrecht, 1989
B. S. Markesinis, M. Coester, G. Alpa, A. Ullstein, *Compensation for personal injury in English, German and Italian law*
B. Soyer, *Warranties in marine insurance*, 2nd ed., Cavendish, 2006
Busnelli and Comande, *Causation under Italian Law* in Unification of Tort Law: Causation (J. Spier eds.), Kluwer, 1996
B. Winiger, H. Koziol, B. A. Cock, R. Zimmerman (eds.) *Essential cases on natural causation (Digest of European tort law, vol. 1)*, Springer, 2007
Carval, *Unification of Tort Law: Damages* in *Principles of European Tort Law* (U. Magnus ed.), Kluwer, 2001
C. Bernasconi, *Civil Laibility and Transfrontier Environmental Damage* in *Hague Yearbook of International Law* (J. G. Lammers ed., 1999) Kluwer
C. Brac, *Operation Tanio*, La Nouvelle Revue Maritime, July 1981
C. C. Nicholls, *Corporate Law*, Ch. 7, Edmond Montgomery, 2005
C. C. A. Voskuil, J. A. Wade, *Carriage of goods by sea: maritime collisions, maritime oil pollution*, T. M. C. Asser Inst., The Hague, 1980
C. E. McDowell, H. M. Gibbs, *Ocean Transportation*, Beardbooks, (1954 Rep. 1999)
CLC, CLC 1992, 1992 Oil Pollution Convention, 1992 Civil Liability Convention in *International Documents on Environmental Liability* (H. Descamps et. al., ed.), Springer, 2008
C. N. Brower, J. D. Brueschke, *The Iran-United States claims tribunal*, Kluwer, 1998
Cohl, Karen, *Tort law*, I. P. I. Publishing, 1982
Colin de La Rue and Charles B. Anderson, "Shipping and the Environment", LLP-London/Hong Kong 1998
Cousy and Vanderspikken, *Causation under Belgian Law* in *Unification of Tort Law: Causation* (J. Spier eds.), Kluwer, 1996
C. Soanes, A. Stevenson eds., *Concise Oxford English Dictionary*, 11th ed. (revised), 2008
C. S. Sangco, *Philippine Law on Torts and Damages, Volume 1*, 1993
C. S. Switzer, L. A. Bulan, *CERCLA: Comprehensive Environmental Response, Compensation, and Liability Act (Superfund)*, American Bar Association, 2002

C. von Bar, *The Common European Law of Torts*, vol. 2, Oxford, 2005
C. Walton, S. E. Wood and R. Cooper, *Charlesworth and Percy on Negligence: Mainwork and Supplement*, Sweet & Maxwell, 2009
Damage from International Disasters in the Light of Tort and Insurance Law, 1990
D. Baker, *Law Made Simple*, 12th ed., Elsevier, 2007
David Freestone, Ellen Hey, *The Precautionary Principle and International Law: The Challenge of Implementation*, Kluwer, 1996
D. Chabas, *1973 Chron.*
D. C. Jackson, *Enforcement of Maritime Claims*, 2 ed., 1996, LLP
D. Delaney, *Law and nature*, Cambridge, 2003
D. Elvin, J. Karas, *Unlawful Interference with Land: Nuisance, Trespass and Covenants and Statutes*, Sweet & Maxwell, 2002
D. Hodgson, *The Law of Intervening Causation*, Ashgate, 2008
D. J. Attard, *The Exclusive Economic Zone in Inetrnational Law*, Oxford University Press, 1986
D. J. Cremean, *Admiralty Jurisdiction: Law and Practice Australia, New Zealand, Singapore and Hong Kong* , The Federation Press, 5th ed., 2008
D. J. Gifford, *Understanding the Australian legal system*, Cavendish, 1997
D. L. A. Barker, *Law Made Simple*, 12th ed., Elsevier, 2007
Douglas and Green on *The Law of Harbours, Coasts and Pilotage*, 5th ed., Informa, 1997
Dr. A. Odeke, *Bareboat charter (ship) registration*, Kluwer, 1998
D. Riesel, *Environmental Enforcement: Civil and Criminal*, 12-54.3 – 12-54.4
Drury (1998)
D. Ulpian, *Essential Cases on Natural Causation* in *Digest of European Tort Law, Vol. 1*, Springer, 2007
D. Ulpian, *Damage caused by multiple tortfeasors* in *Digest of European Tort Law*, vol. 1, Springer, 2007
Edgar Gold, *Gard Handbook on Marine Pollution*, 2nd ed.
E. H. P. Brans, *Liability for damage public natural resources: standing, damage and damage assessment*, Kluwer, 2001
E. J. Weinrib, *Tort Law: cases and materials*, 2nd ed., Emond Montgomery, 2003
E. Louka, *International Environmental Law: Fairness, Effectiveness and World Order*, Cambridge, 2006
E. K. Banakas, *Civil liability for pure economic loss*, Kluwer, 1996
E. Karner, H. Kozoil, *Non-Pecuniary Loss under Austrian Law* in *Damages for non-pecuniary loss in a comparative perspective* (W. V. H. Rogers ed.), Springer, 2001

E. Pattaro, H. Rottleuthner, R. A. Shiner, A. Pec, *A treatise of legal philosophy and general jurisprudence*, vol.1, Springer, 2005

E. R. Anderson, J. S. Stanzler, L. S. Masters, *Insurance Coverage Litigation*, 2nd ed., Aspen, 2008

E. Stuhmcke, *Essential tort law*, 2nd ed., Cavendish, 2001

E. T. Laryea, *Paperless trade: opportunities, challenges and solutions*, Global Trade and Finance Series, Kluwer, 2002

European Group on Tort Law, *Principles of European Tort Law*

E. W. Huffcut, *Elements of the Law of Agency*, BiblioLife LLC

E. W. Vierdag, J. Klabbers, R. Lefeber, *Essays on the law of treaties: a collection of essays in honour of Bert Vierdag*, Kluwer, 1998

F. Cafaggi, H. M. Watt, *The Regulatory Function of European Private Law*, Edward Elgar, 2009

F. B. Cross, R. L. Miller, *The Legal Environment of Business: Text and Cases*, 7th ed., South Western, Cengage (2007 ed., 2009)

F. Lopez-Pena, R. J. Duro, *A Hyperspectral Based Multisensor System for Marine Oil Spill Detection, Analysis and Tracking* in *Knowledge-Based Intelligent Information and Engineering Systems*, Springer, vol. 3213, 2004

F. Maes, *Marine resource damage assessment: liability and compensation for environmental damage*, Springer, 2005

F. Berlingieri, *The Hamburg Rules: a choice for the EEC?*, European Institute of Maritime and Transport Law, Maklu Uitgevers, 1994

F. D. Busneli, G. Comande, *Non-Pecuniary Loss under Italian Law* in *Damages for non-pecuniary loss in a comparative perspective* (W. V. H. Rogers ed.), Springer, 2001

F. D. Busnelli, G. Comande, *Some Remarks on the Borderline between Contract and Tort Liability* in *Unification of Tort Law: Wrongfulness* (H. Koziol, F. D. Busnelli eds.), Kluwer, 1998

F. D.Busnelli, G.Comande, *Unification of Tort Law: Damages*, Kluwer, 2001

F. G. Clarke, *The American ship-master's guide and commercial assistant: being an enlargement of the seaman's manual*, Allen & Co., 1838

F. Wilmer, *The Social Construction of Man, the State, and War: identity, conflict, and violence in former Yugoslavia*, Routledge, 2002

Galand-Carval, *Causation under French Law* in *Unification of Tort Law: Causation* (J. Spier eds.), Kluwer, 1996

Gauci, *Oil Pollutoin at Sea – Civil Liability and Compensation for Damage*, John Wiley & Sons, 1997

G. Betlem, *Liability for Damage to the Environment* in *Towards a European Civil Code* [A. S, Heartkamp et al. (eds.)], Kluwer, 1998

G. E. White, *Tort Law in America: an intellectual history*, Oxford, 1980

G. H. Smith, *The Law of Private Right*, Biblio Bazar LLC, 2008
Gilead, *Unification of Tort Law: Strict Liability*, Kluwer 2002
G. J. de Matons, *La responsabilité de l'Etat pour le fonctionnement de certains services maritimes*
G. Kodilinye, *Commonwealth Caribbean tort law*, 3rd ed., Cavendish, 2003
G. L. Canfield, G. W. Dalzell, *The law of the sea: a manual of the principles of admiralty law for students*
Green, M. Fernando on Spain in *Limitation of Liability for Maritime Claims* (P. Griggs, R. Williams eds.), 3rd ed., LLP, 1998
Griggs and Williams, "*Limitation of Liability for Maritime Claims*", 2ed, 1991
G. Samuel, *Law of obligations and legal remedies*, 2nd ed., Cavendish, 2001
G. Stephenson, *Sourcebook on Torts*, 2nd ed., Cavendish, 2000
G. Sunlai on China in *Limitation of Liability for Maritime Claims* (P. Griggs and R. Williams eds.), 3rd ed., LLP, 1998
G. T. Schwartz, *Causation under US Law* in *Unification of Tort Law: Causation* (J. Spier eds.), Kluwer, 1996
G. V. Scammell, *Shipowning in England c. 1450 – 1550*, Transaction of the Royal Historical Society (1962), 5th Series, Cambridge
G. W. Paulson, *A reassessment of the marine salvage posture of the United States*, National Academy of Sciences, 1994
H. A. Causy, *Risks and Uncertainties in the Law of Tort* in *European Tort Law 2006* (H. Koziol, B. C. Steininger eds.), Springer, vol. 2006, 2008
H. A. Cousy, A.Vanderspikken, *5.6.2., Unification of Tort Law: Damages* in *Principles of European Tort Law* (U. Magnus ed.), Kluwer, 2001
Halisbury, *Laws of England*, Vol. 49
Hall, *International Law*, 8th ed.
H. Carty, *An analysis of the economic torts*, Oxford, (2001 rep. 2003)
H. Cousy, D. Droshout, *Non-Pecuniary Loss under Belgian Law* in *Damages for non-pecuniary loss in a comparative perspective* (W. V. H. Rogers ed.), Springer, 2001
H. G. Beale, W. D. Bishop, M. P. Furmston, *Contract: Cases and Materials*, 5th ed., Oxford, 2008
H. Koziol, B. C. Steininger, *European Tort Law 2002*, Springer, 2003
Hodges, *Law of marine insurance*, Cavendish, 1996
H. Schussler, *Process for monitoring ship traffic at sea while recognizing and oil spills and potential ship collisions*, 1997
H. W. Woolrych, *A practical treatise on the commercial and mercantile law of England*
H. Xue, *Transboundary damage in international law*, Cambridge Studies in International and Comparative Law, Cambridge, 2003

I. Carr, P. Stone, *International Trade Law*, 3rd ed., Cavendish, 2005
I. C. Varotsi, D. A. Pentsov, *The International Safety Management Code* in *Maritime Work Law Fundamentals: Responsible Shipowners, Reliable Seafarers*, Springer, 2008
I. Gilead, *Non-Pecuniary Loss – The Israeli Report*, Appendix 1 in *Damages for non-pecuniary loss in a comparative perspective* (W. V. H. Rogers ed.), Springer, 2001
IMO, MEPC, *Manual on oil pollution*, Ch. 2, IMO, (1983 2nd ed., 1997)
IMO, *Port State Control*, IMO, (1995 ed., 2001)
IMO, *Procedures for Port State Control*, IMO, 2001
I. M. Sinclair, *The Vienna Convention on the Law of Treaties*, Manchester Univ. Press, 1973
J. A. Pontier, E. Burg, *EU Principles on jurisdiction and recognition and enforcement of judgments according to the case law of the European Court of Justice*, T. M. C. Asser, 2004
James Crawford, UN ILC, *The International Law Commission's articles on state responsibility*, Cambridge, 2002
J. Basedow, U. Magnus, *Prevention of Marine Pollution*, Pollution of the Sea – Prevention and Compensation, Hamburg Studies on Maritime Affairs, vol. 10, part II, Springer
J. B. Berryman, *Remedies: Cases and Materials*, 5th ed., Emond Montgomery, 2006
J. B. Loomis, *Contingent Valuation Methodology and the US Institutional Framework* in *Valuing Environmental Preferences: Theory and Practice of the Contingent Valuation Methods in the US, EU, and Developing Countries* (I. J. Bateman et. al. eds.), Oxford, 1999
J. Brunnee, H. M. Kindred, P. M. Saunders, *International Law, Chiefly as Interpreted and Applied in Canada*, 7th Ed, 2006, Emond Montgomery Publications Ltd.
J. C. Care, *Contract law in the South Pacific*, Cavendish, 2001
J. C. D. Clark, *Revolution and Rebellion: state and society in England in the seventeenth and eighteenth centuries*, Cambridge (1986 rep. 1990)
J. D. Tracy, *The Rise of Merchant Empires: long-distance trade in early modern world 1350 – 1750*, Cambridge (19990 rep. 1993)
J. F. Berry, M. S. Dennison, *The environmental law and compliance handbook*, The McGrow Hill, 2000
J. G. Speight, *The Chemistry and Technology of Petroleum*, 4th ed, Taylor & Francis Group, CRC Press, 2007
J. H. Badgley, J. W. Lewis, *Peasant rebellion and Communist revolution in Asia*, Stanford, 1974

J. Hodgson, J. Lewthwaite, *Tort Law Textbook*, 2nd ed., Oxford, 2007
J. L. Mathews, *The Lawsuit Survival Guide: A Client's Companion to Litigation*, 2003
J. L. Sax, "Defending the Environment", 1971
J. Murphy, H. Street, *Street on Torts*, Oxford, 13th ed, 2007
J. Neethling, 5.2.1., *Unification of Tort Law: Damages* in *Principles of European Tort Law* (U. Magnus ed.), Kluwer, 2001
J. O'Brian, *International law*, Cavendish, 2001 rep. 2002
John Kruse, *Bialiff's Law: A Lawful Trespass: v 1: Development of the Common Law to the Present Day*, Wildy, Simmonds and Hill Publishing, 2009
John Spruyt, "Ship Management", Ch.1, 2ed, LLP 1998
J. Rohart on France in P. Griggs and R. Williams, "*Limitation of Liability for Maritime Claims*", 3rd Edition, LLP, 1998
J. Rush, M. Ottley, *Business Law*, Thomson, 2006, pp 200–201 *D*. Kelly, A. E. M. Holmes, R. Heyward, *Business Law*, 5th ed., Cavendish, 2005
J. Spier, *Unification of Tort Law: Causation*, Kluwer, 1996
J. Spier, *Uncertain Partial Causation* and *Uncertain Causes in the Victim's Sphere* in *Principles of European Tort Law: text and commentary* (European Group on Tort Law eds.), Springer, 2005
J. Steele, *Tort Law: Text, Cases and Materials*, Oxford, 2007
J. Teichman, *The Definition of Person*, Philosophy, Cambridge, vol. 60, No. 232 (Apr., 1985)
J. Y. Gotanda, *Supplemental damages in private international law*, Kluwer, 1998
K. D. Kerameus, *Non-Pecuniary Loss under Greek Law* in *Damages for non-pecuniary loss in a comparative perspective* (W. V. H. Rogers ed.), Springer, 2001
Kerameus, *Unification of Tort Law: Damages* in *Principles of European Tort Law* (U. Magnus ed.), Kluwer, 2001
K. J. Goodacre, *Marine Insurance Claims*, 3rd ed., Witherby, 1996
K. J. Holsti, *The State war, and the state of war*, Cambridge Studies in International Relations, Cambridge, 1996
K. J. Rawson, *The carriage of bulk oil and chemicals at sea*, Chameleon, 1994
K. M. Meessen, *Extraterritorial jurisdiction in theory and practice*, Kluwer, 1996
K. N. Waltz, *Man, the State, and War: a theoretical analysis*, Columbia Uni., Press, 2001
Koch, *Contributory Conduct or Activity* in *Principles of European Tort Law*, Springer, 2005
K. Roepke, *Gluabigerschutzregime im europaischen Wettbewerb der Gesselschaftsrecht* in *Schriften zum europaischen unde internationalen privat-, Bank- und wirtschaftsrecht*, De Gruyter , 2007

Law in Environmental Decision-Making: National, European and International Perspectives (T. Jewel, J. Steele eds.,1998), Clarendon Press Oxford
L. Bergkamp, *Liability and Environment*, Kluwer 2001
L. de la Fayette, Achievements in 2001, Environmental Policy and Law, vol. 32, no. 3–4, 2002
Ling Zhu, *Compulsory Insurance and Compensation for Bunker Oil Pollution Damage*, Hamburg Studies on Maritime Affairs 5, Springer, 2007
L. Kurukulasuriya, N. A. Robinson, *Training manual on international environmental law*, UNEP, 2006
L. Moir, *The Law of Internal Armed Conflict*, Cambridge (2002 rep. 2003)
L. Oppenheim, Sir. R. Roxburgh, International law: a treatise, vol. 1, 3rd ed., The Law Book Exchange, 2008
L. S. Seally, S. Worthington, *Cases and materials in company law*, 8th ed., Oxford, 2008
Lucchini, "La pollution des mers par les hydrocarbures : les Conventions de Bruxelles de novembre 1969 ou les fissures du droit international classique" ("Marine oil pollution: the 1969 Brussels Conventions or the cracks in conventional international law"), JDI, 1970
L. Williams, *Navigational Aids*, Cavendish, 2008
Maitland, Frederic, William, *English Law and Renaissance*, Cambridge Univ. Press, 1901 (Rep. The Law Book Exchange Ltd., 2000)
Malcolm Shaw, *International Law*, 6 ed., 2008, Cambridge
Malcolm Willingale, "Ship Management", Ch. 2, 3ed, LLP 1998
M. Anderson, *Frontiers: territory and state formation in the modern world*, Blackwell, 1996 ed. 1997
Marine Environment Law in Yearbook of International Environmental Law, vol. 4 (G, Handl et al., ed.), Lloyds of London Press Ltd., 1994
Martin-Casals, Ribot, Sole, *Unification of Tort Law: Strict Liability*, Kluwer 2002
M. D. Goener-Oezbek, *The Carriage of Dangerous Goods by Sea*, Hamburg Studies on Maritime Affairs 12, Springer, 2008
M. D. Guner-Ozbek, *The Carriage of Dangerous Goods by Sea*, Hamburg Studies on Maritime Affairs, vol. 12, Springer, 2008
M. D. Miller, *Marine War Risks*, 2nd ed., Lloyds 1994
M. Dockray, K. R. Thomas, *Cases & materials on the carriage of goods by sea*, 3rd ed., Cavendish, 2004
M. Faure, J. Boyd, *Deterrence, insurability, and compensation in environmental liability: future developments in the European Union*, Vol. 5, Springer, 2003
M. G. Faure, J. Hu, *Prevention and Compensation of Marine Pollution Damage: Recent Developments in Europe, China and the US*, Kluwer, 2006

M. G. Faure, W. H. van Boom, Concluding Remarks of *Shifts in Compensation Between Private and Public Systems*, Springer, 2007
M. H. Nordquist, S. N. Nandan, S. Rosenne, *United Nations Convention on the Law of the Sea, 1982: a commentary*, Kluwer, 2002
M. H. Wissink, W. H. van Boom, *Non-Pecuniary Loss under Dutch Law* in *Damages for non-pecuniary loss in a comparative perspective* (W. V. H. Rogers ed.), Springer, 2001
M. H.Wissink, W. H van.Boom, 5.6.21., *Unification of Tort Law: Damages* in *Principles of European Tort Law* (U. Magnus ed.), Kluwer, 2001
M. Jacobson, Director, IOPC Fund, *Forward to the First Edition of* Wu Chao, *Pollution from the Carriage of Oil by Sea: Liability and Compensation*, Kluwer, 1996, at xiii
M. L. Busch, *Overlapping Institutions, Forum Shopping, and Dispute Settlement in International Trade*, Int'l Organization (2007)
M. L. Hendrikse, N. H. Margetson, *Aspects of Maritime Law: Claims under Bills of Lading*, Kluwer, 2008
M. Lukas, *Fault Liability, Tort Law of the European Community*, (2008) Springer
M. Lunney, K. Oliphant, *Tort Law Text and Materials*, 3rd ed., Oxford
M. M. Casals, J. Ribot, J. Sole, *Non-Pecuniary Loss under Spanish Law* in *Damages for non-pecuniary loss in a comparative perspective* (W. V. H. Rogers ed.), Springer, 2001
M. Nesterowicz, E. Baginska, *Non-Pecuniary Loss under Polish Law* in *Damages for non-pecuniary loss in a comparative perspective* (W. V. H. Rogers ed.), Springer, 2001
Modern Maritime Law and Risk Management, 2nd ed., 2009, ISBN 978 1 84311 823 7
M. Ottley, *Briefcase on company law*, 2nd ed., Cavendish, 2002
M. Shaw, *International Law*, 5th ed., Cambridge
M. Spadaccini, *Forming an LLC: In Any State*, Entrepreneur Media Inc., 2007
M. Spadaccini, *The Operations Manual for LLCs*, Entrepreneur Media Inc., 2008
M. Stopford, *Maritime Economics*, Routledge, 1997 (Rep. 1999, 2000)
M. W. D. White, *Australian Marine Pollution Laws*, 2nd ed., The Federation Press, 2007
M. W. D. White, *Australian maritime law*, 2nd ed., The Federation Press
National Research Council (U.S.), Committee on Marine Salvage Issues, *Purposeful jettison of petroleum cargo*, National Academy of Sciences, 1994
Neethling, *Element of Causation in South African Law of Delict* in *Unification of Tort Law: Causation* (J. Spier eds.), Kluwer, 1996
N. Hooke, *Maritime Casualties, 1963–1996*, 2nd ed., 1997, LLP Limited, London

NILOS, *International Organizations and the Law of the Sea: Documentary Yearbook*, Kluwer, 1999
N. Schrijver, *Sovereignty over natural resources: balancing rights and duties*, Cambridge, 1997
O. Moreteau, *Revisiting the Grey Zone between Contract and tort: The Role of Estoppel and Reliance in Mapping out the Law of Obligations* in *European Tort Law 2004* (H. Koziol, B. C. Steininger eds.), 2005
O'May, *Marine Insurance*, Sweet & Maxwell, London, 1993
O. Schachter, *International law in theory and practice*, Martinus Nijhoff, Kluwer, 1991
P. Aspden, *Shipping law faces Europe: European policy, competition and environment*, Maklu Uitgevers, 1995
P. A. Symonds, O. Eldhom, J. Muscle, G. F. Moore, *Characteristics of Continental Margins*, Ch. 4, in *Continental Shelf Limits: the scientific and legal interface*, (P. J. Cook, C. Carleton), Oxford, 2000
P. Birts, *Remedies for Trespass*, Sweet & Maxwell, 1990
P. Cane, "The Anatomy of Tort Law", Oxford, Hart Publishing 1997
P. Clarke, *Environmental law toolkit – NSW: a community guide to environmental law*, 5th ed., The Federation Press
Peter Ehlers, *Origin and Compensation of Marine Pollution – A Survey*: Pollution of the Sea – Prevention and Compensation, Hamburg Studies on Maritime Affairs, Part III, 2007, Springer Berlin Heidelberg
P. Griggs, "Limitation of Maritime Claims: The Search for International Uniformity", Lloyds Maritime and Commercial Law…, 1997 – Lloyds of London Press
Philosophy and the Law of Torts (G. J. Postema ed. 2002) Cambridge, 2002
P. I. Blumberg, *The Multinational Challenge to Corporation Law: The Search for a New Corporate Personality*, Oxford, 1993
P. J. Rowe, *The Gulf War 1990–91 in International and English Law*, Sweet & Maxwell, 1993
P. Lehr, *Violence at Sea: Piracy in the age of Global Terrorism*, Taylor and Francis, 2007
Potter, Harold, *An Historical Introduction to English Law and its Institutions*, London, Sweet and Maxwell, 1932 (4th Ed. 1958)
P. P. Craig, G. de Burca, *EU Law: Text, cases, and materials*, 4th ed., Oxford, 2008
P. Rawcliffe, *The challenge of Rio: environmental pressure groups and the UNCED process* in *Environmental pressure groups in transition*, Manchester Univ. Press, 1998
P. R. Barnett, *Res judicata, estoppel, and foreign judgments: the preclusive effects of*, Oxford, 2001

Prof. Staniland, Hilton, on South Africa in *Limitation of Liability for Maritime Claims*, (P. Griggs and R. Williams eds.), 3rd ed., LLP, 1998
P. Sands, *Principles of International Environmental Law*, 2nd ed., Cambridge, 2003
P. Simon, *La réparation civile des dommages causes en mer par les hydrocarbures* (Civil compensation for damage at sea caused by oil), unpublished thesis, 1976
P. Stone, *EU private international law : harmonization of laws*, Edward Elgar, 2006
P. Tercier, *Short comments Concerning Non-Pecuniary Loss Under Swiss Law* in *Damages for non-pecuniary loss in a comparative perspective* (W. V. H. Rogers ed.), Springer, 2001
P. Widmer, W. H. van Boom, *Unification of Tort Law: Fault*, Kluwer, 2005
R. A. Cahill, *Strandings and their causes*, London, Fairplay Publications Inc., (1985)
R. Brillinger, *Canadian Business Law*, Edmond Montgomery, 2007
R. C. Force, A. N. Yiannopoulos, M. Davies, *Admiralty and Maritime Law, Volume 2*, Beard, 2005
R. C. McCullough, *Doctrine of Contributory Negligence and Comparative Negligence*, South Carolina Trial Lawyer Transcript, 1989
R. Force, M. Davies, *Jurisdiction and forum selection in international maritime law*, Kluwer, 2005
R. Force, M. Davies, *Jurisdiction and forum selection in international maritime law: essays in honour of Robert Force*, Kluwer, 2005
R. J. B. Boonekamp, *Schadevergoeding*
R. J. Dupuy, D. Vignes, *A Handbook on the new law of the sea*, vol. 2, Kluwer
R. J. Dupuy, D. Vignes, Hague Academy of International Law, *A Handbook on the new law of the sea*, vol. 2, Kluwer
R. LeRoy Miller, G. A. Jentz, *Fundamentals of Business Law: Summarized Cases*, 8th ed., South Western Cangage, 2007
R. McCracken, G. Johns, J. Pereira, *Statutory Nuisance*, Bloomsbury Publishing PLC, 2008
R. N. Reeve, *Introduction to environmental analysis*, University of Sunderland, 2002
R. Petrow, *In the wake of the Torrey Canyon*, New York, David Mackay Co, (1968)
R. Price, A. Haberbeck, *The maritime laws of the Arabian Gulf Cooperations Council States*, vol. 1, Graham & Trotman, 1986
R. R. Churchill, A.V. Lowe, *The Law of the Sea*, 3 ed., 1999
R. Rodiere, , E. du Pontavice, *Droit Maritime*, 11th ed., Precis Dalloz, Paris, 1991

R. Schaffer, F. Agusti, B. Earle, *International Business Law and its Environment*, South-Western Cengage, (2005 ed., 2009)

R. Tomlinson, *French Peasants in Revolt: the Insurrection of 1851*, (1982) by Ted. W. Margadant. Princeton University Press, 1979

R. V. Randle, *Oil Pollution Deskbook*, Environmental Law Institute, Washington, 1991

R. Wacks, *The New Legal Order in Hong Kong*, Hong Kong University Press, 1999

R. W. Lee, *The Elements of Roman Law*, 4 Ed

R. W. Wright, *Causation in Tort Law*, in *The International Library of Essays in Law and Legal Theory, Tort Law* (E. Weinrib ed.), Dartmouth, 1991

R. Zimmermann, *The law of obligations: Roman foundations of the civilian tradition*, Oxford (1990 ed., 1996)

S. Casey, *Set phases on stun, and other true tales of design, technology, and human error* (1998), Santa Barbara, Aegean Publishing Co., 2nd edition

S. Chertok, T. Coburn, *Jurisdictional Competition in the European Community*, Expresso Preprint Series, 2005, paper 870

Schwartz, *Unification of Tort Law: Damages* in *Principles of European Tort Law* (U. Magnus ed.), Kluwer, 2001

Schwartz, *Unification of Tort Law: Strict Liability*, Kluwer, 2002

S. Deneulin, M. Nebel, N. Sagovsky, *Capabilities and Rights*, in Transorming Unjust Structures The Capability Approach, Springer, 2006

S. G. Carval, *Non-Pecuniary Loss under French Law* in *Damages for non-pecuniary loss in a comparative perspective* (W. V. H. Rogers ed.), Springer, 2001

S. Goulding, *Company Law*, 2nd ed., Cavendish, 1999

S. Harwood, *Shipping finance*, 3rd ed., Euromoney Institutional Investor Plc, 2006 rep. 2008

S. Hodges, *Law of Marine Insurance*, Cavendish, 1996

S. Hodges, R. Carlile, *Case and materials on marine insurance law*, Cavendish, 1999

S. K. Verma, *An Introduction to Public International Law*

S. Mankabady, *The International Maritime Organization*, vol. 2, IMO, 1987

S. P. Baumgartner, *The proposed Hague Convention jurisdiction and foreign judgments: trans*, Paul Siebeck, 2003

S. Schmidt, *Gulf War II (1990/91) – Iraq between United Nations' Diplomacy and United States Policy*, GRIN Verlag, 2006

S. S. Huebner, *Marine Insurance*, Biblio Bazaar, 2008

S. Sime, *A Practical Approach to Civil Procedure*, 10th ed., Oxford, 2007

Stephan Rammeloo, "Corporations in Private International Law, A European Perspective", Oxford, 2001

S. Thacker, H. Ajaib, *Limitation of Liability for Maritime Claims* (P. Griggs and R. Williams eds.), 3rd ed., LLP, 1998
T. Alderton, Seafarers International Research Centre, *The global seafarer: living and working conditions in a globalized industry*, ILO, 2004
T. J. Biersteker, *State, Sovereignty and Territory* in *Handbook of International Relations* (W. Carlsnaes, B. A. Simmons, T. Risse, 2002 ed. 2006), SAGE Publications
Tim Hillier, *Sourcebook on Public International Law*, Routledge – Cavendish, 1998
T. Karatzas, "*The Greek Code of Private Maritime Law*" Kluwer, 1982
T. O'Riordan, J. Cameron, *Interpreting the Precautionary Principle*, Earthscan Publications Ltd., London, 1994
U. Magnus, *Causation under German Law* in *Unification of Tort Law: Causation* (J. Spier eds.), Kluwer, 1996
U. Magnus, J. Fedtke, *Non-Pecuniary Loss under German Law* in *Damages for non-pecuniary loss in a comparative perspective* (W. V. H. Rogers ed.), Springer, 2001
U. Magnus, J. Fedtke, *Unification of Tort Law: Damages* in *Principles of European Tort Law,* Kluwer, 2001
U. Magnus, *Unification of Tort Law: Damages*, Kluwer, 1997
United Nations, Division for Ocean Affairs and the Law of the Sea, *Digest of international cases on the law of the sea, Part 54*, United Nations, 2007
Van Caenegem R. C., *The Birth of the English Common Law*, Cambridge Univ. Press, 1973 (Rep. 1988)
V. Harpwood, *Principles of Tort Law*, 4th ed., Cavendish, 2000
Von Bar, "The Common European Law of Torts", Volume 2, Oxford University Press
W. H. van Boom, H. Koziol, C. A. Witting, L. Bloch, *Pure economic loss* in *Tort and Insurance Law, Volume 9*, Springer, 2004
W. L. Prosser, W. P. Keeton, D. B. Dobbs, R. E. Keeton, D. G. Owen, *Prosser and Keeton on Torts*, 5th ed., West Publishing, 1984
W. P. Stastky, Ch. 2, Foreseeability in Tort Law in *Essentials of Torts*, 2nd ed, West Legal Studies, Thomson Larning, 2001
W. T. Miller, , "A Summary of Intragroup Liability in Light of the Oil Pollution Act of 1990", document contained in documentation of conference 'Oil Pollution – Claims, Liability and environmental Concerns', IBC 23 October 1991
W. van Gerven , *Cases, Materials and Text on National, Supranational and International Tort Law*, Hart, 2000
W. V. H. Rogers, *Causation under English Law* in *Unification of Tort Law: Causation* (J. Spier eds.), Kluwer, 1996

W. V. H. Rogers, *Non-Pecuniary Loss under English Law* in *Damages for non-pecuniary loss in a comparative perspective* (W. V. H. Rogers ed.), Springer, 2001

W. V. H. Rogers, *Unification of Tort Law: Damages* in *Principles of European Tort Law* (U. Magnus ed.), Kluwer, 2001

W. V. H. Rogers, W. H. van Boom, *Unification of Tort Law: Multiple Tortfeasors*, Kluwer, 2004

W. V. H. Rogers, *Winfield and Jolowicz on Tort*, Sweet & Maxwell, 12th Ed, 1984

Wu Chao, *Pollution from the Carriage of Oil by Sea: Liability and compensation*, Kluwer Law International, 1996

X. Chen, *Limitation of liability for maritime claims: a study of U. S. law, Chinese law and International Conventions*, Kluwer, 2001

Z. Wang, S. A. Stout, *Oil spill environmental forensics: fingerprinting and source identification*, Elsevier Inc., 2007

Cases

Aden v. Flood (1891) A.C. 1

Admiralty Commissioners v. SS Volute [1922] 1 AC 129 HL

Amici Curiae Brief, Norfolk & Western Railway Co. v. Ayers, 123 S. Ct. 1210 (2003) (August 19, 2002)

Amoco Cadiz Litigation [1984] 2 Lloyds Report 304, 1992 AMC 913, 954 F.2d 1279 (1992 US App., 7th Cir.)

Athens Maritime Enterprises Corp., v. Hellenic Mutual War Risks Association (Bermuda), [1983] All ER 590

Atlantic Insurance v. King [1919] 1 K.B. 309

Banque Monetaca and Carystuiaka and another v. Motor Union Insurance Co. Ltd., (1923)

Bardeche v. Jonier (Bull. Civ. 1987. II. 86) D 1988. Jur 32

BP Exploration Operating Co. Ltd. v. Chevron Transport (Scotland), [2001] UKHL 50

Bramley Moore Case [1964] 2 Lloyds' Rep. p. 200

Burgess v. M/V Tamano case (1973) AMC 1939

Byrne v. Boadle Court of Exchequer(1863) 2 H. and C. 722, 159 Eng. Rep. 299

Caltex Oil (Australia) Pty. Ltd. v. The Dredge "Willemstad" (1976) 136 CLR 529

Caltex v. BP [1996] 1 Lloyds Rep. 286

Candlewood Navigation Corporation Ltd. v. Mitsui OSK Lines Ltd. [1985] 2 Lloyds' Rep. 303, PC

Centros Case [1999] E. C. R. I-1494

Chorzow Factory (Indemnity) (Jurisdiction) Case, PCIJ Ser. A, No. 8/9, (1927)
Commune de Mesquer v. Total France SA [2008] 3 CMLR 16, [2009]
Dairy Engineering Company of Asia v. American Home Assurance Company [1971] A.M.C. 1001
ED & F Man Sugar Ltd., v. Lendoudis, [2007] EWHC 2268 (Comm), [2008] 1 All ER 952, JIML 14 [2008] 1:17
Exxon Valdez Litigation, 1995 AMC 1440, 1997 AMC 940
Farrell Lines Inc. v. Jones, 530 F.2d 7 (5th Cir. 1976)
Fontaine v. British Columbia (Official Administrator) (1998) 1 S.C.R. 424 (Canada)
Fowler v. Lanning (1959) 2 W.L.R. 249
Froom v. Butcher [(1976) QB 286]
Gilbert v. Corporation of Trinity House 17 QBD 795 (1886)
Golnoy Barge Co. v. M/T Shinoussa, 1994 A.M.C. 1378–1379
Gregory v. Kelly [(1978) RTR 426]
Hashem v. Shayif & Anor [20008] EWHC 2380 (Fam)
Hedley Byrne & C. Ltd. v. Heller & Partners Ltd. [1964] A.C. 465, HL
IMT Shipping and Chartering GmbH v. Chansung Shipping Co. Ltd., (The Zenovia) [2009] EWHC 739 (Comm)
In Re Glacier Bay, (1991 AMC at p. 750)
Joel v. Morison (1934) 6 C & P 501
Jones v. Livox Quarries Ltd., [1952] 2 Q.B. 608
Jose Marti Case 1981
Lennard's Carrying Co. Case [1915] A.C 705
Letang v. Couper (1964) 2 All E.R. 929
Mareva Compania Naviera S.A v. International Bulk Carriers S.A., [1975] 2 Lloyds' Report 509
Maritime Services Board of New South Wales v. Poseidon Navigation Inc. [1982] NSWLR 72
Mulholland & Tedd Ltd., v. Baker (1939) 3 All ER 253
Nance v. British Columbia Electric Ry Co., Ltd., [1951] AC 601, [1951] 2 All ER 448
National Oil Company of Zimbabwe and Other v. Sturge [1991] 2 Lloyd's Rep. 281
Nesbitt v. Luslington (1792)
Nichols v. Marsland (1876) 2 Ex. D. 1
Nissos Amorgos case 1997
Norfolk & Western Ry. Co. v. Ayers, No. 01-963(U.S. Aug. 19, 2002)
Norsk Pacific Steamship Co. Ltd. v. Canadian National Railway Co. 91 DLR 4th 289

Northwestern Utilities Ltd., v. London Guarantee and Accidents Ltd, [1936] A.C. 108
O.P.E Shipping Ltd., and El Povernir Shipping Company Inc. v. Allstate Insurance Company Inc. and Others [1983] A.M.C. 22–30
Pakistan National Shipping Corp., Case [2003] 1 All ER 173 at 180
Pan American World Airways Inc. v. The Aetna Casualty and Surety Co., 505 F.2d 989, [1974] 1 Lloyd's Rep. 207, [1975] 1 Lloyd's Rep. 77
Perry v. Kendricks Transport Ltd., [1956] 1 W.L.R. 87
Pesquerias y Secaderos de Bacalao de Espana v. Beer, (1945) 79 Ll. L. Rep. 417; (1947) 80 Ll. L. Rep.
318; (1948–49) 82 Ll. L. Rep. 501; [1949] 1 All E. R. 845
Piermay Shipping Co. v. A. H. Chester [1978] 1 WLR 411
Polpen v. Commercial Union [1943] 1 K.B. 161
R. v. Caldwell [1982] A. C. at p. 354
R v. Goodwin (2006)1 Lloyd's Rep. 432
R. v. Lawrence Stephen [1982] A. C. 520
Re Piracy Jure Gentium [1934] AC 586
Republic of Bolivia v. Indemnity Mutual Marine Insurance Co. Ltd., [1909] 1 KB 785
Rice v. Harken Exploration Co., [2001] WL 422051 (5th Cir. Tex.)
Robins Dry Dock and Repair Co. v. Flint, 275 US 303, 1928 A.M.C. 61 (1927)
Rylands v. Fletcher (1865) 3 H. & C. 774; 159 ER 737 (Court of Exchequer); (1866) LR 1 Ex. 265 (Court of Exchequer Chamber); (1868) LR 3 HL 330 (House of Lords)
Sanfield Building Contractors Ltd v. Li Kai Cheong (2003) (Hong Kong)
Scott v London & Catherine Dock Co (1865) 3 H&C 596 (Scotland)
Southport Corporation v. Esso Petroleum Co. Ltd. (1956) 2 W.L.R. 93
Spartan Steel & Alloys Ltd. v. Martin & Co. (Constructors) Ltd., (1973) 1 QB 27, CA
Spencer Kellogg & Sons, Inc. v. Hicks (*The Linseed King*), 285 U.S. 502 (1932)
Spinney's (1948) Ltd., and Others v. Royal Insurance [1980] 1 Lloyd's Rep. 406
Stapley v. Gypsum Mines Ltd. [1953] AC 663
States S.S. Co., v. United States, 259 F.2d 458, 466, 1957 AMC 1181 (9th Cir. 1957)
Stevens v. Woodward (1881) 6 QBD 318
Takahashi v. Fish and Games Comm'n, 334 US 410 (1948)
The Capitan San Luis [1994] 1 All ER 1016
The Eurysthenes [1976] 2 Lloyd's Rep. 171
The Irish Stardust case [1977] 1 Lloyd's Rep. 195
The Marel Case [1994] 1 Lloyd's Rep. 624
The Marion [1984] 2 Lloyd's Rep. 1

The Pactolus (1856) Swab.173
The Scaptrade Case [1988] 2 Lloyd's Rep. 253
The Stonedale (No.1) [1956] AC 1 (H.L.)
Uberseering BV v.. Nordic Construction Company Baumanagement GmbH (NCC), [2002] E. C. R. I-0991
Ultramares Corp. v. Touche, [1931] 255 N.Y. 170, 179
United Kingdom v. Albania, I.C.J. Rep., 1949

Electronic Publications

A. Calnan, *Strict Liability and the Liberal Justice Theory of Torts*, March 2, 2007, availableathttp://works.bepress.com/cgi/viewcontent.cgi?article=1000&context=alan_calnan

A. E. Roberts, *Traditional and Modern Approaches to Customary International Law: A Reconciliation*, The American Journal of International Law, vol.95, 757, available at http://www.asil.org/ajil/roberts.pdf

A. Findley, *Practical Considerations in the Recovery of Oil from Sunken and Abandoned Vessels*, available at http://www.oilspillresponse.com/pdf/Articles%20From%20IOSC/Practical%20Considerations%20in%20the%20Recovery%20of%20Oil%20from%20Sunken%20and%20Abandoned%20Vessels.pdf

African Programme and International Law Conference Report, *Piracy and Legal Issues: Reconciling Public and Private Interests*, Oct., 1, 2009, available at http://www.chathamhouse.org.uk/files/15221_011009piracy_law.pdf

A. H. Kassem, *The legal aspects of seaworthiness: current law and development*, 2006, Doctoral thesis, Swansea University, Unpublished, available at http://eprints.ucl.ac.uk/6988/1/6988.pdf

A. H. Lasday, Texaco, Inc,; E. W. Martens, CheyronResearch Co., Fate and Effects of Oil on Marine Life: Progress Report on Research Sponsored by the American Petroleum Institute, Off-Shore Technology Conference, 3–6 May 1976, Huston, Texas, Paper No. 2449-MS, available at http://www.onepetro.org/mslib/servlet/onepetropreview?id=OTC-2449-MS&soc=OTC

A. Honore, *Causation in the Law*, Stanford Encyclopedia of Philosophy, Oct., 12, 2005, available at http://plato.stanford.edu/entries/causation-law/

A. Lewis, F. Merlin, P. Daling, M. Reed, *Applicability of Oil Spill Dispersants Part I: Overview*, European Maritime Safety Agency (EMSA), Aug., 2006, available at http://www.emsa.europa.eu/Docs/opr/emsa_manual_applicability_oil_spill_dispersants_2006.pdf

A. Kiss, D. L. Shelton, *Strict Liability in Environmental Law*, GWU Legal Studies Research Paper No. 345, available at http://papers.ssrn.com/sol3/papers.cfm?abstract_id=1010478

A. O. Sykes, *Transnational Forum Shopping as a Trade and Investment Issue*, http://www.law.virginia.edu/pdf/olin/0708/sykes.pdf

A. P. Y. Ho, *The Organizational and Political Resource of Environmental Pressure Groups in Hong Kong*, Asian Journal of Public Administration, available at http://sunzi1.lib.hku.hk/hkjo/view/50/5000318.pdf

A. Rodriguez, M. C. Hubbard, Fowler, Kingsmill, Flint, Gray, Chalos, LLP, *The International Safety Management (ISM) Code: A New Level of Uniformity*, available at http://www.frc-law.com/files/pub_ISM.pdf

A. W. Ando, et. al., *Natural Resource Damage Assessment: Methods and Cases*, available at http://www.istc.illinois.edu/info/library_docs/RR/RR-108.pdf

B. Grant, *What is a "ship": R v. Goodwin in the Court of Appeal*, [2006] 2 Web JCLI, available at http://webjcli.ncl.ac.uk/2006/issue2/grant2.html

B. H. Oxman, The Third United Nations Conference on the Law of the Sea: The Ninth Session (1980) available at http://www.questia.com/PM.qst?a=o&d=79271683

B. Kathy, *Preparedness to oiled wildlife incidents in Belgium*, available at http://www.interspill.com/PDFs/UK/2006/wildlife_oiledwildlife_doc.pdf

Belize, Registration of Merchant Ships Act, Ch. 236, Rev. ed., 2006, http://www.belizelaw.org/lawadmin/PDF%20files/cap236.pdf

Briefing note: *EU strengthens trials in absentia – Framework Decision could lead to miscarriage of justice*, available at http://www.openeurope.org.uk/research/tia.pdf

Bunkering: Code of Practice, Peel Ports, Oct., 20, 2008, available at http://www.merseydocks.co.uk/assets/pdf/Code-of-Bunkering.pdf

Bunkering Operations, http://www.pka.gov.my/Bunkering%20Checklist%20Appendix%20II.pdf

What is Crude Oil, http://www.opec.org/library/FAQs/CrudeOil/q1.htm

Bunkering Procedure in Singapore, available at http://www.shipowners.co.uk/lossprevent/pdfs/10-08%20BUNKERING%20PROCEDURES%20IN%20SINGAPORE.pdf

C. A. Fleischer, *Liability for Oil Pollution Damage Resulting from Offshore Operations*, available at http://www.cenneth.com/sisl/pdf/20-4.pdf

C. A. Jones, *Compensation for Natural Resource Damages from Oil Spills: A Comparison of USA Law and International Conventions and Pollution*, available at http://papers.ssrn.com/sol3/papers.cfm?abstract_id=1656

C. A. Kontovas, H. N. Psaraftis, *Marine Environment Risk Assessment: A Survey on the Disutility Cost of Oil Spills*, http://www.martrans.org/docs/ws2009/

Kontovas%20Psaraftis%20-%20Disutility%20Cost%20of%20oil%20spills.pdf

C. Connal, *Of Chairs, Trains and Escalators*, The Journal Online, Jun. 15, 2009, available at http://www.journalonline.co.uk/Magazine/54-6/1006655.aspx

C. J. Giaschi, *Carriage of Goods, Bills of Lading and Charterparties*, U. B. C. Maritime Law 332, available at http://www.admiraltylaw.com/UBC%20Law332/carriage_of_goods_outline.pdf

C. J. Giaschi, D. Williams, *The Maritime Liability Act: Parts 2, 3 and 4, Apportionment of Liability, Limitation of Liability and Carriage of Passengers*, Presented at Maritime Conference 2002 (Jan.), Toronto, Ontario, available at http://www.admiraltylaw.com/papers/MLA.htm

C. L. Brooker, *The Clash of Obligations: Exercising Extraterritorial Jurisdiction in Conformance with Transitional Justice*, available at http://works.bepress.com/christen_broecker/2/

Clyde & Co., *Shipping Update: Piracy*, Nov., 2008, available at http://www.clydeco.com/attachments/published/3001/Piracy%20(Nov%2008)%20(2).pdf

C. R. Chase, L. G. Roberts, *Advanced Detection Technology for Early Warning – The Key to Oil Spill Prevention*, available at http://www.slicksleuth.com/pdf/IOSC2008paper.pdf

C. R. Chase, S. van Bibber, *Utilization of Automated Oil Spill Detection Technology for Clean Water Compliance and Spill Discharge Prevention*, available at http://www.epa.gov/OEM/docs/oil/fss/fss06/chase_2.pdf

California Department of Fish and Game National Oceanic and Atmospheric Administration United Nations Fish and Wildlife Service National Park Service, *S. S. J. Luckenbach and Associated Mystery Oil Spills,* Nov., 2006, http://www.dfg.ca.gov/ospr/spill/nrda/luckenbach_final_darp.pdf

Ceccarelli, *The Price for Risk-Taking: Marine Insurance and Probability Calculus in the Late Middle Ages*, Electronic Journ@l for History of Probability and Statistics, vol. 3, no. 1, June 2007, available at http://www.emis.de/journals/JEHPS/Juin2007/Ceccarelli_Risk.pdf

C. J. Giaschi, *Limitation of Liability; Introduction / Case Summaries*, Giaschi & Margolis, available at AdmiraltyLaw.com, http://www.admiraltylaw.com/limitation.htm

C. M. Davis, Punitive Damages in Maritime Law: Exxon Shipping Co. v. Baker; Jun., 24, 2008 available at http://davismarine.com/articles/Exxon%20Shipping%20v%20Baker.pdf

CMI Guidelines on Oil Pollution Damage, Resolution 1994, http://web.uct.ac.za/depts/shiplaw/cmi/cmioil.htm

CMI, *Singapore 1, Implementation and Interpretation of 1976 LLMC Convention: Draft Report*, available at http://www.comitemaritime.org/singapore/implellmc/implemen_draft.html (last visited 01.01.2010)

Colin de la Rue and P. Murray, *Introduction – Compensation for Pollution from Ships' Bunkers* in *Oil Pollution from Ships – Current Legal Issues*, Shanghai International Maritime Forum, 2005, Ince & Co., available at http://www.docstoc.com/docs/12380468/POLLUTION-FROM-SHIPS-BUNKERS

Colin de la Rue, *Bunker Spill Risk*, http://www.incelaw.com/documents/pdf/News-PDF/Bunker-spill-Risk

Int'l Salvage union, *An exchange of view on responder immunity*, Salvage World, June 2007, available at http://www.marine-salvage.com/salvage_world/isu%20salvage%20world%20june%202007.pdf

Commercial Fishing License Update, State of California, The Natural Resources Agency, Department of Fish and Game, Feb., 2009, available at http://www.dfg.ca.gov/licensing/pdffiles/fg1554a.pdf

Cwinya-ai, Robert Ongom, *Implied Warranty of Seaworthiness Under United States Maritime Law: The Differences and Similarities to South African Marine Insurance Law* (December 21, 2008), available at SSRN: http://ssrn.com/abstract=1319064

D. B. Garrow, K. L. Ayre, *The Recovery of Non-Pecuniary Damages in Canada: The Cap on Recovery, Jury Trials, and other Unique Considerations for General Damage Awards*, available at http://monmexique.com/5-Garrow.pdf

D. Ferriere, *Corporate Responsibility: How Ship Owners can prevent Oil Spills*, available at http://www.iosc.org/papers/IOSC%202003%20a161.pdf

D. H. Andersen, H. J. Voth, *The Grapes of War: Neutrality and Mediterranean Shipping under the Danish Flag, 1750–1807*, Uni. of Oxford, available at http://www.nuffield.ox.ac.uk/Economics/History/Paper18/18www.pdf

D. Helton, *Putting Response and Natural Resource Damage Costs in Perspective*, Paper ID #114, 1999 International Oil Spill Conference, available at http://www.darrp.noaa.gov/library/pdf/costsofs.pdf

Directive 2004/35/CE of the European Parliament and of the Council of 21 April 2004 on Environmental Liability with regard to the Prevention and Remedying of Environmental Damage, Notice to Members No. 8, 2009–2010, July 2009, available at http://www.westpandi.com/WestPandI/NoticesToMembers/2009-2010/2009-08+2.htm

D. J. Blackmore, *The Australian Oil Spill Response Program*, AMOSC, available at http://www.pcs.gr.jp/doc/esymposium/12171/97_don-j-blackmore-e.pdf

D. J. Puchala, *Of Pirates and Terrorists: What Experience and History Teach*, Contemporary Security Policy, available at http://www.southchinasea.org/docs/Puchala,%20Pirates%20and%20Terrorists.pdf

D. J. R. White, I. C. White, *Is the oil industry's tiered preparedness and response approach applicable to oiled wild life response planning?*, http://www.oil-spillresponse.com/pdf/Published%20Papers/David%20White/paper_wildlife_tiered.pdf

D. Kanev, *Seaborne trade effects of international terrorism and effectiveness of the anti-terrorist policy*, Nikola Vaptsarov Naval Academy, Varna, Bulgaria, available at http://www.iamu-edu.org/generalassembly/aga6/pdf/s2-kanev.pdf

D. L. C. Miller, *Three of a Kind? Positive Prescription in Sri Lanka, South Africa and Scotland*, Electronic Journal of Comparative Law, vol. 12.1(May 2008), available at http://www.ejcl.org/121/art121-3.pdf

D. Loefgren, *Factual causation in Delict: a look at law of Scotland and Sweden*, European Young Lawyers, Eurodevils, 2002, available at http://www.law.ed.ac.uk/eyl/02repdl.htm

D. M. Engel, M. McCann, Introduction: *Tort Law as Cultural Practice* (July 8, 2009), Buffalo Legal Studies Research Paper No. 2009–15. Available at SSRN: http://ssrn.com/abstract=1431501

D. P. Christodoulou, "*Introduction to the Greek Legal System*" available at http://jurist.law.pitt.edu/world/greececor2.htm

D. S. Etkin, *Estimating Cleanup Costs for Oil Spills*, #168, 1999 Int'l Oil Spill Conf., available at http://www.environmental-research.com/site_files_base/publications/content_pdf/paper6.pdf

D. S. Etkin, *Clean up Costs for Oil Spills in Ports*, Oil Spill Intelligence Report, Arlington, MA, USA, available at http://www.philippines.canurb.com/projectsunrise/files/pdf/paper5.pdf

D. Shelton, A. Kiss, *Judicial Handbook on Environmental Law*, UNEP, 2005, available at http://www.unep.org/Law/PDF/JUDICIAL_HBOOK_ENV_LAW.pdf

D. Stalcup, G. Yoshioka, B. Kaiman, A. Hall, *Effects of OPA 90 Response Planning on Reported Spills*, available at http://www.iosc.org/papers/00640.pdf

Do the provision and work equipment regulations 1998 mean what they say? April Pl Focus vol. 19, issue 6, available at http://www.hughjames.com/pdf/PWER%20-%20Cenric%20Clement-Evans%20-%20July%202009.pdf

Dr. H. D. Nguyen, Eng. M. D. Dao, H. Q. Nguyen, Eng. T. T. Nguyen, *Oil Spills in Vietnam – Facts and Challenges*, http://www.iosc.org/papers/2008%20010.pdf

Dr. I. C. White, *Facilitating the speedy payment of oil spill compensation claims under the CLC and FUND Conventions*, http://www.itopf.com/_assets/documents/tianjin.pdf

Dr. S.C. Derrington, *The nature of the modern action in rem*, Paper presented to the Judges of the Admiralty Panel of the Federal Court of Australia, August

2007, available at http://www.docstoc.com/docs/16056932/The-Nature-of-the-Modern-Action-In-Rem---Federal-Court-of

E. Amato, N. Cardellicchio, R. Giangreco, T. H. Moller, *European Local Authorities role in preparedness and response to oil pollution to safeguard seafood production: state and perspectives*, Report of the European Workshop, 28–31 October 2003, Tarentum, Italy, available at http://ec.europa.eu/environment/civil/marin/reports_publications/taranto_final_report.pdf

EC Law Aspects of Hybrid Entities, Part II, available at http://www.ibfd.org/portal/pdf/Fibbe_Sampleexcerpt.pdf

Edgar Gold, *Liability and Compensation for Ship-Source Marine Pollution: The International System*, available at http://www.fni.no/YBICED/99_02_gold.pdf

E. M. Basse, *Environmental Liability – Modern Developments*, available at http://www.cenneth.com/sisl/pdf/41-2.pdf

EPA Office of Emergency and Remedial Response, *Responding to Oil Spills: The National Response System*, Ch. 7 in Understanding Oil Spills and Oil Response, available at http://www.epa.gov/oem/docs/oil/edu/oilspill_book/chap7.pdf

E. Røsæg, H. Ringbom, *Liability and Compensation with Regard to Places of Refuge*, Final Report, Oct., 2004, Scandinavian Institute of Maritime Law, Uni. of Oslo, pp 20-22, available at http://www.emsa.europa.eu/Docs/other/places%20of%20refuge%20study%20def.pdf

F. Coallier, R. Gerin-Lajoie, *Open Government Architecture: The evolution of De Jure Standards, Consortium Standards and Open Source Software*, Montreal, Feb., 2006, available at http://www.cirano.qc.ca/pdf/publication/2006RP-02.pdf

F. Blobel, *European Tort Jurisdiction and Pure Economic Loss*, The European Legal Forum (E) 3-2004, pp 187–191, available at http://www.simons-law.com/library/pdf/e/507.pdf

Flagship, *Bunker Pollution Convention may fill the last compensation gap*, iss. no. 4, April 2001, available at http://www.liscr.com/liscr/LinkClick.aspx?fileticket=D9Q0EhltD1w=&tabid=154&mid=538 (last visited on 07.01.2010)

Fowler, Rodriguez, Kingsmill, Flint, Gray, Chalos, *The International Convention on Civil Liability for Bunker Oil Pollution Damage, 2001*, available at http://www.frc-law.com/files/pub_international-convention-civil-liability.pdf

G. Calabresi, *Toward A Unified Theory of Torts*, Journal of Tort Law, vol. 1 : iss. 3, Article 1. 2007, available at: http://www.bepress.com/jtl/vol1/iss3/art1

G. de Costa, *Ship Registration in the Cayman Islands*, http://www.conyersdill.com/uploadedfiles/Article%20104%20Ship%20Reg%20in%20Cayman%20Islands.pdf

G. Plant, Third United Nations Conference on the Law of the Sea and the Preparatory Commission: Models for United Nations Law Making available at http://

journals.cambridge.org/action/displayAbstract;jsessionid=00FF63431474917003A43416528AEF9D.tomcat1?fromPage=online&aid=1512780
G. S. Nijar, *Public Interest Litigation: a matter justice an Asian perspective*, available at http://www.aseanlawassociation.org/9GAdocs/Malaysia.pdf
Gaskell, Nicholas and Forrest, *Marine pollution damage in Australia: implementing the Bunker Oil Convention 2001, and the Supplementary Fund Protocol 2003*, Uni. of Queensland Law Journal, 27(2), 103–165, available at http://eprints.soton.ac.uk/66737/01/Marine_Pollution_Damage_in_Australia.pdf
H. J. Peters, *Strengthening Marine Pollution Control the International move towards Statutory Requirements*,http://siteresources.worldbank.org/INTTRANSPORT/Resources/336291-1119275973157/td-ps9.pdf
H. Kim, *Korean Maritime Law Update: 2007 – Focused on the Revised Maritime Law Section in the Korean Commercial Code*, Jour. of Maritime Law and Commerce, vol. 39, no.3, 2008, Available at SSRN: http://ssrn.com/abstract=1284676
H. Schwartz, *Trials in Absentia*, available at http://www.wcl.american.edu/hrbrief/v4i1/schwar41.htm
H. Smith, *What risks dominate the shipping industry from a P&I perspective?*, 5th Int'l Marine Claims Conference, The Standard, Oct., 24, 2008, available at http://www.marineclaimsconference.com/2008/docs/HunterSmith.pdf
Hall, M. Isabel, *After Waterhouse: Vicarious Liability and the Tort of Institutional Abuse*, Journal of Social Welfare and Family Law, vol. 22, no. 2, 2000, pp. 159–173, available at SSRN: http://ssrn.com/abstract=894115
Hess Material Data Sheet, *No. 4 Fuel Oil*, MSDS No. 15054, available at http://www.hess.com/ehs/msds/15054No4FuelOil.pdf
Hess Material Data Sheet, *No. 5 Fuel Oil*, MSDS No. 0332, available at http://www.hess.com/ehs/msds/0332No5FuelOil.pdf
I. C. White, *Factors that determine the cost of oil spills*, available at http://www.itopf.com/_assets/costs03.pdf
IMO: Legal Committee, 80th Session, 11–15 October 1999, http://www.imo.org/Newsroom/mainframe.asp?topic_id=280&doc_id=752
IMO, *Resolutions and other decisions of the 23rd Assembly*, IMO, 2004, pp 210–212; *Pilotage under siege*, http://www.tankeroperator.com/pastissues/2002%20Jul/TO2002Jul%20p1-3%20Pilotage%20under%20siege.pdf
IMO Ship identification number scheme, http://www.imo.org/Facilitation/mainframe.asp?topic_id=388
OECD, Maritime Transport Committee, *Ownership and Control of Ships*, March 2003, available at http://www.oecd.org/dataoecd/53/9/17846120.pdf
IPIECA / ITOPF, *Oil Spill Compensation: a guide to international conventions on liability and compensation for oil pollution damage*, Feb. 2007, pp 10–11,

available at http://www.ipieca.org/activities/oilspill/downloads/publications/reports/english/Compensation.pdf

W. S. Frye, *Oil Spill Prevention Measures: tackling the tough SPCC issues*, http://www.epa.gov/OEM/docs/oil/fss/fss04/frye_04.pdf

IPIECA, *Objectives of Oiled Wild Life Response* in *A Guide to Oiled Wildlife Response Planning*, IPIECA Report Series, vol. 13, at http://www.ipieca.org/activities/oilspill/downloads/publications/reports/english/Vol13_Oiled_Wildlife.pdf

ITOPF, *Dispersants*, available at http://www.itopf.com/spill-response/clean-up-and-response/dispersants/

ITOPF, *The use of Chemical Dispersants to treat oil spills*, Technical Information Paper, available at http://www.itopf.com/_assets/documents/tip4.pdf

J. A. Page, P. Rothstein, *Brief of Amici Curiae of American Law Professors in Support of Respondents, Norfolk & Western Ry. Co. v. Ayers, No. 01-963(U.S. Aug. 19, 2002)*, U. S. Supreme Court Briefs, Georgetown University Law Centre, Year 2002, available at http://scholarship.law.georgetown.edu/cgi/viewcontent.cgi?article=1059&context=scb

J. Coleman, *Theories of Tort Law*, Ch. 2, *The Difference between Fault and Strict Liability*, available at http://plato.stanford.edu/entries/tort-theories/#2

J. F. Baxter Jr., *Oil Spill Detection System*, 1995, US Patent No. US5532679, available at http://www.freepatentsonline.com/5532679.pdf

J. G. Apple, R. P. Deyling, *A Premier on the Civil Law System*, available at http://www.fjc.gov/public/pdf.nsf/lookup/CivilLaw.pdf/$file/CivilLaw.pdf

J. Guyomarch, O. Kerfourn, F. X-Merlin, *Dispersants and Demulsifiers: Studies in the Laboratory, Harbor and Polludrome*, Cedre, France, Paper ID #283, 1999 Int'l Oil Spill Conference, available at http://www.iosc.org/papers/00541.pdf

J. Hare, *Limitation of Liability – A Nigerian Perspective*, available at http://web.uct.ac.za/depts/shiplaw/fulltext/harepapers/limliab-nigeria.pdf

J. Hay, O. Thebaud, J. P. Agundes, *Preventing pollution through the compensation for damage? An appraisal of the European experience in the field of marine oil spills*, http://www.uibcongres.org/imgdb/archivo_dpo4084.pdf

J. L. Gomez Gesteira, J. C. Dauvin, *Impact of the Aegean Sea Oil Spill on the Subtidal fine sand macrobenthic community of the Ares-Betanzos Ria (Northwest Spain)*, http://otvm.uvigo.es/investigacion/informes/documentos/archivos/aegeansea_gomezgesteira.pdf

J. Lauritzen A/S, *Ship Management*, Annual Report 2008 available at http://www.j-lauritzen.com/upload/ship_management__annual_report_2008.pdf

J. Lu, H. Lim, S. C. Liew, M. Bao and L. K. Kwoh, *Oil Pollution Statistics in Southeast Asian Waters compiled from ERS SAR Imagery*, Centre for Remote Imaging, Sensing and Processing, National University of Singapore, available at http://esapub.esrin.esa.it/eoq/eoq61/oil.pdf

J. M. Joseph, *Limitation of Shipowners' Liability in Malaysia*, available at http://www.jbc.com.my/resource/LIMITATION%20OF%20SHIPOWNERS.pdf

J. M. Joseph, *Ship Registration and Ship Mortgages in Malaysia*, http://www.josephandpartners.com/data/SHIP%20REGISTRATION%20AND%20SHIP%20MORTGAGES%20IN%20MALAYSIA.pdf

J. Michel et al, *Potentially Polluting Wrecks in Marine Waters*, An Issue Paper Prepared for the 2005 Int'l Oil Spill Conference, available at http://www.iosc.org/docs/iosc_issue_2005.pdf

J. P. Jones, *The Sky Has Not Yet Fallen on Punitive Damages in Admiralty*, Tulane Law Review, March 24, 2009, Forthcoming. Available at SSRN: http://ssrn.com/abstract=1367909

Jus. R. C. Corona, *Class Action, Public Interest litigation and the Enforcement of Shared Legal Rights and Common Interests in the Environment and Ancestral Lands in the Philippines*, available at http://www.aseanlawassociation.org/9GAdocs/Philipines.pdf

K. Baelz, T. Baldwin, *The End of the Real Seat Theory (Sitztheorie): the European Court of Justice Decision in Uberseering of 5 November 2002 and its impact on German and European Company Law*, German Law Journal, available at http://www.germanlawjournal.de/article.php?id=214

Lloyd's Register, Fairplay, http://www.lrfairplay.com/IMO/imo.html

K. Hofstetter, *The Ecological Liability of Corporate Groups: Comparing US and European Trends*, http://www.hottopos.com/harvard1/ecological_liability_of_corporat.htm

K. Kris Hirst, *Bitumen: A Smelly but Useful Material of Interest*, http://archaeology.about.com/od/bcthroughbl/qt/bitumen.htm

K. Kristl, *Diminishing the Divine: Climate Change and the Act of God Defence*, Widener Law Review, Forthcoming; Widener Law School Legal Studies Research Paper No. 09-21, April 8, 2009, Available at SSRN: http://ssrn.com/abstract=1375115

K. S. Matison, *Comparison of Shipowner's Limitation of Liability Schemes*, Lloyd's Maritime Training Programme, Lane Powell Spears Lubersky LLP, available at http://www.lanepowell.com/wp-content/uploads/2009/04/matisonk_002.pdf

L. Clarke, *Oil Spill Fantasies*, available at http://leeclarke.com/docs/clarke_atlantic_b&w.pdf

L. T. Visscher, *Justifications and Excuses in the Economic Analysis of Tort Law*, RILE Working Paper No. 2008/03, February 21, 2008, published in *Economic Torts*, Hyderabad, (P. S. Khanum ed,), The Icfai University Press 2009, p. 22–57, available at SSRN: http://ssrn.com/abstract=1111547

M. Ashford, *A Jet Ski; a vessel, boat or ship? R v. Goodwin (2006)1 Lloyd's Rep. 432*, https://maritimejournal.murdoch.edu.au/index.php/maritimejournal/article/viewFile/22/43

M. F. Grady (2002), *Proximate Cause Decoded,* UC Los Angeles: UCLA School of Law. Retrieved from: http://escholarship.org/uc/item/2313h061

M. Hasebe, *The International Liability and Compensation Regime*, http://www.cep.unep.org/events-and-meetings/11th-igm/oil-spill-compensation-funds.pdf

M. Hinteregger, Environmental Liability and Ecological Damage, Part I: International and Supranational Systems of Environmental Liability in Europe, Cambridge University Press, 978-0-521-88997-1 available at http://assets.cambridge.org/97805218/89971/excerpt/9780521889971_excerpt.pdf

M. G. Faure, H. Wang, *Financial Caps for Oil Pollution Damage: A Historical Mistake?*, Marine Policy, 2008, available at http://papers.ssrn.com/sol3/papers.cfm?abstract_id=1116173

M. G. Kohn, *Risk Instruments in the Medieval and Early Modern Economy*, Dartmouth College, Department of Economics Working Paper No. 99–07, February 1999., available at SSRN: http://ssrn.com/abstract=151871

M. George, *Choice of Law in Maritime Torts*, Journal of Private International Law, vol. 3, no. 1, 2007, available at http://papers.ssrn.com/sol3/papers.cfm?abstract_id=1117942

M. Jacobson, *The International Oil Pollution Compensation Funds and the International Regime of Compensation for Oil Pollution Damage*, http://www.icao.int/icao/en/anb/atm/SARafrica/docs/Saly_Conf/en/IOPC_Funds_3-2_en.pdf

M. M. Kroll, G. Carpenter, O. Wyman, *Global Marine Practice: Charterparty piracy clause and maritime insurance*, available at http://www.marsh.co.uk/research/documents/CharterpartyPiracy_Technicalpaper_May09.pdf

M. Mason, *Transnational compensation for oil pollution damage: examining changing specialties of environmental liability*, at http://eprints.lse.ac.uk/570/1/RPESA-no69(2002).pdf

M. Mason, *Transnational Compensation for Oil Pollution Damage: Examining Changing Specialties of Environmental Liability*, available at http://www2.lse.ac.uk/geographyAndEnvironment/research/Researchpapers/rp69.pdf

M. Turan, *Turkey's Oil Spill Response Policy: Influences and Implementation*, 2009, available at http://www.un.org/depts/los/nippon/unnff_programme_home/fellows_pages/fellows_papers/turan_0809_turkey.pdf

M. Weitzenbok, *Determining Applicable Law and Jurisdiction in contractual disputes regarding virtual enterprises*, Proceedings of the 8th International Conf. on Concurrent Enterprising: Ubiquitous Engineering in the Collaborative Economy, Rome, Italy, 17–19 June 2000, pp 27–34, available at http://folk.uio.no/emilyw/documents/S01_p04_Weitzenboeck.pdf

Marcos, Francisco, S. Graells, Albert, *Towards a European Tort Law? Damages Actions for Breach of the EC Antitrust Rules: Harmonising Tort Law Through the Back Door?* , October 26, 2006, available at SSRN: http://ssrn.com/abstract=1028963

M. Bussani, V. V. Palmer, *Pure Economic Loss in Europe*, Cambridge, 2003, available at http://assets.cambridge.org/97805218/24644/sample/9780521824644ws.pdf

Merchant Shipping (Registration of Ships) Regulations 2009, Mauritius, http://supremecourt.intnet.mu/Entry/dyn/GuestGetDoc.Asp?Doc_Idx=2334775&Mode=Html&Search=No

M. Ravin, *ITLOS and Dispute Settlement Mechanisms of the United Nations Convention on the Law of the Sea*, 2005, available at http://www.un.org/Depts/los/nippon/unnff_programme_home/fellows_pages/fellows_papers/mom_0506_cambodia_itlos.pdf

N. P. Ventikos, I. D. Genikomsidou, E. D. Grigoriou, *Oil Pollution in Greek Seas: Critical Review of Confrontation and Waste Management*, Submission of Greece to the 9th Meeting of the Focal Points of the Regional Marine Pollution Emergency Response Centre for the Mediterranean Sea (REMPEC), Malta, 21–24 April 2009, available at http://www.rempec.org/admin/upload/publications/FPM09WG30INF9(E)Consolidated.pdf;

Nanaimo Ports Authority, *Practices and Procedures to be followed by Ships in the Port of Nanaimo*, updated May 7, 009, available at http://www.npa.ca/PandPmay72009.pdf

NOAA/ Hazardous Materials Response and Assessment Division, Seattle, Washington, *Oil Spill Case Histories: Summaries of Significant US and International Spills*, available at http://response.restoration.noaa.gov/book_shelf/26_spilldb.pdf

O. Thebaud, D. Bailly, J. Hay, J. Perez, *The cost of oil pollution at sea: an analysis of the process of damage valuation and compensation following oil spills*, http://otvm.uvigo.es/investigacion/informes/documentos/archivos/Prestige_Hayetal.pdf

Oil Pollution Claims in Australia: The Legal Remedies available at http://www.au.findlaw.com/article/4410.htm

Oil Companies International Marine Forum: OCIMF's comments following the EP Transport Committee meeting on 28th and 29th May 2001, p. 1–2, available at http://www.ocimf.com/view_document.cfm?id=322

Oil Spill and Compensation available at

http://www.unescobkk.org/fileadmin/user_upload/westpac/documents/News/Marine_oil_spill_Apr_09/Invited_lecturers/20_Apr/04_Liability___Compensation.pdf

P. Bennett, *Mutuality at a distance ? Risk and regulation in marine insurance clubs*, Environment & Planning A 2000, vol. 32, pp 147–163, available at http://www.envplan.com/epa/fulltext/a32/a3215.pdf

P. Cane, *Taking Disagreement Seriously: Courts, Legislatures and the Reform of Tort Law*, Oxford Journal of Legal Studies, vol. 25, no. 3, pp 393–417, 2005, available at SSRN: http://ssrn.com/abstract=821686

P. D. Boehm et. al., *Advances Chemical Fingerprinting for Oil Spill Identification and Natural Resource Damage Assessments*, available at http://www.iosc.org/papers/00033.pdf

P. Griggs, *International Convention on Civil Liability for Bunker Oil Pollution Damage 2001*, p. 2, http://www.bmla.org.uk/documents/imo-bunker-convention.doc

P. J. S. Griggs, *Obstacles to Uniformity of Maritime Law*, The Nicholas J. Healy Lecture, available at http://www.comitemaritime.org/year/pdfiles/obst_uniform.pdf

P. Pelle, *Companies crossing borders within Europe*, Utrecht Law Review, Igitur, vol. 4, Issue 1, (March), 2008, available at http://www.utrechtlawreview.org/publish/articles/000056/article.pdf

P. Sarcevic, Yearbook of Private International Law, Vol. IX, Staempfli, 2007, available at http://www.sellier.de/pages/downloads/9783866530713_leseprobe.pdf?code=28fa0049d3f20d01a9cbe47eabd022b2

P. Saundry, *Exxon Valdez Oil Spill*, http://www.eoearth.org/article/exxon_valdez_oil_spill

Peter Morgan, *News from Intergovernmental and International Organizations: News from IMO: IMO Legal Committee – 80th Session*, CMI News Letter, No. 4-1999, p 7, available at http://www.comitemaritime.org/news/pdfiles/nl99_4.pdf

Petroleum, http://en.wikipedia.org/wiki/Petroleum

Lubricating Oil for Small Scale Live Steam Locomotives, http://www.southernsteamtrains.com/notes/lubeoil.htm

Prof. B. H. Oxman, *The Development of the Modern Law of the Sea* in *Law of the Sea*, UN/UNITAR Fellowship Programme in International Law, The Hague, 2008, http://www.law.miami.edu/news/images/999images/Oxman%20UNITAR%20Outline.pdf

Prof. S. Jayakumar, Deputy Prime Minister, Keynote Address at the Singapore Conference on Freedom of Seas, Passage Rights and the 1982 Law of the Sea Convention (January 9, 2008) available at http://app.mfa.gov.sg/pr/read_content.asp?View,10946

R. C. Helm, R. G. Ford, H. R. Carter, *The Oil Pollution Act of 1990 and Natural Resource Damage Assessment*, available at http://www.marineornithology.org/PDF/34_2/34_2_99-108.pdf

R. Gillich, *Practical issues arising from the Arrest of Ships*, available at https://maritimejournal.murdoch.edu.au/index.php/maritimejournal/article/viewFile/96/140
R. Martin, *Recovery of Pure Economic Loss in Tort*, available at http://sunzi1.lib.hku.hk/hkjo/view/14/1400126.pdf
R. S. Neyland, *Sovereign Immunity and the Management of United States Naval Shipwrecks*, http://www.history.navy.mil/branches/org12-7h.htm
R. Wolfrum, C. Langenfeld, P. Minnerop, *Environmental Liability in International Law: Toward a Coherent Conception*, Max Planck Institute for Comparative Public and Law and International Law, Research Paper for the Federal Environmental Agency, Nov. 2004 available at http://www.umweltdaten.de/publikationen/fpdf-k/k2942.pdf
R. W. Wright, *The Vitality of Joint and Several Liability: Amici Curiae Brief, Norfolk & Western Railway Co. v. Ayers*, 123 S. Ct. 1210 (2003) (August 19, 2002), available at SSRN: http://papers.ssrn.com/sol3/papers.cfm?abstract_id=392786
Report of the Transport and Industrial Relations Committee, House of Representatives, New Zealand, *International Treaty Examination of the international Convention on Civil Liability for Bunker Oil Pollution Damage*, http://www.parliament.nz/NR/rdonlyres/C901FD34-146E-43A8-A8CD-3EE643C-293CB/94261/DBSCH_SCR_4183_6372.pdf
S. Bloodworth, *Death on the High Seas: the demise of TOVALOP and CRISTAL*, Florida States University Journal of Land Use and Environmental Law, 1998, available at http://www.law.fsu.edu/journals/landuse/Vol132/Bloo.htm
S. C. Derrington, J. M. Turner, *The Law and Practice of Admiralty Matters*, Oxford, Ch. 10, available at www.gbv.de/dms/spk/sbb/recht/toc/527702471.pdf
S. D. Sugarman, *Tort Reform Through Damages Law Reform: An American Perspective*, available at http://works.bepress.com/cgi/viewcontent.cgi?article=1009&context=stephen_sugarman
S. Flogeland, *DNV Classification of mobile Offshore Units*, http://www.dnv.ru/Binaries/dnv_classification_mobile_offshore_units_rev2_tcm83-10777.pdf
S. M. Hope, *Recent Private International Law Developments Before the Supreme Court of Canada*, The Globetrotter, Ontario Bar Association – International Law Section Newsletter, vol. 13, no.2, 2009, available at http://papers.ssrn.com/sol3/papers.cfm?abstract_id=1342413
Sandra Spears, *Probing the Limits of Maritime Regimes*, llyodslist.com available at http://folk.uio.no/erikro/WWW/corrgr/presseklipp/probing.pdf
Ships' Bunkers Convention Conference proposed for 2000–2001, LEG/80, Oct., 11–15, 1999 available at http://www.imo.org/Newsroom/mainframe.asp?topic_id=280&doc_id=752

Status of Multilateral Conventions and Instruments in Respect of which the International Maritime Organization or its Secretary General Performs Depository or Other Functions, as at December 31, 2008 , L:\LED\LO\MS-Depositary\STATUS OF MULTILATERAL CONVENTIONS\2008 & 2009\Status-2008.doc, p 445 available at http://www.imo.org/includes/blastDataOnly.asp/data_id%3D25891/Status-2008.pdf

T. Nast, *Weathering the Storm: How the Lessons of History will enable the Marine Insurance Industry to survive the current economic downturn*, available at http://works.bepress.com/cgi/viewcontent.cgi?article=1000&context=timothy_nast

T. Tietenberg (Lead Author); S. Niggol Seo (Topic Editor), 2006, *Open access resources* in *Encyclopedia of Earth*. Eds. Cuttler J. Cleverland (Washington D. C.: Environmental Information Coalition, National Council for Science and the Environment) published in the Encyclopedia of Earth, Dec., 11, 2006 [Retrieved Nov., 18, 2009], http://www.eoearth.org/article/Open_access_resources

The Belgian Ships Register, http://www.brv.be/eng/download/BELGIAN%20REGISTER%20-%20Recommendations.pdf

The Charterers P&I Club Newsletter, *The Charterer*, July 2009, available at http://www.else.co.uk/Documents/The%20Charterer%20July%202009.pdf

The Falkland Island – A History of the 1982 Conflict: Preface to a Conflict, available at http://www.raf.mod.uk/falklands/preface.html

The Gulf War 1990/1991, available at http://www.rickard.karoo.net/articles/wars_gulf1990.html

Transport and Housing Bureau of Hong Kong, discussion paper No. CB(1)1344/08-09(03), *Legislative Council Panel on Economic Development Bunker Oil Pollution (Liability and Compensation) Bill*, April 27, 2009 available at http://www.legco.gov.hk/yr08-09/english/panels/edev/papers/edev0427cb1-1344-3-e.pdf

Transport Canada, Canada's Government takes action to protect the environment with changes to the Marine Liability Act, June 25, 2009, news report available at http://www.tc.gc.ca/eng/mediaroom/releases-nat-2009-09-h096e-3293.htm (last visited 05.01.2010)

UNCTAD, *Review of Maritime Transport* (2008) available at http://www.unctad.org/en/docs/rmt2008_en.pdf

United States' Environmental Protection Agency (EPA), *In-Use Marine Diesel Fuel*, Air and Radiation, EPA420-R-99-027, available at http://www.epa.gov/oms/regs/nonroad/marine/ci/fr/dfuelrpt.pdf

U. S. Environmental Protection Agency, Regulatory Information for the Petroleum Sector, available at http://www.epa.gov/regulations/bizsector/petroleum.html

U. S. Fish & Wild Life Services, *Effects of Oil Spills on Wildlife and Habitat: Alaska Region*, Dec. 2004, available at http://alaska.fws.gov/media/unalaska/Oil%20Spill%20Fact%20Sheet.pdf

W. Anderson, *Environmental Law Enforcement: the role of the judiciary*, available at http://www.inece.org/conf/proceedings2/58%20-%20Enviro%20Lw%20Enforcement.pdf

Waltons & Morse LLP, *Limitation of Liability for Maritime Claims: New Limits Come into Force*, available at http://www.waltonsandmorse.com/Bulletins2.jsp?bulletinID=10 (last visited 01.01.2010)

W. L. Peck, *Responder Immunity and Salvage*, http://www.iosc.org/papers/01852.pdf

W. M. Sullivan, M. J. Mulhare, *Oil Spill Prevention, Administration and Response (OSPAR) Fund, Annual Report FY 2006*, Rhode Island Department of Environmental Management, available at http://www.dem.ri.gov/pubs/ospar06.pdf

Whale Oil, http://en.wikipedia.org/wiki/Whale_oil

Finnish Environment Institute, *The Impacts of Oil on the Marine Environment*, Doc. 9/14/2009, available at http://www.ymparisto.fi/print.asp?contentid=334186&lan=en&clan=en

Wreck Oil Removal Programme (WORP), *2009 Demonstration Project Overview*, available at http://www.mlaus.org/content/ItemContent/10431Events.pdf

Y. K. Lee, A Brief Introduction to the International Convention on Liability and Compensation for Bunker Oil Pollution Damage (The Bunkers Convention), available at http://www.seatransport.org/seaview_doc/SV_84/ICLC%20for%20Bunker%20Oil%20Pollution%20Damage.pdf

IMO Documents

71 FUND/EXC.54/10
71 FUND/EXC.54/INF.3
FUND 1971/EXC. 50/17
FUND/A.4/10
FUND/A/ES.1/13
FUND/EXC. 35/5/Add. 1
FUND/EXC 35.10
FUND/EXC. 37.3, 8 October 1993
FUND/EXC. 37/3, 18 October 1993
FUND/EXC. 38.9, 11 February 1994
FUND/EXC. 50/17, 23 October 1996

FUND/WGR.7/3
FUND/WGR.7/21
LEG 73 agenda item 12, 12 October 1995
LEG 74/4/1, 9 August 1996
LEG 74/4/2, 9 August 1996
LEG 74/4/4, 13 September 1996
LEG 75/5/1 17 February 1997
LEG 75/11, 29 April 1997
LEG 75/WP.1 24 April 1997
LEG 76/3
LEG 76/WP.3, 13 October 1997
LEG 77/WP.3, 22 April 1998
LEG 78/5/2
LEG 78/11, 2 November 1998
LEG 79, 12 February 1999
LEG 79/6/1, 12 February 1999
LEG 79/6/2, 12 February 1999
LEG 79/6/3, 18 March 1999
LEG/80, Oct., 11–15, 1999
LEG 80/4/1, 13 August 1999
LEG 80/4/2, 9 September 1999
LEG 81/4, 21 January 2000
LEG 81/4/3, 21 February 2000
LEG 81/WP. 2, 30 March 2000
LEG 82/3, 30 June 2000
LEG 83/11, 6 September 2001
LEG 94/11/2
LEG/95, Mar., 30 – Apr., 4, 2009
Leg/Conf.11/6, 27 Sep. 2000
MEPC.67(37)
O.R. 1969, LEG/CONF/C.2/SR.8
O.R., 1969, LEG/CONF/C.2/WP.9
O.R., 1969, LEG/CONF/4, September 1969
O.R , 1971, LEG/CONF.2/C.1/SR.11
O.R. 1984, LEG/CONF.6/C.2/SR.5

International, European Conventions, Codes and Arrangements

Brussels Convention on Jurisdiction and the Enforcement of Judgments in Civil and Commercial Matters (1968)

Contract Regarding an Interim Supplement to Tanker Liability for Oil Pollution (1971)

Convention (IV) respecting the Laws and Customs of War on Land and its annex: Regulations concerning the Laws and Customs of War on Land, The Hague (1907)

Convention for the Protection of the Marine Environment of the North-East Atlantic (1992)

Convention on International Civil Aviation, Chicago (1944)

Convention on International Liability for Damage Caused by Space Objects (1972)

Convention on Civil Liability for Damage Resulting from Activities Dangerous to the Environment, Lugano (1993)

Convention on Jurisdiction and the Enforcement of Judgments in Civil and Commercial Matters, Lugano (1968)

Convention on Limitation of Liability for Maritime Claims (1976)

Convention on the International Hydrographic Organization (1967)

Convention on the Organization for Economic Co-operation and Development (1960)

Convention on the Protection of the Marine Environment of the Baltic Sea Area (1992)

Draft Treaty of the Hague Conference on Mutual Recognition of the Legal Personality of Companies (1956)

European Community Convention on Jurisdiction and Enforcement of Judgments in Civil and Commercial Matters (EC Regulation No. 44/2001)

European Directive 2004/35/CE

European Directive No. 90/314/EEC

European Treaty Series No. 150

Geneva Convention on the Territorial Sea and the Contiguous Zone (1958)

Hague Convention on Territorial Sea and Contiguous Zone (1930)

International Convention for the Prevention of Oil Pollution from Ships (1973)

International Convention for the Prevention of Pollution of the Sea by Oil (1954)

International Convention for the Safety of Life at Sea (1974)

International Convention for the Unification of Certain Rules relating to the Limitation of Liability of the Owners of Sea-going Vessels, Brussels (1924)

International Convention on Civil Liability for Oil Pollution Damage (1969)

The International Conference on Liability and Compensation for Damage in Connection with the Carriage of Certain Substances by Sea (1984)

International Convention on Liability and Compensation for Damage in connection with the Carriage of Hazardous and Noxious Substances by Sea (1996)
International Convention on the Establishment of an International Fund for Compensation for Oil Pollution Damage (1971)
International Convention Relating to Intervention on the High Seas in Cases of Oil Pollution Casualties (1969)
International Convention relating to the Limitation of Liability of Owners of Sea-going Ships, Brussels (1957)
International Convention Relating to the Arrest of Sea-Going Ships (1952)
International Maritime Dangerous Goods Code (2008)
International Convention on Civil Liability for Bunker Oil Pollution Damage (2001)
International Convention on Maritime Liens and Mortgages (1993)
International Convention on Tonnage Measurement of Ships (1969)
International Safety Management Code (2002)
International Tanker Owners Pollution Federation Limited
International Union for Conservation of Nature
Nairobi International Convention on the Removal of Wrecks (2007)
Protocol to Amend the Convention on Limitation of Liability for Maritime Claims, 1976 (1996)
Protocol to the International Convention for the Prevention of Oil Pollution from Ships (1978)
Protocol to the International Convention on Civil Liability for Oil Pollution Damage (1984)
Protocol to the International Convention on Civil Liability for Oil Pollution Damage (1992)
Protocol to the International Convention on the Establishment of an International Fund for Compensation for Oil Pollution Damage (1992)
Protocol to the International Convention on the Establishment of an International Fund for Compensation for Oil Pollution Damage (1984)
Protocol to the International Convention on the Establishment of an International Fund for Compensation for Oil Pollution Damage (2000)
Tanker Owners Voluntary Agreement Concerning Liability for Oil Pollution (1969)
Treaty on European Union (1992)
United Nations Conference on Environment and Development (1992)
United Nations Conference on Trade and Development (1964)
United Nations Convention on the Law of the Sea (1982)
Vienna Convention on the Law of Treaties between States and International Organizations or between International Organizations (1986)

International Organizational Resolutions, Reports and Documents

Organization of Economic Coorporation and Development (OECD), Recommendation of 14 November 1974 on the Implementation of the Polluter Pays Principle, C (74) 223

Principle 16 of the Rio Declaration on United Nations Conference on Environment and Development (UNCED)

Communication from the Commission to the Council and Parliament and the Economic and Social Committee, "Green Paper on Remedying Environmental Damage", Brussels, 14 May 1993. COM (93) 47 Final

Official Record (O.R.), International Conference on Liability and Compensation for Damage in Connection with the Carriage of Certain Substances by Sea, 1984 and the International Conference on the revision of the 1969 Civil Liability Convention and the 1971 Fund Convention, 1992, Vol.1, pages 296/297

Internet Sources

http://conventions.coe.int/Treaty/en/Treaties/Word/150.doc

http://en.wikipedia.org/wiki/Diesel_fuel

http://en.wikipedia.org/wiki/Fuel_oil

http://en.wikipedia.org/wiki/Hydrocarbon

http://eur-lex.europa.eu/en/treaties/dat/11992M/htm/11992M.html#0001000001 http://untreaty.un.org/ilc/texts/instruments/english/conventions/1_1_1969.pdf

http://fletcher.tufts.edu/multi/texts/BH595.txt

http://nudeal.com/selling-lead/ywc-type-bilge-water-treatment-plant-used-ship

http://restoration.doi.gov/

http://restoration.doi.gov/laws.html

http://sedac.ciesin.org/entri/texts/intervention.high.seas.casualties.1969.html

http://untreaty.un.org/ilc/texts/instruments/english/conventions/1_2_1986.pdf

http://untreaty.un.org/ilc/texts/instruments/english/conventions/8_1_1958_territorial_sea.pdf

http://www.lboro.ac.uk/departments/hu/ergsinhu/aboutergs/lasttrip.htm

http://www.admiraltylawguide.com/conven/arrest1952.html

http://www.admiraltylawguide.com/conven/passengers1974.html

http://www.admiraltylawguide.com/conven/tonnage1969.html http://www.admiraltylaw.com/statutes/hamburg.html

http://www.amsa.gov.au/Marine_Environment_Protection/Major_oil_spills_in_australia/Iron_Baron/index.asp

http://www.amsa.gov.au/Marine_Environment_Protection/Revision_of_Annexes_I_and_II_of_MARPOL_73-78/index.asp

http://www.aph.gov.au/House/committee/jsct/28march2006/treaties/bunkers_ris.pdf
http://www.archive.org/stream/merchantshipping00grearich#page/n3/mode/2up http://www.asl.ie/fleet/tonnage_pdf/t_rover.pdf https://www.ccaimo.mar.mil.br/secimo/convencoes/Prot2000.pdf
http://www.comitemaritime.org/
http://www.comitemaritime.org/histo/his.html
http://www.dfg.ca.gov/ospr/spill/nrda/nrda_kure.html
http://www.ecolex.org/ecolex/ledge/view/RecordDetails;jsessionid=813C42F0F5707D9A343DF010BBEE4A1E?id=TRE-000117&index=treaties
http://www.edinformatics.com/math_science/c_atom.htm
http://www.england-legislation.hmso.gov.uk/acts/acts1983/pdf/ukpga_19830021_en.pdf
http://www.greenpeace.org/international
http://www.handytankers.com/handy/certificates/fileShow.asp?intFileId=3552
http://www.helcom.fi
http://www.historyoftheuniverse.com/molecule.html
http://www.igpandi.org
http://www.igpandi.org/downloadables/submissions/imo/IMO%20LEG%2094%20Paper%20-%20October%202008.pdf
http://www.igpandi.org/Home/Sitemap http://www.imo.org
http://www.iho-ohi.net/english/home/
http://www.imo.org/Conventions/contents.asp?doc_id=660&topic_id=256
http://www.imo.org/Conventions/contents.asp?doc_id=664&topic_id=256
http://www.imo.org/Conventions/contents.asp?doc_id=678&topic_id=258#3 http://www.imo.org/Conventions/contents.asp?topic_id=256&doc_id=665
http://www.imo.org/Conventions/contents.asp?topic_id=256&doc_id=666 http://www.imo.org/Conventions/mainframe.asp?topic_id=256&doc_id=660
http://www.imo.org/Conventions/mainframe.asp?topic_id=256&doc_id=661
http://www.imo.org/HumanElement/mainframe.asp?topic_id=287
http://www.imo.org/includes/blastDataOnly.asp/data_id%3D23754/InternationalShippingandWorldTrade-factsandfigures.pdf
http://www.imo.org/newsroom/mainframe.asp?topic_id=1772&doc_id=11233
http://www.incidentnews.gov/incident/6241
http://www.iopcfund.org
http://www.itlos.org/
http://www.itopf.com/information-services/data-and-statistics/case-histories/tlist.html#TANIO
http://www.itopf.com/marine-spills/fate/persistence-of-oil/
http://www.itutorials.co.uk/freedom_of_establishment.php

http://www.iucn.org
http://www.jus.uio.no/lm/brussels.jurisdiction.and.enforcement.of.judgments.in.civil.and.commercial.matters.convention.1968/doc.html
http://www.marisec.org
http://www.mcga.gov.uk/c4mca/mcga07-home/emergencyresponse/mcga-pollutionresponse/mcga-dops_cp_environmental-counter-pollution.htm
http://www.miamicharters.com/2002/vessels/Fastrack%20Bareboat%20Contract.doc
http://www.mms.gov/ooc/Assets/KatrinaAndRita/BackgrounderMODU.pdf
http://www.oecd.org
http://www.ofj.admin.ch/etc/medialib/data/wirtschaft/ipr.Par.0022.File.tmp/260307_entw_lugano_convention-e.pdf
http://www.opsi.gov.uk/ACTS/acts1995/ukpga_19950021_en_1
http://www.opsi.gov.uk/RevisedStatutes/Acts/ukpga/1945/cukpga_19450028_en_1
http://www.ospar.org
http://www.simsl.com/Articles/Wreck%20Removal%20Convention.pdf
http://www.simsl.com/Shipman98.html
http://www.ukpandi.com/ukpandi/infopool.nsf/html/LP_Init_AnalyMajorClaims
http://www.un.org/Depts/los/convention_agreements/convention_overview_convention.htm http://www.un.org/Depts/los/convention_agreements/texts/unclos/unclos_e.pdf
http://www.unep.org/regionalseas
http://www.upv.edu.ph/oilspill/compensation.php
http://www.usmma.edu/waterfront/kingspointer/General%20Documents/Int%20Tonnage%20Certificate.pdf http://www.windsofireland.net/files/contract.pdf

Journals and Articles

A. A. Ehrenzweig, *Releases of Concurrent Tort-Feasors in the Conflict of Laws: Law and Reason versus the Restatement*, Virginia Law Review, vol. 46, no. 4 (May, 1960), pp 712–716

A. Alberini, D. Austin, *On and Off the Liability Bandwagon: Explaining State Adoptions of Strict Liability in Hazardous Waste Programs*, Journal of Regulatory Economics, Springer, vol. 15, No. 1/Jan. 1999, pp 41–64

A. Bianchi, *The Harmonization of Laws on Liability for Environmental Damage in Europe: An Italian Perspective*, Journal of Environmental Law, 1994, 6(1): 21–42, Oxford

A. Calnan, *Anomalies in International Tort Law*, Tennessee Jour., of Law and Policy, vol.1, 2005, p 187

A. C. Gross, N. E. Scott, *Comparative Environmental Legislation and Action*, Int'l and Comparative Law Quarterly (1980), 29: 619–663, Cambridge

A. C. Homan, T. Steiner, *OPA 90s impact at reducing oil spills*, Marine Policy, vol. 32, issue 4, July 2008, pp 711–718

A. Dandal, P. Gillies, *Characterizing the Action in rem in Australia and the Implications on International Commercial Arbitration*, 40 J. Mar. L. & Com. 271, Apr., 2009

A. Dahdal, P. S. Gillies, *Characterizing the Action in Rem in Australia and the Implications on International Commercial Arbitration*, Journal of Maritime Law and Commerce, vol. 40, no. 2, 2009

Admiralty, *Federal Court Applies Maritime rather than State Law in Diversity Suit for Injury sustained within State's Territorial Waters*, Harvard Law Review, vol. 63, no. 3 (Jan., 1950) pp 525–526

A. E. Boyle, *Globalizing Environmental Liability: The Interplay of National and International Law*, Journal of Environmental Law, 2005, 17(1): 3-26, Oxford

A. E. Richard, Defenses and Subsequent Pleas in a System of Strict Liability, The Journal of Legal Studies, vol. 3, no. 1 (Jan., 1974), pp. 165–215

A. Geddes, *Locus standi and EEC environmental measures*, Journal of Environmental Law (1992), 4(1), pp 29–39

A. Gray, *Flexibility in conflict of laws multistate tort cases: the way forward in Australia*, University of Queensland Law Journal, 2004, vol. 23, issue 2, pp 435–463

A. G. S. Moldan, et al, *Some Aspects of the Castillo de Bellver oil spill*, Marine Pollution Bulletin (1985), vol. 16, Issue 3, pp 97–102

A.H.E. Popp, *Limitation of Liability in Maritime Law – An Assessment of Its Viability from a Canadian Perspective,* 24 J. Mar. L. & Com. 335 (1993)

A. J. Rodriguez, P. A. C. Jaffe, *The Oil Pollution Act of 1990*, 15 Tul. Mar. L. J. (1990–1991)

A. Lebrun, *Aménagement, environnement, urbanisme et droit frontière (1990)*, Brussels, 2 Nov., 1989, p.42

A. M. Tettenborn, Reviewed work(s): *Limitation of Shipowners' Liability: The New Law* by Nicholas Gaskell, The Cambridge Law Journal, vol. 46, no. 3 (Nov., 1987), pp. 535–536

A. N. Ince and E. Topuz, *Modeling and Simulation for Safe and Efficient Navigation in Narrow Waterways*. The Journal of Navigation, 57 , 2004, pp 53–71

A. Nussbaum, *Proving the Law of Foreign Countries*, The American Journal of Comparative Law, vol. 3, no. 1 (Winter, 1954), pp 60–67

A. P. Higgins, *War and Law*, the Cambridge Law Journal, vol. 1 1922, pp 177–184

A. R. B. Amerasinghe, *Rylands v. Fletcher in the Law of Ceylon*, International and Comparative Law Quarterly (1962), vol. 11

A. Rein, *International Variations on Concepts of Limitation of Liability*, 53 Tul. L. Rev., 1259 (1978–1979)

A. V. Cicourel, Review: D'Antonio and Pike's „Rebellion, Revolution, and Reform", Reviewed work(s): Rebellion, Revolution, and Reform: New Forces for Change in Latin America by W. V. D'Antonio; F. B. Pike, Economic Development and Cultural Change, vol. 16, no. 2, Part 1 (Jan., 1968), pp. 315–327

A. V. Y. Ong, *Issues in the Application of Depecage in Chinese Private International Law*, Chinese Jour., of Int'l Law, vol. 8, no. 3, 2009, Oxford, pp 637–655

A. W. McThenia, J. E. Ulrich, *A Return to Principles of Corrective Justice in Deciding Economic Loss Cases*, Virgenia Law Rev., vol. 69, no. 8 (Nov., 1983), pp 1517–1535

B. A. Chokor, *Environmental pressure groups and habitat protection in the developing world: The case of Nigeria*, The Environmentalist, Springer, vol. 12, no. 3, Sep., 1992

B. E. Bengtsson, T. L. Coombs and M. Waldichuk, *Biological Variables, Especially Skeletal Deformities in Fish, for Monitoring Marine Pollution [and Discussion], Philosophical* Transactions of the Royal Society of London. Series B, Biological Sciences, vol. 286, no. 1015, The Assessment of Sublethal Effects of Pollutants in the Sea (Aug. 8, 1979), pp. 457–464

Bernauer, L. Caduff, *In whose interest? Pressure group Politics, Economic Competition and Environmental Regulation*, Journal of Public Policy, 24, pp 99–126

B. G. Rabe, *Federalism and Entrepreneurship: Explaining American and Canadian Innovation in Pollution Prevention and Regulatory Integration*, Policy Studies Journal, vol. 27, 1999

B. Jenkins, *International Terrorism – a new mode of conflict*, NCJRS, 1975; Y. Sheffi, *Supply Chain Management under the Threat of International Terrorism*, Int'l Journal of Logistics Management, vol. 12, iss. 2, 2001, pp 1–11

C. A. Hopkins. *Vicarious Liability—Negligence on the Way to Work*. The Cambridge Law Journal, 48, 1989, pp 368–370

C. A. Jones, *Compensation for natural resource damages from oil spills: a comparison of US law and international conventions*, Int'l Jour. of Environment and Pollution, vol. 11, no. 1, 1999

Capt. T. Hughes (2009). *When is a VTS not a VTS?*. Journal of Navigation, 62 , pp 439–442

Case Law Analysis, *Waste and Liability in Environmental Law*, vol. 14, iss. 1, Oxford, 2002

C. Y. C. Chu, Y. Qian, *Vicarious Liability under the Negligence Rule*, Int'l Review of Law and Economics, vol. 15, Issue 3, Sep. 1995

C. C. Turpin, *Roman-Dutch Law Remoteness of Damage. Wagon Mound Not Followed*, The Cambridge Law Journal, vol. 23, no. 1 (Apr., 1965)
C. Fairman, *A Note on re Piracy Jure Gentium*, The American Journal of International Law, vol. 29, No. 3 (Jul., 1935), pp. 508–512
C. J. Colombus, *Ships and Their Owners in the Prize Courts of France, Great Britain, Italy, and China*, Journal of Comparative Legislation and International Law, Cambridge, 3rd Series, vol. 2, no. 3 (1920)
C. J. Strobel, A. H. Brenowitz, *Effects of Bunker Coil on Juvenile Horseshoe Crabs (Limulus polyphemus)*, Estuaries and Costs Journal, vol. No. 2/ June 1981, Springer
C. L. Cochran, *De Facto and De Jure Recognition: Is there a Difference?*, The American Journal of International Law, vol. 62, no. 2 (Apr., 1968), pp. 457–460
C. M. Vazquez, *Sovereign Immunity, Due Process, and the Alden Trilogy*, Yale Law Journal, vol. 109, No. 1927, June 2000
Contributory Negligence. Doctrine of Last Clear Chance Retained by Comparative Negligence Statute, Harvard Law Review, vol. 52, no. 7 (May, 1939), pp 1187–1188
C. R. Dunlop, *Immunity of State Ships*, Journal of Comparative Legislation and International Law, Third Series, vol. 6, No. 4 (1924), Cambridge
C. Y. C. Chu, Y. Qian, *Vicarious liability under a negligence rule*, Int'l Review of Law and Economics, vol. 15, iss. 3, Sep., 1995, pp 305–322
D. Anderson, *The Effect of Port State Control on Substandard Shipping*, Maritime Studies, Issue 125, July/Aug 2002
D. Dion, *Canadian Approach to Limiting the Liability of pleasure craft owners*, 24 J. Mar. L. & Com. 561 (1993)
D. G. S. Chong, *Security in Actions in Rem*, Singapore Academy of Law Journal, vol. 1, 1989, p 81
D. Harris, *Tort Law Reform in the United States*, Oxford Journal of Legal Studies (1991), vol. 11, no. 3, pp 407–415
D. Jielong, *Statement on Diplomatic Protection, International Liability for Injurious Consequences Arising out of Acts not Prohibited by International Law*, Chinese Journal of Int'l Law 2007, 6(1): 189–194
D. Kretzmer, *Transformation of Tort Liability in the Nineteenth Century: The Visible Hand*, Oxford Jour., of Legal Studies, vol. 4, iss. 1, 1984, pp 46–87
D. Lasok, *The Polish Merchant Shipping Code*, Int'l and Comparative Law Quarterly (1964), 13:1087-1093
D. L. C. Miller, *Three of a Kind? Positive Prescription in Sri Lanka, South Africa and Scotland*, Electronic Journal of Comparative Law, vol. 12.1(May 2008)
D. L. Kahn, R. H. Steeves, *The experience of suffering: conceptual clarification and theoretical definition*, J. Adv. Nurs., 1986 Nov., 11(6):623–31

D. M. Collins, *Comments on the American Rule of in rem Liability*, 10 Tul. Mar. L. J. 71, Spring 1985

D. Ozonoff, *Legal Causation and Responsibility for Causing Harm*, American Journal of Public Health, vol. 95, no. S1, July 2005

D. Pastor, J. Sanchez, C. Porte, J. Albaiges, *The Aegean Oil Spill in the Galicia Coast (NW Spain). I. Distribution and Fate of the Crude Oil and Combustion Products in Subtidal Sediments*, Marine Pollution Bulletin, vol.42, Issue 10, Oct. 2001, pp 895–904

D. R. Cooley, *Strict Joint and Several Liability and Justice*, Journal of Business Ethics, Springer, vol. 47, issue 3, Oct. 2003

Dr. V. J. G. Power, *EU competition law on sharing information between competitors*, JIML 15 [2009] 3: 224

Dr. A. D. Lazaraki, *Third party rights of suit in contracts for the carriage of goods by sea and the Contracts (Rights of Third Parties) Act 1999*, JIML [2008] 3, 208

D. R. Dillon, *Maritime Piracy: Defining the problem*, SAIS Review – Volume 25, Number 1, Winter-Spring 2005, pp. 155–165

D. R. Vieites, S. Nieto-Roman, A. Palanca, X. Ferrer and M. Vences, *European Atlantic: the hottest oil spill hotspot Worldwide*, Naturwissenchaften, Springer, vol. 91, No. 11/Nov. 2004, pp 535–538

D. Wilkinson, *Moving the boundaries of compensable environmental damage caused by marine oil spills: The effect of two new international protocols*, Jour., of Environmental Law, vol. 5, iss. 1, 1993, pp 71–90

D. Yang, *A Comparative Analysis of Maritime Lien Priority under United States and Chinese Maritime Law*, 23 Tul. Mar. L. J. 465 (1998–1999)

E. Alkema, *Seaplane a "Vessel" for Purpose of Salvage*, Michigan Law Review, vol. 52, No. 8 (Jun., 1954), pp 1229–1230

E. C. German, *Seller Beware – Strict Liability not Absolute Liability*, 37 Ins. Counsel J. 44 (1970)

E. Chadwick (2007), *The 'Impossibility' of Maritime Neutrality during World War 1.* Netherlands International Law Review, 54 , pp 337–360

E. Lauterpacht, *International Law Reports*, vol. 42, Cambridge, 1971 rep. 1994, p 249

E. Lauterpacht, *International Law – Sovereign Immunity – Proof of Interest*, The Cambridge Law Journal (1955), 13: 19–23

E. L. Johnson, *Government, Law and Courts in the Soviet Union and Eastern Europe.* General Editors, Vladimir Gsovski and Kazimerz Grzybowski, Stevens & Sons, Ltd. Mouton & Co., N.V. 1959. 2 Vols. xxxii and 2068, 1960, *International & Comparative Law Quarterly,* 9 , pp 716–720

E. L. Sherwin, K. M. Clermont, *A Comparative View of Standards of Proof*, American Journal of Comparative Law, vol. 50, 2002, p. 243

E. S. Knutsen, *The Ultimate Limitation Period: Updating the Limitation Act*, Canadian Business Law Journal, vol. 39, no. 149, 2003

E. S. Tzannatos, *A decision support system for the promotion of security in shipping*, Disaster Prevention and Management, vol. 12, iss. 3, MCB UP Ltd., 2003, pp 222–229

E. W. O'Brian, *The Hague Convention on Jurisdiction and Judgments: The Way Forward*, The Modern Law Review, vol. 66, no. 4 (Jul., 2003), pp 491–509

E. W. Vierdag, *The Law Governing Treaty Relations between Parties to the Vienna Convention on the Law of Treaties and States not party to the Convention*, American Journal of International Law, vol. 76, 1982

F. Douever, F. Maes, A. Vanhulle, J. Schrijvers, *The role of marine spatial planning in sea use management: The Belgian case*, Marine Policy, vol. 31, iss. 2, Mar., 2007

F. G. Harmon, *Discharge and Waiver of Maritime Liens*, 47 Tul. L. Rev., 786, Apr., 1973

F. Gomez, H. B. Schafer, *The law and economics of pure economic loss: Introduction to the special issue of the International Review of Law and Economics*, Int'l Rev. of Law and Economics, vol. 27, iss. 1, March 2007, pp 1–7

Ganten, "The International Oil Pollution Compensation Fund", Environmental. Policy and Law, vol. 12, February 1984

G. Berens, C. van Riel, *Corporate Associations in Academic Literature: Three Main Streams of thought in the Reputation Measurement Literature*, Corporate Reputation Review, Palgrave McMimillan, vol. 7, No. 2 (July 2004), pp 161–178(18)

G. Doeker, T. Gehring, *Private or International Liability for Transnational Environmental Damage – The Precedent of Conventional Liability Regimes*, JEL 1990, Oxford, 2(1):1–16

G. F. Birch, *Marine Pollution in Australia, with special emphasis on central New South Wales estuaries and adjacent continental margin*, Int'l Journal of Environment and Pollution, vol. 13, Nos. 1–6/2000, pp 573–607

G. Gauci, *Limitation of Liability in maritime Law: an anachronism?*, Marine Policy, vol. 19, Issue 1, Jan. 1995

G. Gunasekara, A. Sims, *Statutory Trends and the Genetic Modification' of the Common Law: Company Law as a Paradigm*, Statue Law Review (2005), 26(2), 82–102

G. G. Wilson, *Armed Merchant Vessels and Submarines*, The American Journal of International Law, vol. 24, no. 2 (Apr., 1930), pp. 337–339

G. J. Mulkins-Phillips, J. E. Stewart, *Effect of Environmental Parameters on Bacterial Degradation of Bunker C Oil, Crude Oils, and Hydrocarbons*, Appl Microbiol, 1974 Dec.; 28(6): 915–922 corrected by Appl Microbiol, 1975 April; 29(4): 569
G. Plant, *The Convention for the Suppression of Unlawful Acts against the Safety of Maritime Navigation*, Int'l and Comparative Law Quarterly, 39, 1990, pp 27–56
G. Plant, The Third United Nations Conference on the Law of the Sea and the Preparatory Commission: Models for United Nations Law Making, Int'l and Comparative Law Quarterly (1987), 36:525–558 Cam. Uni. Press
G. Stessens, *Corporate Criminal Liability: A Comparative Perspective*, Int'l and Comparative Law Quarterly (1994), 43:493–520
G. Williams, *Vicarious Liability and the Master's Indemnity (Continued)*, The Modern Law Review, vol. 20, No. 5 (Sept., 1957), pp 437–446
H. Bernstein, *Civil Liability for Pure Economic Loss under American Tort Law*, The American Journal of Comparative Law (1998), vol. 46, p 111
H. J. H. W., *Tort—Volenti Non Fit Injuria—Knowledge of Risk*, The Cambridge Law Journal, 3 , 1927, pp 98–99
H. Karan, *Revising Liability Limits in International Maritime Conventions: A Turkish Perspective*, 34 J. Mar. L. & Com. 615
H. Ohtsubo, Y. Kawamoto, T. Kuroiwa, *Experimental and numerical research on ship collision grounding oil tankers*, Nuclear Engineering and Design, 1994, vol. 150, no. 2–3, pp 385–396
H. T. Chester, *Round Table VI – Marine Insurance Law: Some Aspects of the Free of Capture and Seizure Clause*, 1943 A.B.A. Sec. Ins. Negl. & Comp. L. Proc. 240 (1943)
H. Tiberg, *Oil Pollution of the Sea and the Swedish Tsesis decision*, 1984 LMCLQ, pp. 218–226
H. Xanthaki, *Centros: Is this really the end for the theory of Siege Reel*, Company Law, vol. 22, no. 1, 2001, pp 2–8
I. d'Almeida Pires-Filho, *Priority of Maritime Liens in the Western Hemisphere: How Secure Is Your Claim?*, The University of Miami Inter-American Law Review, Vol. 16, No. 3 (Spring, 1985), pp. 505–537
I. Englard, *Can Strict Liability be Generalized?*, Oxford Journal of Legal Studies, vol. 2, no. 2, 1982, pp 245–256
I. Kim, *A comparison between the international and US regimes regulating oil pollution liability and compensation*, Marine Policy, vol. 27, issue 3, May 2003, pp 265–279
J. A. Schwartz, B. M. Aldridge, J. L. Stott and F. C. Mohr, *Immunophenotypic and functional effects bunker C fuel oil on the immune system of American*

mink (Mustela vison), Veterinary Immunology and Immunopathology, vol. 101, Issues 3–4, Oct. 2004, pp. 179–190

J. Bywater, *Government response to marine pollution from ships*, Marine Policy, vol. 19, iss. 6, Nov., 1995, pp 487–496

J. Borgo, *Casual Paradigms in Tort Law*, The Journal of Legal Studies, vol. 8, no. 3 (June 1979)

J. Cartwright, *Remoteness of Damage in Contract and Tort: A Reconsideration*, The Cambridge Law Journal, vol. 55, 1996, pp 488–514

J. C. Knobel, *The Feasibility of the co-existence of concrete negligence and legal causation*, Electronic Publishing (Online Journal), vol. 124, issue 3, 2007

J. C. McHorse, *Admiralty Jurisdiction and Limitation of Liability in Single Claim Cases*, California Law Review, vol. 22, no. 5 (Jul., 1934), pp 526–534

J. C. Macintosh (1922). Contributory Negligence. The Cambridge Law Journal, 1 , pp 185–191

J. Dewey, *The Historic Background of Corporate Legal Personality*, The Yale Law Journal, vol. 35, No. 6 (Apr., 1926), pp 655–673

J. E. Thomson, *State Sovereignty in International Relations: Bridging the Gap Between Theory and Empirical Research*, International Studies Quarterly (1995), Blackwell Publishers, 39, 213–233

J. G. B. Hutchins, *The Effect of the Civil War and the Two World Wars on American Transportation*, The American Economic Review, vol. 42, no. 2, Papers and Proceedings of the Sixty-fourth Annual Meeting of the American Economic Association (May, 1952), pp. 626–638

J. H. Mathias Jr,, J. D. Shugrue, T. A. Marrinson, *Insurance Coverage Disputes*, ALM Properties Inc., Law Journal Press, 2006, p 9

J. Hogendorn, *Review: Blaming the British*, Reviewed work(s): *Britain and Nigeria: Exploitation or Development?* by Toyin Falola, The Journal of African History, vol. 29, no. 1, Special Issue in Honour of Roland Oliver (1988), pp. 133–134

J. J. O. Harwood, *Lessons in Lien – The Separate Nature of in personam and in rem Claims*, The Arbitrator, Society of Maritime Arbitrators, vol. 40, no. 3, April 2009

J. J. White, *Corporate judgment proofing: a response to Lynn LoPucki's 'The Death of Liability'*, Yale Law Journal, vol. 107, 1998

J. Kasum, *Updating Sea Charts and Navigational Publications*, The Journal of Navigation, vol. 56, 2003, pp 497–505

J. Murphy, *The Merits of Rylands v. Fletcher*, Oxford Journal of Legal Studies, vol. 24, no. 4, 2004

J. O'Brian, R. Smith, *Conflict of Laws*, 2nd ed., Cavendish, 1999, pp 351–360; W. Tetley, *New Development in Private International Law: Tolofson v. Jensen and Ganon v. Lucas*, The American Jour., of Comparative Law, vol. 44, no. 4, Autumn 1996, pp 647–668

J. Payne, *Lifting the Corporate Veil: A Reassessment of the Fraud Exception*, The Cambridge Law Journal (1997), 56:284–290

J. P. Brennan , M. B. Gordillo, *Working class protest, popular revolt, and urban insurrection in Argentina: the 1969 Cordobazo*, Journal of Social History, vol. 27, 1994

J. P. Pederson, L. G. Seljelv, T. Bauna, O. A. Follum, J. H. Anderson, T. Whal, A. Skoelv, *Towards an Operational Oil Spill Detection Service in the Mediterranean? The Norwegian Experience: A Pre-operational Early Warning Detection Service using ERS SAR Data*, Spill Science and Technology Bulletin, vol. 3, no. 1, 1996, pp 41–46(6)

J. R. Mario, *Law amid Flux: The Economics of Negligence and Strict Liability in Tort*, The Journal of Legal Studies, vol. 9, no. 2, Change in the Common Law: Legal and Economic Perspectives (Mar., 1980), pp. 291–318

J. S. Gray, D. Calamari, R. Duce, J. E Portmann, P. G Wells, *Scientifically based strategies for Marine Environmental Protection and Management*, Marine Pollution Bulletin MPNBAZ, vol. 22, no. 9, Sept., 1991, p 432–440

J. S. Johnston, *Punitive Liability: A New Paradigm of Efficiency in Tort Law*, 87 Colum. L. Rev. 1385 (1987)

J. Spier, , *Sluipende schade and Mon*, Nieuw Burgerlijk Wetbeok – 36, nr.52

J. T. Biezup, T. J. Abeel, *The Limitation Fund and its Distribution*, 1979, 53 Tul. L. Rev. 1185

J. Thompson, *Remoteness of Damages – Extending the Doctrine?*, Edinburg Law Rev., vol. 8, iss. 3, 2004

J. Wardley-Smith, *The Castillo de Bellver*, Oil and Petrochemical Pollution (1983), vol. 4, Issue 1, pp 291–293

J. W. Boulton, *The Modern International Law of Piracy: Content and Contemporary Relevance*, Int'l Relations, vol. 7, No. 6, 1983, pp 2493–2511

J. Wren, *Overview of the Compensation and Liability Regimes Under the International Oil Pollution Compensation Fund (IOPC)*, Spill Science and Technology Bulletin, vol. 6, issue 1, Feb., 2000, pp 45–58

K. E. Goodpaster, *The concept of corporate responsibility*, Journal of Business Ethics, Springer, vol. 2, No. 1 (Feb., 1993), pp 1–22

K. Iwai, *Persons, things and Corporations: The Corporate Personality Controversy and Comparative Corporate Governance*, The American Journal of Comparative Law, vol. 47, No. 4 (Autumn, 1999), pp 583–632

K. J. Rawson, *The carriage of bulk oil and chemicals at sea*, Inst.., of Chemical Engineers, 1994, pp 28–31; J. Wren, *Overview of the Compensation and Liability Regimes under the International Oil Pollution Compensation Fund (IOPC)*, Spill Science & Technology Bulletin, vol.6, iss. 1, Feb., 2000, pp 45–58

K. Lipstein, *Conflict of Laws – Jurisdiction – Mareva Injunction as Basis for Service Aboard – R. S. C., Ord. 11, R.1(1)(i)*, The Cambridge Law Journal, 37, pp 241–244

K. Mitroussi, *The role of organizational characteristics of ship owning firms in the use of third party ship management*, Marine Policy, vol. 28, issue 4, July 2004, pp 325–333

K. Mitroussi, *Third party ship management: the case of separation of ownership and management in the shipping context*, Marine Policy and Management, vol. 30, no. 1, Jan. 2003

K. N. Hylton, *Missing Markets Theory of Tort Law*, 90 Nw. U. L. Rev. 977 (1995–1996)

K. S. Abraham, *Environmental Liability and the Limits of Insurance*, Columbia L. Rev., 1988, p 942

K. Segerson, *Government regulation and compensations for environmental quality and natural resource use*, Contemporary Economic Policy, col. 15, 1997

K. Takahashi, *II. External Competence Implications of the EC Regulation on Jurisdiction and Judgments*, Int'l and Comparative Law Quarterly (2003), 52:529–534

K. W. Ketkar, *The Oil Pollution Act of 1990: A Decade Later*, Spill Science & Technology Bulletin, vol. 7, Issues 1–2, June 2002, pp 45–52

L. B. Burlington, *A Update on Implementation of Natural Resource Damage Assessment and Restoration under OPA*, Spill Science and Technology Bulletin, vol. 7, issues 1–2, June 2002

L. Bender, *Frontier of Legal Thought III: Feminist (Re) Torts: Thoughts on the Liability Crisis, Mars Torts, Power and Responsibilities*, 1990 Duke L. J. 848

L. Burglass, *Limitation of Liability from a Marine Insurance Viewpoint*, Tul. Law Preview, vol. 53, 1979, p. 1364

L. de la Fayette, *Achievements in 2001*, Environmental Policy and Law, vol. 32, no. 3–4, IOS Press, 2002, pp 146–151

L. F. E. Goldie, *Liability for Damage and the Progressive Development of International Law*, Int'l & Comparative Law Quarterly (1965), Cambridge University Press, 14:1189–1264

L. Gundling, "*The Status in International Law of the Principle of Precautionary Action*", IJECL (1990)

L. I. Kiern, *Liability, Compensation and Financial Responsibility Under the Oil Pollution Act of 1990: A Review of the First Decade*, Spring 2000, 24 Tul. Mar. L. J. 481

L. I. De. Winter, *Exclusive Jurisdiction in Private International Law*, Int'l and Comparative Law Quarterly (1968), 17, pp 706–720

Ling Zhu, *Can the Bunkers Convention Ensure Adequate Compensation for Pollution Victims*, Journal of Maritime Law and Commerce, April 2009

L. J. Kirmayer, *Psychotherapy and the Cultural Concept of the Person*, Transcultural Psychiatry, vol. 44, No. 2, 232–257 (007)

L. Orlando, *'Piracy' provisions under the Enforcement Directive and patent infringement*, Jour. of Intellectual Property Law & Practice Advance Access (originally published online on August 31, 2007), available in Jour. of Intellectual Property Law & Practice, vol. 2, no. 10, 2007, pp 642–643

L. Tamiotti, M. Finger, *Environmental Organizations: Changing Roles and Functions in Global Politics*, Global Environmental Politics, vol. 1, no. 1, Feb, 2001

M. Boile, S. Theofanis, F. Mikiki, *Oil Spills in Maritime Transport*, Transportation Research Record: Journal of the Transportation Research Board, vol. 1909, 2005, pp 100–107

M. D. French, R. J. Jennings, W. Nicole, S. Sankar, E. D. Schmidt, *Estimation of potential impacts and natural resource damages of oil*, Journal of Hazardous Materials, Elsevier – Amsterdam, 2004, vol. 107, pp 11–25

M. George, *Choice of Law in Maritime Torts*, Journal of Private International Law, vol. 3, no. 1, 2007

M. Grey, *Why heavy weather is not like Act of God*, Lloyd's List, March 2, 1989

M. G. Ross, *Concurrent liability in tort and contract*, Structural Survey, 1987, vol.5, issue 3, pp 288–299

M. Jacobson, *Bunkers Convention in force*, JIML 15 [2009] 1:21

M. Johnson, *Roll on the 14th Directive – Case law fails to solve the problems of corporate mobility within the EU – again*, Hertfordshire Law Journal, vol. 2, no. 2, 2004, pp 9–18

M. L. Honkasalo, *Grips and Ties: Agency, Uncertainty, and the Problem of Suffering in North Karelia*, Medical Anthropology Quarterly, Blackwell, vol. 23, no. 1, Mar., 2009, pp 51–69(19)

M. Mason, *Civil liability for oil pollution damage: examining the evolving scope for environmental compensation in the international regime*, Marine Policy, vol. 27, issue 1, Elsevier Science Ltd., Jan. 2003, pp 1–12

M. O. Tsavaris, *Calming Troubled Waters for Cruise Ship Owners and Their Passengers: Carlisle V. Carnival Corp*, The University of Miami Inter-American Law Review, vol. 35, no. 1, Symposium Edition: Forum Non Conveniens: Development and Issues over the Past Seven Years (Fall, 2003 – Winter, 2004), pp. 153–181

M. R. Badykov, *The Russian Civil Code and the Rome Convention: Applicable Law in the Absence of Choice by the Parties*, Jour., of Private Int'l Law, vol. 1, no. 2, Hart, Oct., 2005, pp 269–289

M. Sole, C. Porte, X. Biosca, C. L. Mitchelmore, J. K. Chipman, D. R. Livingston, J. Albaiges, *Effects of the "Aegean Sea" oil spill on bio transformation enzymes, oxidative stress and DNA-adducts in digestive gland of the mussel (Mytilus edulus L.)*, Comparative Biochemistry and Physiology Part C: Pharmacology, Toxicology and Endocrinology, vol. 113, Issue 2, Feb. 1996, pp 257–265

M. T. Katzman, *Pollution Liability Insurance and Catastrophic Environmenatl Risk*, The Journal of Risk and Insurance, vo. 55, No. 1 (Mar., 1998)

N. B. Richards, *Maritime Liens in Tort, General Average, and Salvage*, 47 Tul. L. Rev., 569 (1972–1973)

Priorities of Maritime Liens, Harvard Law Rev., vol. 69, no. 3 (Jan., 1956), pp 525–528

N. Campbell, J. Armour, *Demystifying the Civil Liability of Corporate Agents*, The Cambridge Law Journal (2003), 62:290–304

N. de Sadeleer, *Liability for Oil Pollution Damage versus Liability for Waste Management: The Polluter Pays Principle at the Rescue of the Victims: Case C-188/07, Commune de Mesquer v. Total France SA [2008] 3 CMLR 16, [2009] Env LR 9*, Journal of Environmental Law 2009, 21(2):299–307

N. Emmott, N. Haigh, *Integrated Pollution Prevention and Control: UK and EC Approaches and Possible Next Steps*, Journal of Environmental Law, 1996, 8(2): 301–311, Oxford

N. Klein (2008), *The Rules of the International Tribunal for the Law of the Sea: A Commentary* by P Chandrasekhara Rao and Ph. Gautier (eds) (Martinus Nijhoff Publishers, 2006; ISBN-13: 978-9004152403, 521 pp, hb)]. *International and Comparative Law Quarterly,* 57 , pp 999–1004

N. M. Hunnings and Research Officers of the British Institute of International and Comparative Law (1970), *Pollution of the Seas*. International & Comparative Law Quarterly, 19 , pp 341–343

N. S. Marsh, *The Ethiopian Law of Extra-Contractual Liability*, by G. Krzeczunowicz. [Addis Ababa: Faculty of Law. 1970. xv (preface and table of contents), 140 and 41 (Appendices)]. *International & Comparative Law Quarterly,* 20 , 1971, pp 788–789

Oil Pollution Liability of Colliding Vessels in China, Lloyd's Maritime Law Quarterly, [2001] L. M. L. C. Q. 19

P. Bennett, *Environmental governance and private actors: enrolling insurers in international maritime regulation*, Political Geography, vol. 19, iss. 7, Sep., 2000, pp 875–899

P. Bennet, *Mutual risk: P&I insurance clubs and maritime safety and environmental performance*, Marine Policy, vol. 25, issue 1, Jan 2001

P. Giliker, *Codifying Tort Law: Lessons from the Proposals for Reform of the French Civil Code*, Int'l & Comparative Law Quarterly (2008), Cambridge University Press, 57:561–582

P. Hardouin, R. Weichhardt, *Terrorist fund raising through criminal activities*, Journal of Money Laundering Control, Emerald Group Publishing, vol. 9, iss. 3, 2006, pp 303–308

P. Hassan, A. Azfar, *Comment: Securing environmental rights through public interest litigation in South Asia*, 22 Virginia Envtl. L. J. 215

P. K. Shin, *Effects of a spill of bunker oil on the marine biological communities in Hong Kong*, Environment Int'l (USA), vol. 14:6, pp 545–552

P. McEleavy, E. Micheler, *Current Developments – Private International Law*, International and Comparative Law Quarterly, vol. 52, no, 2, 2003

P. Salvador-Coderch, N. Garoupa, C. Gomez-Liguerre, *Scope of Liability: The vanishing distinction between negligence and strict liability*, European Jour., of Law and Economics, Springer, vol. 28, no. 3, Dec., 2009

P. T. Pederson, S. Zhang, *Effects of ship structure and size on grounding and collision damage distributions*, Ocean Engineer, vol. 27, Issue 11, Nov. 2000

P. Zweifel, J. R. Tyran, *Environment Impairment Liability as an Instrument of Environmental Policy*, Ecological Economics, vol. 11, iss. 1, Sep., 1994, pp 43–56

R. Asariotis, *Implementation of the ISPS Code: an overview of recent developments*, JIML 11[2005] 4: 266

R. B. Grantham, *The Limited Liability of Company Directors*, Univ. of Queensland TC Beirne School of Law Research Paper No. 07-03, Maritime and Commercial Law Quarterly, 2007

R. Cooke, *Remoteness of Damages and Judicial Discretion*, The Cambridge Law Journal, vol. 37, 1978, pp 288–300

R. D. Thomas, *Bunker oil pollution – a new liability convention*, JIML 14 [2008] 5:375

R. F. Blomquist, *Re-Enchanting Torts*, 56, South Carolina Law Review, 481 (2005)

R. G. Hansen, R. S. Thomas, *The efficiency of sharing liability for hazardous waste: effects of uncertainty over damages*, Int'l Review of Law and Economics, vol. 19, Issue 1, Mar. 1999

R. Gottschalk, *Problems in Israeli Maritime Law*, Int'l and Comparative Law Quarterly (1965), 14:302–308

R. Grantham, C. Rickett), *A Normative Account of Defences to Restitutionary Liability*, The Cambridge Law Journal, vol. 67, 2008, pp 92–125

R. H. Ganten, *The Tanio Spill: A case history illustrating the work of the International Oil Pollution Compensation Fund*, Proceedings of the 1985 Oil Spill

Conference, 135–139, API Publication No. 4385, American Petroleum Institute, Washington DC, USA
R. H. Stern, Vicarious liability for infringement, *IEEE Micro*, vol. 24, no. 5, Sep./Oct. 2004, pp. 6, 79–82
An Efficiency Analysis of Vicarious Liability under the Law of Agency, The Yale Law Journal, vol. 91, no. 1, Nov., 1981, pp. 168–206
R. Lefeber, *Frontiers of International Law: Counteracting the Exercise of Extraterritorial Jurisdiction*, Leiden Journal of Int'l Law (1997), 10:1:1–7
R. L. V. Eisma, P. Christie, M. Hershman, *Legal issues affecting sustainability of integrated coastal management in the Philippines*, Ocean and Coastal Management, vol. 48, iss. 3–6, 2005, pp 336–359
Robert Gay, H. Dickinson, *Time charterparty – whether implied term that charterers would not undertake additional voyage – whether promissory estoppel against charterers*, JIML 15[2009] 3: 213
R. Price, *The Carriage of Hazardous Cargo by Sea: A UAE Law Perspective*, Arab Law Quarterly, vol. 10, no. 4, 1995, pp. 310–330
R. R. W. Brooks, *Liability and Organizational Choice*, Journal of Law and Economics, vol. 45, No. 1, (Apr., 2002), pp 91–125
R. Shaw, *What is a ship in maritime law*, JIML 11(2005)
R. Wolfson, *The English and French Carriage of Goods by Sea Enactments*, Int'l & Comparative Law Quarterly, vol. 4, 1955, pp 508–532
R. W. Right, *Causation in Tort Law*, Dec. 1985, 73 Calif. L. Rev. 1735
R. Y. Jenings, *The Limits of Continental Shelf Jurisdiction: Some Possible Implications of the North Sea Case Judgment*, Int'l and Comparative Law Quarterly, (1969), 18:819–832, Cambridge
S. Bloodworth, *Death on the High Seas: the demise of TOVALOP and CRISTAL*, Florida State Univ., Jour., of Land Use & Environmental Law, 1998
S. Ebert, *European Union: Company Law – Freedom of Establishment*, International Company Commercial Law Review, vol. 14, no. 5, 2003, pp 51–52
S. Ebert, *The law applicable to groups of companies involving European companies (Societas Euopea)*, Company Lawyer, vol. 24, no. 4, 2004, pp 108–113
Seismic Shipping Inc and another v Total E & P UK plc (The Western Regent): Limitation of liability – 1976 Convention – jurisdiction to grant limitation decree – limitation fund – foreign litigation of limitable claim – anti-suit injunction, JIML 11 [2005] 4, 263
S. Knapp, M. van de Velden, *Visualization of differences across port state control regimes by means of correspondence analysis*, Econometric Institute Report, Erasmus University Rotterdam, EI 2007-32, Revision, 2009-07-29
S. N. Vlachos, "*Greece: New Code of Private Maritime Law*", The American Journal of Comparative Law, vol. 7, No. 4 (Autumn 1958), pp 599–601

S. R. Pezeshki, M. W. Hester, Q. Lin, J. A. Nyman, *The effects of oil spill and clean up on dominant US Gulf coast marsh macrophytes: review*, Environmental Pollution, vol. 108, Issue 2, pp 129–139
S. Shavell, *An Analysis of Causation and the Scope of Liability in the Law of Torts*, Journal of Legal Studies, 1980, The University of Chicago
S. S. Karic, *NOTE: Admiralty – Procedure for Maritime Action in Rem Found Unconditional*, 55 Tul. L. Rev. 936 (1981)
T. A. Mensah, *The Place of the ISPS Code in the International Regime – for the Security of International Shipping*, WMU Jour. of Maritime Affairs, vol. 3, no. 1, April 2004, pp 17–30
T. J. Schoenbaum (1998), *Oil Pollution at Sea*. By Ganci Potthard [Chichester: John Wiley 1997, xxx + 313 pp. ISBN 0–471–97066–2. No price given], International & Comparative Law Quarterly, 47, pp 966–966
T. J. Wagner, *Recoverable Damages under the Oil Pollution Act 1990*, S. U. S. F. Mar. L. J. 283 (1992–1993)
T. Koller, *The English Limited Company ready to invade Germany*', International Company and Commercial Law Review, vol. 15, no. 11, 2004, pp 334–343
T. M. Bos (2007). P. van der Grinten; T. Heukels, eds., *Crossing Borders: Essays in European and Private International Law in Honour of Frans van der Velden*, Kluwer, Deventer 2006, xvi + 262 pp., ISBN 90-13-03040-8.. Netherlands International Law Review, 54 , pp 168–173
T. Nunes, Charterer's Liabilities under the Ship Time Charter, Houston Journal of International Law, vol.26, 2004
T. V. van Flein, *Allocation of Fault and Products Liability: A comment on safety products and human error*, 19 Alaska Law Rev., 141
United States Code Congressional and Administrative News (101st Congress, 2nd Session 1990) Volume5, p. 780
V. Petaccio, *Water Pollution and the Future Law of the Sea*, Int'l and Comparative Law Quarterly (1972), 21, pp 15–42
V. P. Goldberg, *Recovery for Economic Loss following the Exxon „Valdez" Oil Spill, The* Journal of Legal Studies, vol. 23, no. 1 (Jan., 1994), pp. 1–39
V. Tomljenovic, *New Conflicts Approach: Is It Necessary*, Maritime Torts, Yearbook of Private International Law, vol. 1 (1999), pp 249–298
W Chao, *Liability and Compensation for Oil Pollution Damage: Some Current Threats to International Convention System*, Spill Science and Technology Bulletin, vol. 7, Issues 1–2, Jun 2002
W. J. Ronan, *English and American Courts and the Definition of War*, The American Journal of International Law, vol. 31, no. 4 (Oct., 1937), pp. 642–658
W. K. Tally, *Determinants of Ship Accident Seaworthiness*, Int'l Jour., of Maritime Economics, vol. 1, 1999, pp 1–14

W. L. Walker, *Recognition of Belligerency and Grant of Belligerent Rights*, Transactions of the Grotius Society, Vol. 23, Problems of Peace and War, Papers Read before the Society in the Year 1937 (1937), pp. 177–210

W. M. Landes, R. A. Posner, *Causation in Tort Law: An Economic Approach*, Journal of Legal Studies, vol. XII, Jan. 1983, The University of Chicago

W. Oosterveen, *Some Recent Developments Regarding Liability for Damage Resulting from Oil Pollution – from the Perspective of and EU Member State*, Environmental Law Review, vol. 6, issue 4, 2005, pp 223–239

W. Poor, *A Shipowner's Right to Limit Liability in Cases of Personal Contracts*, The Yale Law Journal, vol. 31, no. 5 (Mar., 1922), pp. 505–513

W. Tetley, *Uniformity of International Private Maritime Law – the pros, cons and alternatives to international Conventions – How to Adopt an International Convention*, (2000) 24 Tul. Mar. L. J. 775–856

Wu Chao, *Liability and Compensation for Bunker Pollution*, 33 J. Mar. L. & Com. 553 (2002)

W. W. Eyer, *Shipowners' Limitation of Liability. New Directions for an Old Doctrine*, Stanford Law Review, vol. 16, no. 2 (Mar., 1964)

X. Liu and K. W. Wirtz, *The economy of oil spills: Direct and indirect costs as a function of spill size*, Journal of Hazardous Materials, Volume 171, Issues 1–3, 15 November 2009, pp 471–477

Yearbook of International Cooperation on Environment and Development 2003–2004 (O. S. Stokke, O. B. Thommessen, Int'l Environmental Governance Set, Oct. 2009

Z. Keyuan, *New Developments in the International Law of Piracy*, Chinese Jour. of Int'l Law, vol. 8, no. 2, 2009, pp 323–345

Z. Wang, M. Finags, L. Sigouin, *Characterization and identification of a "mystery" oil spill from Quebec (1999)*, A Journal of Chromotography, Elsevier, vol. 909, no. 2, Feb. 2001, pp 155–169(15)

Z. Wang, M. Fingas, D. S. Page, *Oil spill identification*, Journal of Chromatography A vol. 843, issues 1–2, 28 May 1999, pp 369–411

National Statutes

Animals Act, United Kingdom (1971)

Civil Procedure Rules, United Kingdom [(1998) S. I. 1998/3132]

Civil Procedure Rules (Admiralty Claims), United Kingdom (1998)

Comprehensive Environmental Response, Compensation, and Liability Act, United States (42 USC 9601)

Contracts (Rights of Third Parties) Act, United Kingdom (1999)

Cons. Stato, France (1991, II)
Corte Cost., Spain, (22 November 1991)
Dutch Civil Code, Netherlands (1980)
Limitation Act, Australia (1969)
Limitation Act, Canada (2002)
Limitation Act, India (1963)
Limitation Act, Malaysia (1953)
Limitation Act, New Zealand (1950)
Limitation Act, Samoa (1975)
Limitation Act, Singapore (Ordinance No. 57 of 1959)
Limitation Act, United Kingdom (1980)
Marine Insurance Act, United Kingdom (1906)
Merchant Shipping Act, India (1958)
Merchant Shipping Act, South Africa (57 of 1951)
Merchant Shipping Act, South Africa [(1951) as amended by Act of 1997]
Merchant Shipping Act, United Kingdom (1894)
Merchant Shipping Act, United Kingdom (1995)
Merchant Shipping (Oil Pollution) Act, United Kingdom (1971)
Merchant Shipping (Oil Pollution) (Bunkers Convention) Regulation 2006 (SI 2006/1244), United Kingdom
Oil Pollution Act, United States (1990)
Oil Spill Prevention and Response Act, United States R.S. 30
Pilotage Act, United Kingdom (1983)
Ports and Merchant Marine Act, Spain (Law of November, 24, 1992)
Practice Direction [(61) Admiralty Claims], United Kingdom
Prescription and Limitation (Scotland) Act (1984)
Prescription Ordinance, Sri Lanka (No. 22 of 1871)
Rules of Court (Admiralty Proceedings), United Kingdom (1996)
Supreme Court Act, United Kingdom (1981)
UmweltHG, Germany
Wasserhaushaltsgesetz (Water Resources Act), Germany (1986)

Reports and Statistics

Comite Maritime International Guidelines on Oil Pollution (Part III, 1994)
Hess Material Data Sheet, *No. 4 Fuel Oil*, MSDS No. 15054
Hess Material Data Sheet, *No. 5 Fuel Oil*, MSDS No. 0332
International Maritime Organization, Maritime Knowledge Centre, International Shipping and World International Oil Spill Statistics: 1994 – Oil Spill Intelligence Report

IOPC Fund Annual Report 1987
IOPC Fund Annual Report 1988
IOPC Fund Annual Report 1989
IOPC Fund Annual Report 1990
IOPC Fund Annual Report 1991
IOPC Fund Annual Report 1994
IOPC Fund Annual Report 1995
IOPC Fund Claims Manual (5th Edition)
Lloyds' Casualty Reports (21 July – 11 August, and December 1995)
Newcastle P&I Association Rulebook (1997)
Trade, Facts and Figures, (updated November 2008)
Trade Winds (3 April 1998)
United Kingdom Clubs Analysis of Major Claims (1993)
West of England Ship Owners' Mutual Insurance Association (Luxemburg) Rulebook (1997)

INTERNATIONALRECHTLICHE STUDIEN
Beiträge zum Internationalen Privatrecht,
zum Einheitsrecht und zur Rechtsvergleichung

Herausgegeben von Ulrich Magnus und Peter Mankowski

Band 1 Birgit Reimers-Zocher: Beweislastfragen im Haager und Wiener Kaufrecht. 1995.

Band 2 Lenja Vespermann: Schadensersatz bei Beeinträchtigung rechts- oder sittenwidriger Interessen im deutschen und französischen Recht. 1996.

Band 3 Peter Cypra: Die Rechtsbehelfe im Verfahren der Vollstreckbarerklärung nach dem EuGVÜ unter besonderer Berücksichtigung der Ausgestaltung in Deutschland und Frankreich. 1996.

Band 4 Brigitta Lurger: Regulierung und Deregulierung im europäischen Privatrecht. 1997.

Band 5 Uta Gutknecht: Das Nacherfüllungsrecht des Verkäufers bei Kauf- und Werklieferungsverträgen. Rechtsvergleichende Untersuchung zum CISG, zum US-amerikanischen Uniform Commercial Code, zum deutschen Recht und zu dem Vorschlag der Kommission zur Überarbeitung des deutschen Schuldrechts. 1996.

Band 6 Stefan Abel: Die Qualifikation der Schenkung. Zur Methode der Qualifikation im Internationalen Schuldvertragsrecht. 1997.

Band 7 Daniel Kötz: Verkäuferpflichten und Rechtsbehelfe des Käufers im neuen norwegischen Kaufrecht vom 13. Mai 1988 Nr. 27 im Vergleich zum UN-Kaufrecht vom 11. April 1980 (CISG). 1997.

Band 8 Olaf Beier: Grundsätze eines europäischen transportmittelübergreifenden Schadensrechts für den Gütertransport. Eine rechtsvergleichende Untersuchung unter besonderer Berücksichtigung der CMR, der ER/CIM 1990, des WA/HP und der HVR. 1999.

Band 9 Ulrich Krüger: Modifizierte Erfolgshaftung im UN-Kaufrecht. Die Haftungsbefreiung bei Lieferung vertragswidriger Ware gemäß Art. 79 CISG. 1999.

Band 10 Wolf-Henning Hammer: Das Zurückhaltungsrecht gemäß Art. 71 CISG im Vergleich zu den Kaufgesetzen der nordischen Staaten unter Einbeziehung transportrechtlicher Aspekte. Eine rechtsvergleichende Studie. 1999.

Band 11 Ulrich Quay: Der Consulting-Vertrag im Internationalen Privatrecht. 2000.

Band 12 Axel Halfmeier: Die Veröffentlichung privater Tatsachen als unerlaubte Handlung. Eine rechtsvergleichende Untersuchung. 2000.

Band 13 Stephanie Scholz: Lohngleichheit durch Verfahren. Eine Studie zur Förderung der Lohngleichheit durch verfahrensrechtliche Ansätze. 2000.

Band 14 Tarık Cebecioglu: Stellung des Ausländers im Zivilprozeß. Eine rechtsvergleichende Untersuchung des türkischen und deutschen Rechts. 2000.

Band 15 Gunnar Straube: Sozialrechtliche Eingriffsnormen im Internationalen Privatrecht. 2001.

Band 16 Daniela de Lukowicz: Divergenzen in der Rechtsprechung zum CISG. Auf dem Weg zu einer einheitlichen Auslegung und Anwendung? 2001.

Band 17 Mohamed Hassan Fadlalla Ali: Das islamische Ehe- und Kindschaftsrecht im Sudan. Mit Hinweisen zu den Lehren der Islamischen Rechtsschule, der anwendbaren Familienrechtsvorschriften für Nichtmuslime im Sudan und im deutschen Familienrecht. 2001.

Band 18 Ole Lando/Ulrich Magnus/Monika Novak-Stief (Hrsg./eds.): Angleichung des materiellen und des internationalen Privatrechts in der EU. Harmonisation of Substantive and International Private Law. 2003.

Band 19 Florian Schulz: Der Ersatzlieferungs- und Nachbesserungsanspruch des Käufers im internen deutschen Recht, im UCC und im CISG. Eine rechtsvergleichende Untersuchung unter Berücksichtigung der Schuldrechtskommission, der EG-Richtlinie 1999/44/EG und des Diskussionsentwurfs eines Schuldrechtsmodernisierungsgesetzes. 2002.

Band 20 Ingo Janert: Die Aufrechnung im internationalen Vertragsrecht. 2002.

Band 21 Armin Jentsch: Die Erhaltungspflichten des Verkäufers und des Käufers im UN-Kaufrecht im Vergleich zum US-amerikanischen Uniform Commercial Code und zum deutschen Recht. 2002.

Band 22 Martin Taschner: Arbeitsvertragsstatut und zwingende Bestimmungen nach dem Europäischen Schuldvertragsübereinkommen. Einheitliche Auslegung. 2003.

Band 23 Klaus Bitterich: Die Neuregelung des Internationalen Verbrauchervertragsrechts in Art. 29a EGBGB. 2003.

Band 24 Sascha J. Morgenroth: Umsetzung der Verbrauchsgüterkaufrichtlinie in Spanien und Deutschland. Umsetzung der Richtlinie 1999/44/EG des Europäischen Parlaments und des Rates vom 25. Mai 1999 zu bestimmten Aspekten des Verbrauchsgüterkaufs und der Garantien für Verbrauchsgüter in Spanien und Deutschland. 2003.

Band 25 Markus Renfert: Über die Europäisierung der ordre public Klausel. 2003.

Band 26 Henning Mordhorst: Die Behandlung vertraglicher Mobiliarsicherheiten im US-amerikanischen und deutschen Insolvenzrecht. 2003.

Band 27 Hans Markus Wulf: UN-Kaufrecht und eCommerce. Problembereiche bei der Anwendung des Wiener Übereinkommens auf Internet-Verträge. 2003.

Band 28 Julia Kaiser: Freiheit oder Schutz? Die Scheidungsfolgenvereinbarung im deutschen und im englischen Recht. 2003.

Band 29 Astrid Hillebrenner: Die private Zustimmung zu Rechtsgeschäften Dritter im englischen, dänischen und deutschen Recht. 2004.

Band 30 Irene von der Heyde: Das deutsche und italienische internationale Kindschaftsrecht im Rechtsvergleich. Favor filii oder favor filiationis? 2004.

Band 31 Sven-Michael Werner: Recht im Systemwandel. *Ein Land, zwei Systeme* und das Kontinuitätsproblem in Hongkong. 2004.

Band 32 Holger Ellenberger: Wirtschaftsrelevante Kollisionsnormen im japanischen internationalen Privatrecht. 2004.

Band 33 Sebastian Kaufmann: Parol Evidence Rule und Merger Clauses im internationalen Einheitsrecht. 2004.

Band 34 Britta Christina Weber: Eigentumsübertragung und gutgläubiger Erwerb beim Kauf von beweglichen Sachen im vietnamesischen Zivilgesetzbuch von 1995. Eine rechtsvergleichende Analyse unter Berücksichtigung des Eigentumsbegriffs im vietnamesischen Recht. 2004.

Band 35 Alexander von Saucken: Die Reform des österreichischen Schiedsverfahrensrechts auf der Basis des UNCITRAL-Modellgesetzes über die internationale Handelsschiedsgerichtsbarkeit. Ein Diskussionsbeitrag. 2004.

Band 36 Jasnica Garašić: Anerkennung ausländischer Insolvenzverfahren. Teil 1 und Teil 2. Ein Vergleich des kroatischen, des deutschen und des schweizerischen Rechts sowie der Europäischen Verordnung über Insolvenzverfahren, des Istanbuler Übereinkommens und des UNCITRAL-Modellgesetzes. 2005.

Band 37 Eugenia Kurzynsky-Singer: Anknüpfung und Reichweite des Vollmachtsstatuts. 2005.

Band 38 Timo Utermark: Rechtsgeschichte und Rechtsvergleichung bei Ernst Rabel. 2005.

Band 39 Mohamed Lakmes: Die IPR-Anknüpfung der Schuldverträge. Ein Vergleich zwischen dem deutschen und dem syrischen Recht. 2005.

Band 40 Christian Haas: Vertragsstrukturen privatfinanzierter Infrastrukturprojekte. Nationale und internationale Perspektiven. 2005.

Band 41 Tanja Herzog-Engels: Gesellschaftsrechtliche Formvorschriften und Internationales Privatrecht. 2005.

Band 42 Hassan Siahpoosh: Das Familien- und Erbrecht im Iran. 2006.

Band 43 Carola Reifenrath: Das Vertragsrecht Hongkongs und dessen zukünftige Entwicklung. Unter besonderer Berücksichtigung des englischen Common Law. 2006.

Band 44 Arkadiusz Wudarski: Der Pfandbrief und sein Treuhänder. Eine rechtsvergleichende Untersuchung zum polnischen und deutschen Recht. 2006.

Band 45 Ulla Liukkunen: Cross-Border Services and Choice of Law. A Comparative Study of the European Approach. 2006.

Band 46 Claire Anna Reifner: Die Umsetzung der Verbrauchsgüterkaufrichtlinie in Italien und Deutschland. 2007.

Band 47 Shahriar Rabbanifar: Vergleich des Rechts der öffentlichen Unternehmen in Deutschland, Iran und Polen. 2007.

Band 48 Judith Farokhmanesh: Der Schutz des Urhebers im internationalen Vertragsrecht. Funktion, Reichweite und Effektivität zwingender Bestimmungen im deutschen, französischen und englischen Recht. 2007.

Band 49 Christoph Hellriegel: Schuldnerschutz in Polen. Rechtsvergleichende Untersuchung eines universalen Regelungsproblems. 2007.

Band 50 Ingemar Carl: Einstweiliger Rechtsschutz bei *Torpedoklagen*. 2007.

Band 51 Gabriele Wunsch: Möglichkeiten und Grenzen eines europäischen Hypothekarkredits. Problemanalyse und Lösungswege für die Schaffung eines Binnenmarktes für Grundpfandrechte unter Berücksichtigung des deutschen, englischen, schwedischen und spanischen Rechts. 2009.

Band 52 Sven Kuchmann: Die Haftung des Geschäftsführers einer deutschen GmbH und des Gerente einer argentinischen SRL im Rechtsvergleich. 2008.

Band 53 María Elena Alvarez de Pfeifle: Der Ordre Public-Vorbehalt als Versagungsgrund der Anerkennung und Vollstreckbarerklärung internationaler Schiedssprüche. Unter Berücksichtigung des deutschen, schweizerischen, französischen und englischen Rechts sowie des UNCITRAL-Modellgesetzes. 2009.

Band 54 Amir-Said Ghassabeh: Die Zustellung einer *punitive-damages*-Sammelklage an beklagte deutsche Unternehmen. Zugleich ein Beitrag zum „unnötigen“ transatlantischen Justizkonflikt. 2009.

Band 55 Gunnar Lennart Schmüser: Das Zusammenspiel zwischen Haupt- und Sekundärinsolvenzverfahren nach der EuInsVO. 2009.

Band 56 Johanna Siemonsen: Die deutschen Ausführungsvorschriften zur Europäischen Insolvenzverordnung. 2009.

Band 57 Dan Malika Gunasekera: Civil Liability for Bunker Oil Pollution Damage. 2010.

Band 58 Sarah Mukaddes Oguz: Das türkische und das europäische Verbraucherrecht. Eine vergleichende Analyse mit Blick auf die Beitrittsbestrebungen der Türkei in die Europäische Union. 2010.